D1168225

Safety of Irradiated Foods

FOOD SCIENCE AND TECHNOLOGY

A Series of Monographs, Textbooks, and Reference Books

1. Flavor Research: Principles and Techniques, *R. Teranishi, I. Hornstein, P. Issenberg, and E. L. Wick (out of print)*
2. Principles of Enzymology for the Food Sciences, *John R. Whitaker*
3. Low-Temperature Preservation of Foods and Living Matter, *Owen R. Fennema, William D. Powrie, and Elmer H. Marth*
4. Principles of Food Science
   Part I: Food Chemistry, *edited by Owen R. Fennema*
   Part II: Physical Methods of Food Preservation, *Marcus Karel, Owen R. Fennema, and Daryl B. Lund*
5. Food Emulsions, *edited by Stig Friberg*
6. Nutritional and Safety Aspects of Food Processing, *edited by Steven R. Tannenbaum*
7. Flavor Research: Recent Advances, *edited by R. Teranishi, Robert A. Flath, and Hiroshi Sugisawa*
8. Computer-Aided Techniques in Food Technology, *edited by Israel Saguy*
9. Handbook of Tropical Foods, *edited by Harvey T. Chan*
10. Antimicrobials in Foods, *edited by Alfred Larry Branen and P. Michael Davidson*
11. Food Constituents and Food Residues: Their Chromatographic Determination, *edited by James F. Lawrence*
12. Aspartame: Physiology and Biochemistry, *edited by Lewis D. Stegink and L. J. Filer, Jr.*
13. Handbook of Vitamins: Nutritional, Biochemical, and Clinical Aspects, *edited by Lawrence J. Machlin*
14. Starch Conversion Technology, *edited by G. M. A. van Beynum and J. A. Roels*

15. Food Chemistry: Second Edition, Revised and Expanded, *edited by Owen R. Fennema*
16. Sensory Evaluation of Food: Statistical Methods and Procedures, *Michael O'Mahony*
17. Alternative Sweeteners, *edited by Lyn O'Brien Nabors and Robert C. Gelardi*
18. Citrus Fruits and Their Products: Analysis and Technology, *S. V. Ting and Russell L. Rouseff*
19. Engineering Properties of Foods, *edited by M. A. Rao and S. S. H. Rizvi*
20. Umami: A Basic Taste, *edited by Yojiro Kawamura and Morley R. Kare*
21. Food Biotechnology, *edited by Dietrich Knorr*
22. Food Texture: Instrumental and Sensory Measurement, *edited by Howard R. Moskowitz*
23. Seafoods and Fish Oils in Human Health and Disease, *John E. Kinsella*
24. Postharvest Physiology of Vegetables, *edited by J. Weichmann*
25. Handbook of Dietary Fiber: An Applied Approach, *Mark L. Dreher*
26. Food Toxicology, Parts A and B, *Jose M. Concon*
27. Modern Carbohydrate Chemistry, *Roger W. Binkley*
28. Trace Minerals in Foods, *edited by Kenneth T. Smith*
29. Protein Quality and the Effects of Processing, *edited by R. Dixon Phillips and John W. Finley*
30. Adulteration of Fruit Juice Beverages, *edited by Steven Nagy, John A. Attaway, and Martha E. Rhodes*
31. Foodborne Bacterial Pathogens, *edited by Michael P. Doyle*
32. Legumes: Chemistry, Technology, and Human Nutrition, *edited by Ruth H. Matthews*
33. Industrialization of Indigenous Fermented Foods, *edited by Keith H. Steinkraus*
34. International Food Regulation Handbook: Policy • Science • Law, *edited by Roger D. Middlekauff and Philippe Shubik*
35. Food Additives, *edited by A. Larry Branen, P. Michael Davidson, and Seppo Salminen*
36. Safety of Irradiated Foods, *J. F. Diehl*

Other Volumes in Preparation

Omega-3 Fatty Acids in Health and Disease, *edited by Robert Lees and Marcus Karel*

Food Emulsions, Second Edition, Revised and Expanded, *edited by Kare Larsson and Stig Friberg*

Safety of Irradiated Foods

J. F. Diehl
Karlsruhe
Federal Republic of Germany

MARCEL DEKKER, INC. New York and Basel

Library of Congress Cataloging-in-Publication Data

Diehl, Johannes Friedrich
Safety of irradiated foods /J.F. Diehl
p. cm.--(Food science and technology ; 36)
ISBN 0-8247-8137-6 (alk. paper)
1. Irradiated foods--Health aspects. I. Title. II. Series: Food science and technology (Marcel Dekker, Inc.) ; 36.
[DNLM: 1. Food Irradiation--adverse effects. W1 FO509P v. 36 / WA 710 D559s]
RA 1258.D54 1990
363.19'2--dc20
DNLM/DLC
for Library of Congress 89-23277
CIP

This book is printed on acid-free paper.

MARCEL DEKKER, INC.
270 Madison Avenue, New York, New York 10016

Current printing (last digit):
10 9 8 7 6 5 4 3 2 1

PRINTED IN THE UNITED STATES OF AMERICA

To the memory of

Professor Johann Kuprianoff

(December 7, 1904—January 31, 1971),
Director of the Federal Research Center for
Food Preservation, Karlsruhe, from 1948 until 1969,
who pioneered food irradiation research in Germany.

Preface

In recent months food irradiation has been hotly debated in the United States Congress, in the state legislatures of Hawaii, Illinois, Maine, Massachusetts, Pennsylvania, New Jersey, New York, Vermont, and Oregon, in the European Parliament in Strasbourg, in the British House of Commons, and in several other parliaments. Some consumer organizations have vigorously opposed this new method of food preservation, others have approved it in principle but insisted on appropriate labeling of irradiated foods. While some proponents of food irradiation have described it as the greatest invention since Nicolas Appert developed food canning, opponents claim that its introduction could lead to widespread concealment of food contamination, a lowering of food quality standards, and an increased risk to public health.

Many of these claims and debates are characterized by exaggerations, misunderstandings, and muddled terminology. Radiation, irradiation, and radioactivity are frequently confused. Results of research carried out here years ago or in other countries are often completely ignored. Several good texts on food irradiation exist, but they treat the question of safe consumption of irradiated foods, which is of central interest in the public debate, briefly or not at all.

In this book I have sought to present a balanced view of the subject written for the nonspecialist, for individuals who have a limited background in chemistry, physics, and biology. A special effort has been made to explain the problems in simple and, wherever possible, nontechnical language without sacrificing accuracy (a glossary of terms is provided in Appendix III). The book is aimed at the educated citizen who wants to base his own political and economic decisions on rational choice. It should be of interest to scientists from neighboring disciplines who are looking for an up-to-date overview and for facilitated access to the vast literature on food irradiation, to decision-makers in the food industry, to government officials involved in regulating and/or controlling food manufacture and food trade, and to nutritionists, dieticians, and physicians who seek answers to questions asked by their clients. Perhaps it would be expecting too much if I hoped that one or the other parliamentarian will find the time to read this book before entering legislative debate on the subject.

The first four chapters can be regarded as being the introduction to the main topic, providing some basics of radiation technology, radiation chemistry, and radiation biology. Chapters 5 to 8 present the central issue: results of studies on the wholesomeness of irradiated foods and their evaluation. The final four chapters deal with practical aspects of food irradiation, regulation, and consumer acceptance.

Although it is impossible to list all those who have in some way contributed to this book and to whom I have become deeply indebted, there are several whose contributions were essential. First among these is my wife, Eva, who provided the understanding and encouragement needed for the completion of a task that, for a considerable period of time, has occupied me intensely. My secretary, Diana Inkster, worked tirelessly to type the many revisions of the manuscript; I would not miss this opportunity to applaud her skill and perseverance. Mrs. Edith Haller has prepared the illustrations expertly and has patiently accepted the many modifications that were required. Thanks are also due to Mrs. Isolde Lang for providing the photographs.

It is my pleasure to record the debt of gratitude that I owe to my colleagues, Dr. Henry Delincee, Prof. Peter Elias, Mr. Dieter Ehlermann and Dr. Theo Gruenewald, for counsel and advice, and for their critical review of the text.

J. F. Diehl

Contents

1

Introduction: How It All Began

The first documented proposal to use ionizing radiation "to bring about an improvement in the condition of foodstuffs" and in "their general keeping quality" was made in the United Kingdom over 80 years ago in British Patent No. 1609 (1905) issued to J. Appleby, miller, and A. J. Banks, analytical chemist. The inventors proposed the treatment of foods, especially cereals and their products, with alpha, beta, or gamma rays from radium or other radioactive substances. Remarkably, they stressed "the exceptionally marked advantage of an entire absence of the direct use or employment of chemical compounds" in this process. They suggested that the effects of radiation treatment were due to "chemical changes similar to those which occur in nature." However, the radium preparations suggested by these inventors as sources of ionizing radiation were not available in sufficient quantity to irradiate food commercially. Similar difficulties still existed in 1921, when B. Schwartz of the U.S. Department of Agriculture's Bureau of Animal Industry suggested the use of X-rays for inactivating trichinae in pork [1]; the X-ray machines available at that time were not powerful enough to treat pork in commercially interesting quantities.

The food laws of many countries apply also to tobacco products, and it is perhaps not too far-fetched to mention irradiation of a tobacco product in this context. Cigars can be attacked and badly damaged by the tobacco beetle, *Lasioderma serricorne*. This used to be a serious problem for the cigar industry. Many shipments of cigars had to be discarded because the product was criss-crossed by the feeding tunnels of the insect. G. A. Runner of USDA's Bureau of Entomology had demonstrated in 1916 that eggs, larvae, and adults of the tobacco beetle could be killed in cigars by X-rays

[2]. At the request of American Tobacco Co., an X-ray machine with a conveyor system for the irradiation of cigar boxes was built by American Machine and Foundry Co. in New York City and put into operation in 1929. A water-cooled X-ray tube with a maximal power of 30 mA at 200 kV was the radiation source. (The author is obliged to F. J. Bartlewski, the engineer who designed this facility, for communicating this information.) Although the treatment effectively prevented damage to the cigars, the machine turned out to be unsuitable for continuous use. Details can no longer be reconstructed, but it appears that the X-ray tubes then available were built for intermittent use in medical diagnosis and therapy, not for continuous use on a production line. At any rate, chemical fumigation later replaced this first industrial application of radiation processing.

A French patent was granted in 1930 to O. Wüst for an invention described by the words: "Foods of all kinds which are packed in sealed matallic containers are submitted to the action of hard (high-voltage) X-rays to kill all bacteria." However, the patent never led to a practical application.

New interest was stimulated in 1947 by a publication [3] of two expatriate German scientists, Arno Brasch and Wolfgang Huber, co-inventors of a pulsed electron accelerator, the Capacitron, and founders of Electronized Chemicals Corporation in Brooklyn, New York. They reported that meats and some other foodstuffs could be sterilized by high-energy electron pulses, that some foodstuffs, particularly milk and other dairy products, were susceptible to radiation and developed off-flavors, and that these undesirable radiation effects could be avoided by irradiation in the absence of oxygen and at low temperatures. With regard to cost efficiency, they concluded that irradiation "will not materially increase the final price of the treated product." At about the same time, J. G. Trump and R. J. van de Graaff of the Massachusetts Institute of Technology, who had developed another type of electron accelerator, also studied effects of irradiation on foods and other biological materials [4]. They collaborated in these studies with MIT's Department of Food Technology. The foundations of food irradiation research had been laid when B. E. Proctor and S. A. Goldblith reviewed these early studies in 1951 [5]. Surveying the available radiation, these authors concluded that neutron radiation could not be used because it would produce radioactivity in the irradiated food; alpha particles and ultraviolet light were ruled out because of their low penetration, as were X-rays because of insufficient power of available X-ray machines. The sterilization of one medium-sized can of food by X-rays would require 10 to 20 minutes, and that excluded X-rays from commercial application. This left only accelerated electrons, then called cathode rays. Gamma rays of radioactive isotopes were not even mentioned, presumably because suitable isotopes were not yet available on a sufficiently large scale.

At about this time, the United States Atomic Energy Commission (USAEC) initiated a coordinated research program on the use of ionizing radiation for food preservation and began to provide gamma radiation from spent-fuel rods of nuclear reactors as another source of ionizing radiation. Most of the food irradiation experiments carried out in the United States during the mid- and late 1950s made use of the spent-fuel rod source at the Argonne National Laboratory in Lemont, Illinois. However, the limitations of spent-fuel rod sources, especially with regard to exact dosimetry, became increasingly apparent, and ^{60}Co sources were provided instead by USAEC to several academic institutions. Experimental food irradiators with a source strength of approximately 30,000 curies of ^{60}Co were installed at the Massachusetts Institute of Technology; the University of California, Davis; the University of Washington, Seattle; and the University of Florida, Gainesville, in 1961–62. The Marine Products Development Irradiator (MPDI) with 235,000 curies of ^{60}Co was built for the National Marine Fisheries Service's research laboratory at Gloucester, Massachusetts, in 1964, and a Grain Products Irradiator with 35,000 curies ^{60}Co for the USDA's Entomological Research Center, Savannah, Georgia, in 1965. (Radioactivity should now be expressed in becquerel (Bq). One radioactive disintegration per second equals 1 Bq. However, the older unit curie (Ci) is still used to describe the source strength of gamma irradiators. One Ci equals 3.7×10^{10} Bq.)

The U.S. Army supported research on both low- and high-dose food irradiation from 1953 to 1960. After 1960 the Army concentrated its research efforts on developing radiation-sterilized meat products to substitute for canned or frozen military rations. During 1961–62 a large food irradiation laboratory was constructed at the U.S. Army Natick Laboratories in Natick, Massachusetts. It was equipped with a 1.3 million Ci ^{60}Co source and an 18 kW electron linear accelerator. The ready availability of gamma sources and electron accelerators in many parts of the United States then provided opportunities for food irradiation research and development work of which the earlier advocates of the use of X-ray machines could have only dreamt.

Reports from the United States about successful experiments on food irradiation stimulated similar efforts in other countries. In the United Kingdom investigation of the effects of radiation on food began in 1950 at the Low Temperature Research Station at Cambridge and somewhat later at the Wantage Research Laboratories of the Atomic Energy Research Establishment. By the mid- or late 1950s national research programs on food irradiation were also underway in Belgium, Canada, France, the Netherlands, Poland, the Soviet Union, and the Federal Republic of Germany. This early history of food irradiation has been reviewed by Goldblith [6], Goresline [7], and Josephson [8].

The first commercial use of food irradiation occurred in 1957 in the Federal Republic of Germany, when a spice manufacturer in Stuttgart began to improve the hygienic quality of his products by irradiating them with electrons using a Van de Graaff generator [9]. The machine had to be dismantled in 1959 when a new food law prohibited the treatment of foods with ionizing radiation, and the company turned to fumigation with ethylene oxide instead.

In Canada irradiation of potatoes for inhibition of sprouting was allowed in 1960, and a private company, Newfield Products Ltd., began irradiating potatoes at Mont St. Hilaire near Montreal, in September 1965. The plant used a ^{60}Co source and was designed to process some 15,000 t of potatoes a month. It closed after only one season, when the company ran into financial difficulties [10].

In spite of these setbacks, interest in food irradiation grew worldwide. At the first International Symposium on Food Irradiation at Karlsruhe, organized by the International Atomic Energy Agency (IAEA), representatives from 28 countries reviewed the progress made in research laboratories [11]. However, health authorities in these countries still hesitated to grant permissions for marketing irradiated foods. At that time only three countries, Canada, the United States, and the Soviet Union, had given clearance for human consumption of a total of five irradiated foods, all treated with low radiation doses. The food industry had not yet made use of the permissions. Irradiated foods were still not marketed anywhere.

Questions about the safety for human consumption of irradiated foods were still hotly debated, and this was recognized as the major obstacle to commercial utilization of the new process. As a result of this recognition the International Project in the Field of Food Irradiation was created in 1970, with the specific aim of sponsoring a worldwide research program on the wholesomeness of irradiated foods. Under the sponsorship of the IAEA in Vienna, the Food and Agriculture Organization (FAO) in Rome, and the Organization for Economic Cooperation and Development (OECD) in Paris, 19 countries joined their resources, this number later growing to 24. The World Health Organization (WHO) in Geneva became associated with the International Project in an advisory capacity. Animal feeding studies contracted by the International Project with various laboratories, mostly in the United States, United Kingdom, and France, involved irradiated wheat flour, potatoes, rice, iced ocean fish, mangoes, spices, dried dates, cocoa beans, and legumes. The research programs also included short-term screening tests and evaluation of chemical changes in irradiated foods.

The results obtained in the framework of the International Project and in numerous national testing programs were repeatedly evaluated by the FAO/IAEA/WHO Joint Expert Committee on Irradiated Foods,

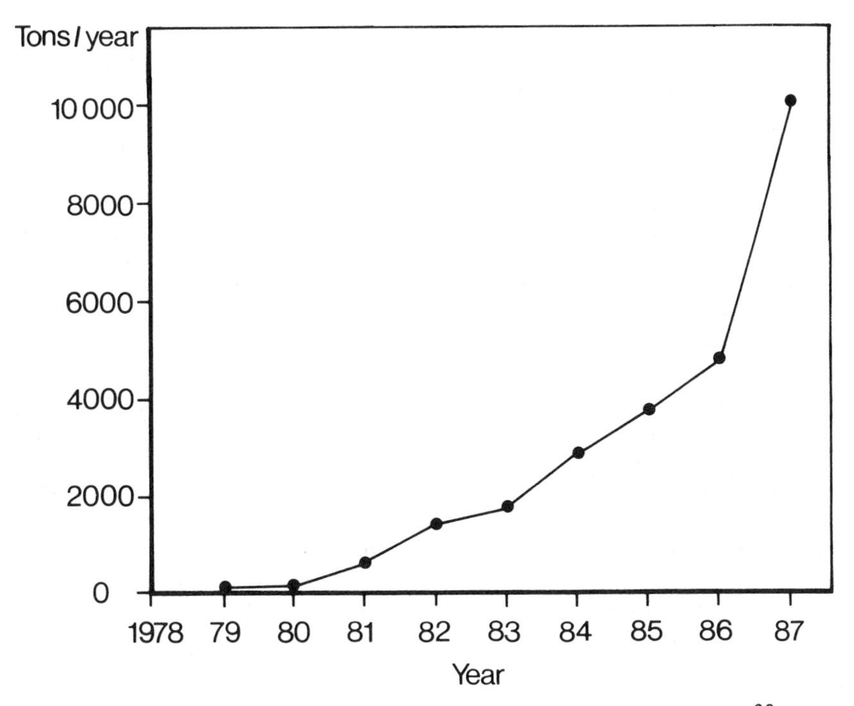

Figure 1 Quantity of foodstuffs irradiated per year in the ^{60}Co-facility GAMMIR II in Belgium. (Adapted from Ref. 13.)

the internationally recognized arbiters in this field. At its session in November 1980 this Committee concluded "that the irradiation of any food commodity up to an overall average dose of 10 kGy presents no toxicological hazard; hence, toxicological testing of foods so treated is no longer required" [12].

As a result of this decision, many national governments have approved the marketing of a growing number of irradiated foods. Not in all of these countries has this led to commercial use of the process. In other countries many thousand tons of foods are irradiated every year, and the trend is increasing. This is demonstrated for Belgium in Figure 1.

In other countries some political and consumer organizations are vigorously opposing the introduction of this new method of food processing, for various reasons. Unfortunately, the public debate about the presumed virtues or disadvantages of food irradiation is often characterized by a lack of solid information on this subject. Both opponents and proponents of food irradiation have been sources of misinformation or valid information presented in a misleading way

[14]. In this book an attempt will be made to provide factual data as a basis for a more rational approach to these controversies.

There is voluminous scientific literature on food irradiation, but it is not easy to come by because contributions have come from so many disciplines. Relevant reports have been published in journals of food technology, microbiology, analytical chemistry, food chemistry, radiation chemistry, radiation physics, toxicology, health physics, and some other fields. In order to facilitate access to this literature a computerized irradiation information database called IRREFCO (Irradiation Reference Collection) has been installed recently at the National Agricultural Library in the United States. A bibliography on food irradiation has been prepared since 1955 by the Federal Research Center for Nutrition, Karlsruhe, Federal Republic of Germany, and now contains close to 9000 documents. The whole database is processed and stored on computer and is also available in printed form. In recent years one issue of the printed bibliography has been published annually, each with 400 to 600 references [15]. In the following chapters only a small fraction of these documents can be mentioned. The author endeavors to quote primarily those studies that will guide the reader to key issues, to review articles, and to other works showing a path to the remaining literature. Some of the references are mentioned primarily for documentation, for instance, to give proper credit to the sources from which data in tables and figures were taken.

REFERENCES

1. Schwartz, B., Effects of X-rays on trichinae, *J. Agric. Res.*, 20:845 (1921).
2. Runner, G. A., Effect of Roentgen rays on the tobacco or cigarette beetle and the results with a new form of Roentgen tube, *J. Agric. Res.*, 6:383 (1916).
3. Brasch, A., and W. Huber, Ultrashort application time of penetrating electrons: a tool for sterilization and preservation of food in the raw state, *Science*, 105:112 (1947).
4. Trump, J. G., and R. J. van de Graaff, Irradiation of biological materials by high energy roentgen rays and cathode rays, *J. Appl. Physics*, 19:599 (1948).
5. Proctor, B. E., and S. A. Goldblith, Food processing with ionizing radiation, *Food Technol.* (Chicago), 5:376 (1951).
6. Goldblith, S. A., Historical development of food irradiation, in *Food Irradiation*, Proceedings of a Symposium, Karlsruhe, June 1966, Internat. Atomic Energy Agency, Vienna, 1966, p. 3.
7. Goresline, H. E., Historical aspects of the radiation preservation of food, in *Preservation of Food by Ionizing Radiation*,

E. S. Josephson and M. S. Peterson, eds., CRC Press, Boca Raton, FL, 1982, vol. 1, p. 2.

8. Josephson, E. S., An historical review of food irradiation, *J. Food Safety*, 5:161 (1983).
9. Maurer, K. F., On sterilization of spices (in German), *Ernährungswirtschaft*, 5:45 (1958).
10. Masefield, J., and G. R. Dietz, Food irradiation: The evaluation of commercialization opportunities, *Crit. Revs. Food Sci. Nutrit.*, 19:259 (1983).
11. IAEA, *Food Irradiation*, Proceedings of a Symposium, Karlsruhe, June 1966, Internat. Atomic Energy Agency, Vienna, 1966.
12. WHO, *Wholesomeness of Irradiated Foods*, Technical Report Series 659, Geneva, 1981.
13. Constant, R., and J. P. Lacroix, Irradiation in the service of mankind (in French), *Revue IRE Tijdschrift*, 11:(1) 2 (1987).
14. Pauli, G. H., and C. A. Takeguchi, Irradiation of foods—an FDA perspective, *Food Revs. Internat.*, 2:79 (1986).
15. Diehl, J. F., ed., Bibliography on Irradiation of Foods, no. 32, Federal Research Centre for Nutrition, Karlsruhe, 1988.

2

Radiation Sources and Process Control

I. TYPES OF RADIATION

The types of radiation discussed in this book are called ionizing radiations because they are capable of converting atoms and molecules to ions by removing electrons. Ionizing radiations can be energetic charged particles, such as electrons, or high-energy photons, such as X-rays or gamma rays. Not all types of ionizing radiation are suitable for irradiation of foods, either because they do not penetrate deep enough into the irradiated material (this applies, for instance, to alpha particle radiation) or because they make the irradiation material radioactive (this applies to electrons and X-rays of very high energy).

The FAO/IAEA/WHO Joint Expert Committee on Irradiated Foods considered only the following types of ionizing radiation as appropriate for irradiation of foods, and the Codex General Standard for Irradiated Foods (see Appendix I) has followed that recommendation:

gamma rays from the radionuclides ^{60}Co or ^{137}Cs
X-rays generated from machine sources operated at or below an energy level of 5 MeV
electrons generated from machine sources operated at or below an energy level of 10 MeV

The eV (electronvolt) is the unit of energy used to measure and describe the energy of electrons and of other types of radiation. The energy of 1 eV is equivalent to the kinetic energy acquired by an electron on being accelerated through a potential difference of 1 volt. The eV is a very small unit of energy. It is therefore more common to speak of keV (kiloelectronvolt = 1000 eV) or MeV (mega-

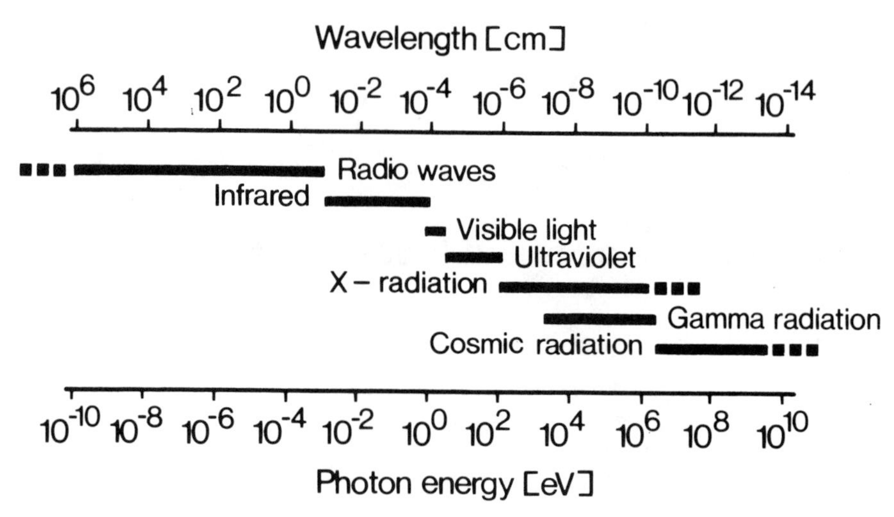

Figure 1 The electromagnetic spectrum.

electronvolt = 1 million eV). To convert eV to units of energy, one can use the conversion 1 MeV = 1.602×10^{-13} J (joule).

Gamma rays and X-rays are part of the electromagnetic spectrum (Fig. 1), which reaches from the low-energy, long wavelength radio waves to the high-energy, short wavelength cosmic rays. Radio waves, IR waves, and visible light are nonionizing radiations. UV light can ionize only certain types of molecule under specific conditions and is generally not considered as ionizing radiation. X-rays and gamma rays are identical in their physical properties and in their effect on matter; they differ only in their origin, X-rays being produced by machines, gamma rays coming from radioactive isotopes (radionuclides).

When ionizing radiation penetrates into a medium (for instance, the irradiated food), all or part of the radiation energy is absorbed by the medium. This is called the *absorbed dose*. The unit in which the absorbed dose is measured is the gray (Gy); it is equal to the absorption of 1 J/kg. One kGy(kilogray) = 1000 Gy. Formerly the dose unit rad was used. It was defined as 100 erg/g. The conversion of old to new units is based on the relationship 100 rad = 1 Gy, or 1 krad = 10 Gy, or 1 Mrad = 10 kGy.

The energy absorbed per unit of time is called the *dose rate*. Gamma ray sources provide a relatively low dose rate (typically 100 to 10,000 Gy/hr), whereas electron accelerators provide a high dose rate (10^4 to 10^9 Gy/sec). As a consequence, to achieve a specified absorbed dose, irradiation with a gamma source may take many hours, while irradiation with an electron accelerator may take only seconds or minutes.

Table 1 Units of Radiation Dose and Radioactivity

	Absorbed dose	Radioactivity
Unit	gray (Gy)	becquerel (Bq)
Definition	1 Gy = 1 J/kg	1 Bq = 1 disintegration/sec
Old unit	rad	curie (Ci)
Conversion	1 rad = 0.01 Gy	1 Ci = 3.7×10^{10}Bq = 37 GBq
	1 krad = 10 Gy	1 kCi = 37 TBq
	1 Mrad = 10 kGy	1 MCi = 37 PBq

As mentioned in Chapter 1, radioactivity is measured in Bq (becquerel), formerly in Ci (curie). These units are summarized in Table 1. Because the Bq is an extermely small unit, the following larger units of radioactivity are also used.

10^3 Bq = 1 kBq (kilobecquerel)

10^6 Bq = 1 MBq (megabecquerel)

10^9 Bq = 1 GBq (gigabecquerel)

10^{12} Bq = 1 TBq (terabecquerel)

10^{15} Bq = 1 PBq (petabecquerel)

(The "new units" described here and used throughout this book are those of the International System now recommended by all international scientific bodies. The change from Ci to Bq or from rad to Gy corresponds to that from the calorie to the joule.)

II. GAMMA SOURCES

Before more is said about radioactive elements and the radiations they give off, a short introduction to *atomic structure* may be useful. All chemical elements consist of submicroscopic particles, called atoms. Atoms consist of a nucleus containing positively charged protons and uncharged neutrons, surrounded by negatively charged electrons. The electrons are arranged in shells or orbits. In general, it is the number and arrangement of electrons in the outermost shell that determine how an element behaves chemically. Normally the number of electrons in an atom is equal to the number of protons. If an electron is removed from the outer shell, the atom has lost one negative charge. It is now a positively charged ion, or cation. In contrast, if an electron is added to the outer shell, the atom becomes a negatively charged ion, or anion.

For the atom to be stable, the protons and neutrons in the nucleus must be present in a certain ratio, which is 1:1 for the lighter elements and closer to 1:1.5 in heavy elements. The *atomic number* is determined by the number of protons, the *mass number* by the sum of protons and neutrons. Helium, for instance, having 2 protons and 2 neutrons, has the atomic number 2 and the mass number 4. (The *atomic weight* is approximately equal to the mass number.)

An element can exist as different *isotopes,* which all have the same number of protons (i.e., the same atomic number) but different numbers of neutrons (i.e., different atomic weights). Some of these isotopes, particularly among the heavy elements, are unstable or radioactive. A number of such *radioisotopes* exist, and have always existed, in nature. An example among the lighter elements is potassium, an important element in the human body and in all biological systems. It consists mostly of the stable isotope $^{39}_{19}K$, which contains 19 protons and 20 neutrons in its nucleus. But natural potassium also contains 0.0118% $^{40}_{19}K$, which has one more neutron in its nucleus and is unstable. An example of an unstable heavy isotope in nature is $^{238}_{92}U$, or uranium-238. It has 92 protons and 146 neutrons.

The disintegration of radioisotopes is associated with the release of one or more of the following types of radiation:

α *particles*: fast-moving helium nuclei consisting of 2 protons and 2 neutrons

β^+ *and* β^- *particles*: high-speed positrons or electrons

γ *ray photons*: electromagnetic wave packets moving at the speed of light

neutrons: particles with no charge and a mass close to that of protons

The rate of radioactive decay is characterized by the radioactive *half-life*, the time required for disintegration of one-half the atoms of a given radioactive substance. Potassium-40 has a half-life of 1.83×10^9 years. Giving off gamma radiation with an energy of 1.46 MeV and beta radiation with a maximum energy of 1.3 MeV, it is transformed to the stable elements $^{40}_{18}Ar$ (argon) and $^{40}_{20}Ca$ (calcium).

With a half-life of 4.5×10^9 years and emission of alpha particles, $^{238}_{92}U$ is converted to $^{234}_{90}Th$ (thorium), which is itself unstable. Through a series of further steps, $^{206}_{82}Pb$ (lead) is formed as the stable end product.

In addition to these naturally occurring radionuclides, atomic bomb explosions and nuclear energy production have created anthropogenic (man-made) radioactive materials on earth. When a uranium nucleus is split in a nuclear fission reaction, many lighter elements (*fission products*) are formed. Most of these have too many neutrons and are therefore unstable.

Typical fission products are ^{89}Sr and ^{90}Sr (strontium), ^{89}Kr (krypton), ^{133}X (xenon), and ^{137}Cs (cesium), but there are many others. Some of these have half-lives of only seconds, others have half-lives of many years.

As mentioned in the previous chapter, many experiments on food irradiation in the 1950s were carried out with spent-fuel rods from nuclear reactors. Such fuel rods contain a mixture of many fission products, with greatly differing half-lives, emitting different types of radiation with different energies. The composition of fuel rods changes all the time because the radionuclides with short half-lives disappear quickly, those with longer half-lives remain. Although fuel rods are primarily a source of gamma radiation (the less penetrating alpha and beta radiation are absorbed by the steel hull of the rods), they do give off some neutrons. Since the latter can produce radioactivity when they interact with matter such as food, fuel rods have not been used for irradiation of foods since the early 1960s. Because of their constantly varying composition, fuel rods also make dosimetry difficult, and this was another reason for abandoning their use.

Individual constituents of fuel rods can be separated in reprocessing plants by chemical methods. One of the radionuclides obtainable in this way is ^{137}Cs. With a half-life of 30 years and emission of gamma radiation (0.66 MeV) and beta radiation (0.51 MeV and 1.18 MeV), $^{137}_{55}Cs$ decays to stable $^{137}_{56}Ba$ (barium). After the ^{137}Cs is separated from the other constituents of the fission waste in the form of CsCl, it is triply encapsulated in stainless steel containers because CsCl is soluble in water. If it leaked out it could cause contamination of the environment. As provided by the Waste Encapsulation and Storage Facility (WESF) at Hanford, Washington, the ^{137}Cs capsule is 400 mm in active length (500 mm in total length) and 67 mm in diameter.

There are few reprocessing plants in the world, and the capacity for extracting ^{137}Cs from spent fuel rods is very limited. Plans for building several commercial reprocessing facilities in the United States were cancelled by President Carter's 1977 decision that the United States would not engage in commercial reprocessing of spent nuclear fuel. As a consequence, not much ^{137}Cs is available, and there are not many gamma radiation facilities that use ^{137}Cs.

Instead, ^{60}Co has become the choice for gamma radiation sources. It is not a fission product. In order to obtain ^{60}Co, slugs or pellets of nonradioactive ^{59}Co are bombarded with neutrons in nuclear reactors for one to one and a half years. This can be done economically in only certain types of nuclear reactors, such as the CANDU reactors in Canada. Over 80% of the ^{60}Co available on the world market is being produced in Canada. For use as a radiation source, the activated cobalt pellets are encapsulated in a stainless steel liner in the form of a pin or pencil, 450 mm in length and 12.5 mm in diameter, as provided by AECL (Atomic Energy of Canada, Ltd.). The reason for this long, thin design of the source pins is the desire to minimize self-absorption of the cobalt and to minimize heat buildup. The ^{60}Co is present in a water-insoluble form and thus presents very little risk of environmental contamination. With a half-life of 5.27 years and with emission of gamma radiation (1.17 and 1.33 MeV) and beta radiation (0.31 MeV), $^{60}_{27}$Co disintegrates to stable $^{60}_{28}$Ni (nickel). An annual replenishment of 12.4% is needed if a ^{60}Co source is to be kept at its original strength. The basic design of a gamma irradiation facility, viewed from above, is shown in Figure 2. Goods to be irradiated enter the concrete-shielded irradiation chamber through a labyrinth, which prevents radiation from reaching the work area where the goods are loaded onto and unloaded from the conveyor system.

A commercial ^{60}Co irradiation facility designed by AECL is shown in Figure 3. When the facility is not in operation, the ^{60}Co is stored under water in a depth of about 6 m ("source storage pool"). Personnel can now safely enter the irradiation chamber, for instance, to carry out repairs on the transport mechanism. Only when the

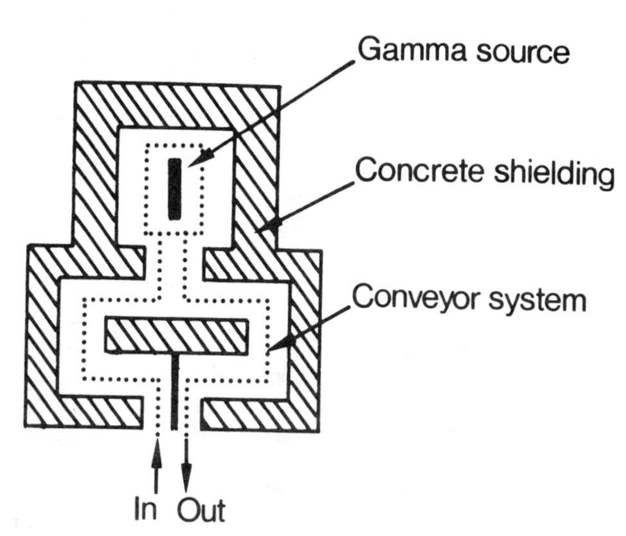

Figure 2 Basic design of a gamma irradiation facility, viewed from above.

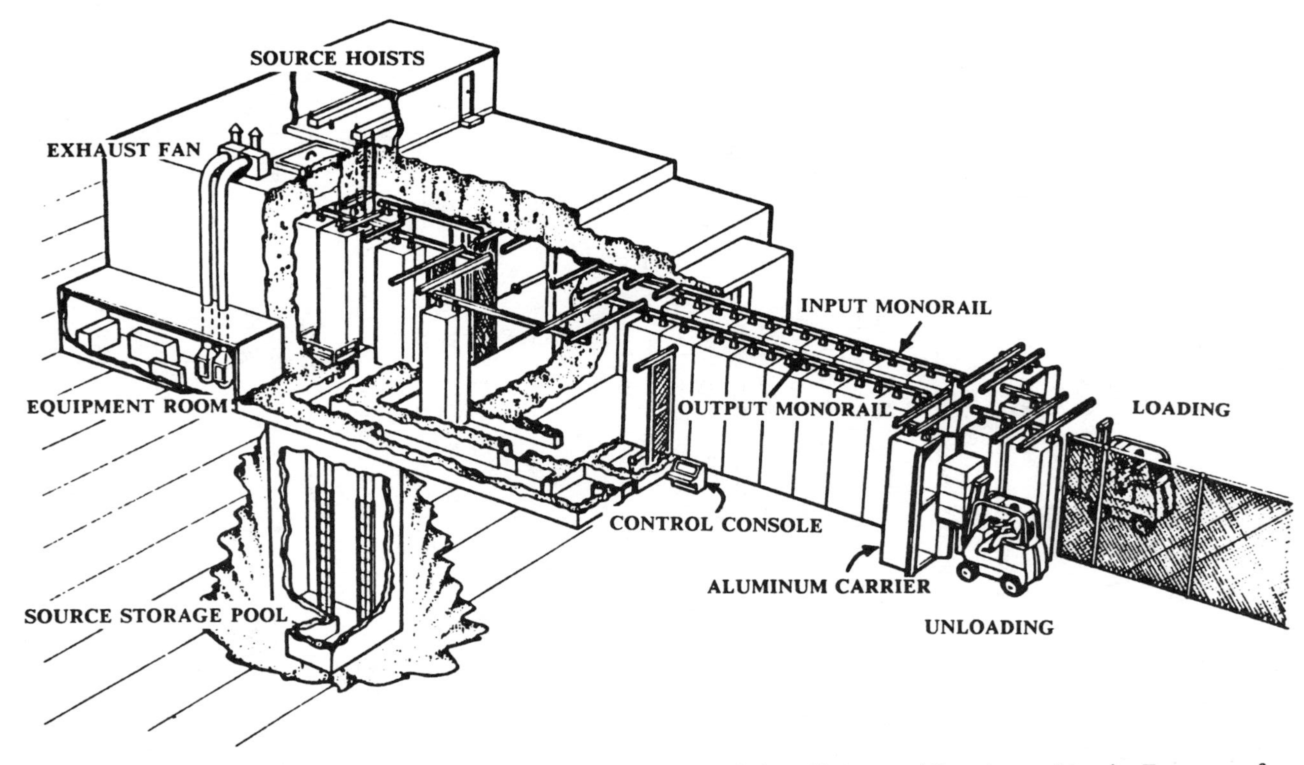

Figure 3 Drawing of an automatic carrier type cobalt-60 irradiator. (Courtesy Atomic Energy of Canada Limited, Ottawa.)

personnel has left the irradiation chamber and all safety devices have been activated can the source hoists lift the rack with the ^{60}Co pins into the irradiation position above the water level. Material to be irradiated is placed into the aluminium carriers and moved into the concrete-shielded irradiation chamber. The thickness of the concrete shield is usually about 1.5–1.8 m. The carriers move around the ^{60}Co rack, and they turn around their own axis so that their contents are irradiated equally from two sides. The radiation dose absorbed by the material contained in the carrier is determined by the strength of the cobalt source, i.e., by the number of curies or becquerels, and by the dwell time of the carrier in the irradiation position. In most irradiators the source strength cannot easily be modified, and the radiation dose is only determined by the dwell time of the carrier. The length of time that each carrier spends in the irradiation position is preset at the control console and is registered by an automated printout system. The printout also contains a description and batch number of the material contained in the carrier. It is thus possible at any time to verify which material has been irradiated when and for how long. Together with the information on source strength this allows an estimation of the absorbed dose. Additionally, the absorbed dose must be measured by placing dose meters into the carriers, as described in Sec. V.

If all of the gamma radiation coming from the ^{60}Co source were absorbed by the irradiated goods, the radiation facility would have 100% *irradiation efficiency*. This is of course impossible because some of the radiation will be absorbed by the concrete shielding, by the carrier metal, and by the water below the source rack. Well-designed gamma facilities have an efficiency of about 30%. Designs different from that shown in Figure 3 may use tote boxes on a conveyor belt instead of carriers hanging from a monorail, or the carriers may move around the source in a double loop instead of a single loop. Storage of the radioactive source in air in a concrete-shielded or lead-shielded cask instead of the water pool is also possible but is not often practiced.

In principle, a ^{137}Cs irradiator operates exactly like a ^{60}Co irradiator. Somewhat less concrete shielding (1.2 m) is needed because the 0.66 MeV gamma rays of ^{137}Cs are less penetrating than the 1.33 MeV of ^{60}Co. The longer half-life of ^{137}Cs means that, to keep the source at its original strength, only 2.3% per year have to be replaced instead of the 12.4% in the case of ^{60}Co. Due to the required triple encapsulation in steel, which absorbs some of the radiation, and the relatively large diameter of the capsules, which causes considerable self-absorption, only about 70% of the gamma radiation emitted by the ^{137}Cs escapes from the source geometry and is available for irradiation. In contrast, the thinner ^{60}Co source pins make 95% of their gamma radiation available. The lower efficiency of ^{137}Cs partly offsets the advantage of the lower annual replacement.

A ^{60}Co irradiator as shown in Figure 3 may be loaded with 3 MCi of ^{60}Co (or 111 PBq) or even more. How much product can be irradiated with such a facility per hour or per year? As mentioned earlier, each radioactive decay of a ^{60}Co atom produces one gamma photon of 1.17 MeV and another one of 1.33 MeV. Together that is 2.5 MeV. A 3 MCi or 111 PBq source produces $2.5 \times 111 \times 10^{15}$ MeV/sec = 2.775×10^{17} MeV/sec.

The energy dose absorbed by the irradiated product is measured in Gy. The different energy units can be converted using the equation 1 Gy = 6.242×10^{12} MeV/kg. If we assume that all of the gamma energy emitted from the 3 MCi cobalt source were absorbed by the irradiated product (100% efficiency), this would mean 10 Gy in 4.44 t in 1 sec or 10 kGy in 16 t in 1 hr. Realistically we may assume 25% efficiency, and that means 4 t irradiated with a dose of 10 kGy in 1 hr. If this facility was in operation 10 months per year, 24 hr per day, it could irradiate about 29,000 t at the dose level of 10 kGy, or 290,000 t at the dose level of 1 kGy. In actual practice one would have to consider factors such as the maximum speed of the conveyor system and the logistics of transporting the products to the irradiator and removing them after irradiation. The purpose of this exercise was only to show how one can estimate the throughput of a cobalt facility for a given source strength and radiation dose.

In the case of ^{137}Cs, the gamma energy per decay is only 0.66 MeV—less than a quarter of the ^{60}Co decay. Moreover, the higher self-absorption of the cesium capsules makes only about 70% of the cesium activity available. As a result, a ^{137}Cs irradiator having the same source strength as a ^{60}Co irradiator provides only about a seventh of the dose rate or throughput.

A survey of large-scale gamma irradiators in the United States carried out in 1981 showed 32 million installed curies of ^{60}Co and 2 million installed curies of ^{137}Cs. All available information indicates that the share of ^{137}Cs has not significantly increased since that time. In other countries the use of ^{137}Cs is also low in comparison with the use of ^{60}Co.

Opposition to food irradiation is often based on the notion that the existence of gamma irradiation facilities would cause a growing accumulation of radioactive waste material. This is a misconception. Cobalt-60 is converted to nonradioactive nickel, ^{137}Cs to nonradioactive barium. The sale of ^{60}Co or ^{137}Cs is based on a contract which commits the supplier to take back the cobalt pins or cesium capsules when the radioactivity has fallen below a certain level. This makes space in the source rack for installing new full-strength ^{60}Co pins or ^{137}Cs capsules. Neither the concrete shielding nor the parts of the conveyor system nor the water of the source storage pool becomes radioactive. For the operator of a gamma irradiation facility there is no radioactive waste problem. If for any reason a gamma facility should be closed, no radioactivity would remain in the building once the supplier had taken out the gamma source. The building could be safely razed without any need for special precautions. Descriptions of gamma irradiators as something akin to a nuclear reactor, as are often heard in the public debate about food irradiation, are quite misleading.

The task of handling the radioactive waste returned by the users thus remains for the supplier. With the few ^{137}Cs facilities in existence this has not yet become a problem because of the long half-life (30 years) of this radionuclide. When used cesium capsules are eventually returned, the remaining ^{137}Cs can be separated from the nonradioactive barium in a reprocessing plant and can be reused.

^{60}Co pins may be returned after three or four half-lives, i.e., after 16 to 21 years, when they have decayed to about 6–12% of their original strength. Reactivation of depleted ^{60}Co sources by inserting them into a reactor for the second time is technically feasible as they still consist primarily of cobalt-59, mixed with some remaining ^{60}Co and some ^{60}Ni. If, however, reactivation is not the preferred alternative, the depleted source can be disposed of. Atomic Energy of Canada Ltd. (AECL) estimated in March 1988 that approximately 100 MCi of ^{60}Co they had supplied was in service, representing more than 22,000 source pencils. If all of this ^{60}Co was gathered in one stack, it would occupy a space of 1.25m^3, the size of a small office desk. AECL deposits radioactive waste material at its Chalk River disposal site.

For transport from the supplier to the user (and back to the supplier when most of the source strength is exhausted), ^{60}Co and ^{137}Cs are shipped in special steel casks. These are designed to survive severe traffic accidents, fires, or other types of disasters.

In order to minimize corrosion of cobalt pins or cesium capsules in the irradiation facility, the water of the source storage pool is constantly recirculated through a deionizer. A sensitive radiation monitor close to the deionizer is used to detect any radioactivity in case the water should be contaminated by a leak in the source.

Some 150 commercial gamma irradiators exist all over the world, and many of these have been in operation for more than 20 years. They are mostly used for radiation-sterilization of medical disposables, and only in a few countries they are also used for irradiation of foodstuffs. All of these facilities are regulated by strict radiation protection laws, and they have a safety record that compares very favorably with that of other industries.

Details on the design of various types of gamma irradiation facilities can be found in a number of publications [1–5].

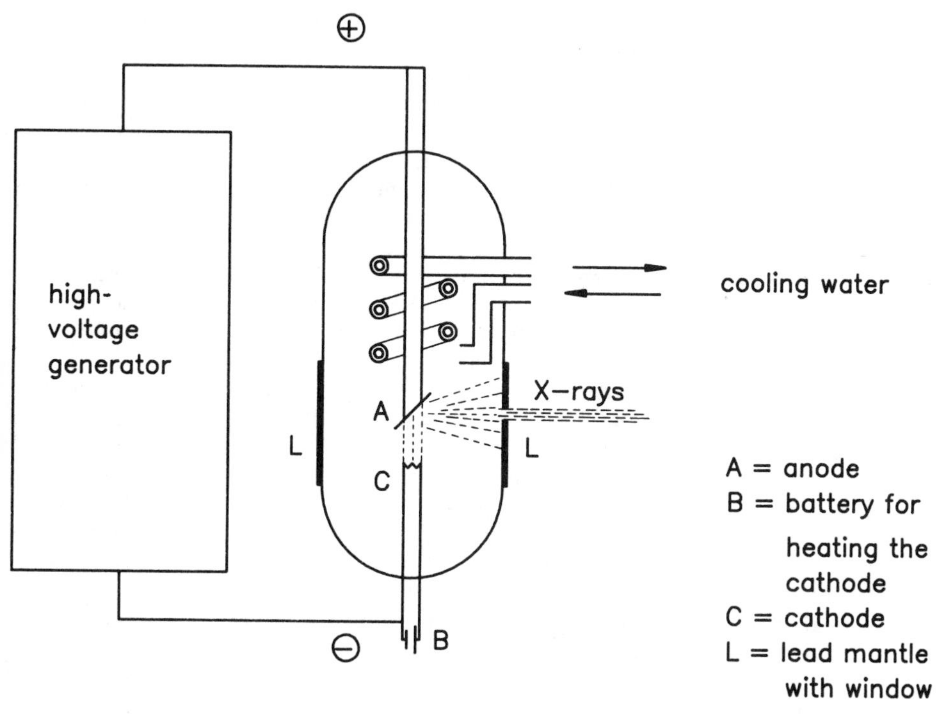

Figure 4 Principal parts of an X-ray machine.

III. MACHINE SOURCES

Some of the earliest systematic studies on food irradiation were carried out with *X-ray machines*. X-rays are produced when matter is bombarded by electrons of sufficiently high kinetic energy. An X-ray tube (Fig. 4) consists basically of an evacuated glass bulb containing a wire filament, which can be heated and opposite it a piece of metal, usually tungsten, which can be cooled. The filament is given a negative charge (cathode) and the metal a positive charge (anode, or anticathode). When the cathode filament is heated by passing an electric current through it, it emits electrons. When these hit the anode, X-rays are produced. The anode must be cooled because in a typical X-ray tube about 99% of the energy carried by the electrons may be converted to heat, and only about 1% appears in the form of X-rays. The energy or wave length of the X-rays (see Fig. 1) is determined by the potential difference between cathode and anode. The X-ray tube together with the high-voltage generator that produces this potential difference are the essential parts of an X-ray machine. For laboratory-scale experiments on food irradiation, X-ray machines are still used occasionally, but they have never found use in commercial food irradiation.

Electron accelerators capable of commercial-scale irradiation became available in the 1950s and have been much improved since then. Various designs are now available which are used for the sterilization of medical supplies and packaging materials, radiotherapy, curing of wire and cable insulations, removal of toxic components from exhaust gases, and many other applications.

The effective penetration range of an electron beam depends on its energy level. Low-energy (up to 300 keV) and medium-energy (300 keV to 1 MeV) electron beam accelerators can be used only for treating thin-layer materials, because these electrons penetrate a thickness of not more than a few mm. With regard to food processing, they are only of interest for the sterilization of films or sheets of packaging material.

High-energy electron beam accelerators produce electrons with energies above 1 MeV. For purposes of food irradiation, 10 MeV is the upper limit. As a rule of thumb, the depth of penetration of an electron beam in most foodstuffs is 5 mm per MeV. A 10 MeV electron beam can thus be used for irradiation of thicknesses up to 5 cm if irradiated from one side or 10 cm if irradiated from two sides.

There are two basic designs of high-energy electron accelerators. One is the DC (direct current) accelerator, the other the microwave or RF (radio frequency) linear accelerator (linac) [6]. In both types, electrons are accelerated to very high speed, close to the speed of light, in an evacuated tube. In DC accelerators this is achieved by applying a high voltage across the terminals to this tube. Electrons emitted form an electron source, a heated wire or an "electron gun," are pushed away from the negative end of the tube and attracted by the positive end. The higher the potential difference, the higher the speed attained by the electrons.

Various makes of DC accelerators, such as the ICT® (Insulating Core Transformer) or the Dynamitron®, differ primarily in the way the high voltage is produced. The Dynamitron utilizes a cascade rectifier system in which all rectifiers are driven in parallel from a high frequency oscillator. For optimal insulation the high voltage generator is contained in a pressurized tank filled with SF_6 (sulfurhexafluoride) gas. The direct current from the rectifiers establishes a DC potential of up to about 5 MV. The monoenergetic electron beam thus developed is essentially of constant current. Energy and current determine the output capacity of an electron accelerator. For example, a 4 MeV beam of 25 mA will produce 100 kW of beam power. The manufacturer of the Dynamitron, Radiation Dynamics Inc., Melville, NY, offers various models, such as a 150 kW/3 MV, a 200 kW/4 MV, and a 100 kW/5 MV [7]. Higher voltage means

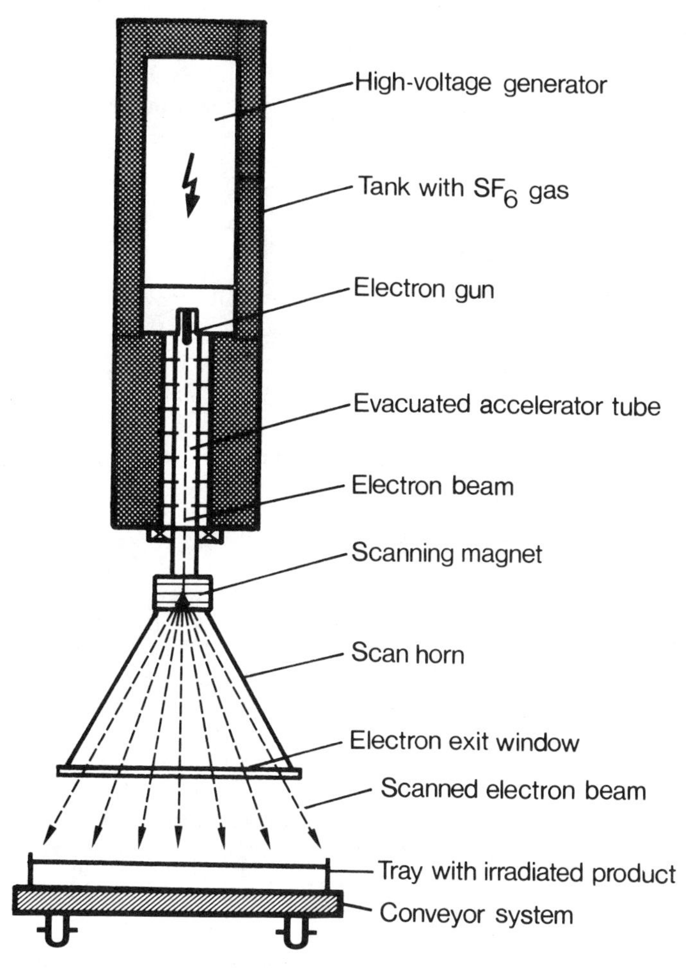

Figure 5 Basic design of a DC electron accelerator.

greater depth of penetration; higher beam power means higher throughput of material to be irradiated.

As shown schematically in Figure 5, the electron beam is rythmically deflected by a scanning magnet attached to the end of the accelerator tube. With a frequency of 100–200 Hz the beam is moved to and fro like a pendulum. The reason for this is that the material to be irradiated must, as nearly as possible, receive the same radiation dose in all its parts. The electron beam has a diameter of only a few mm or cm. Without scanning, the electron beam would concentrate the whole beam on a very small area of the ir-

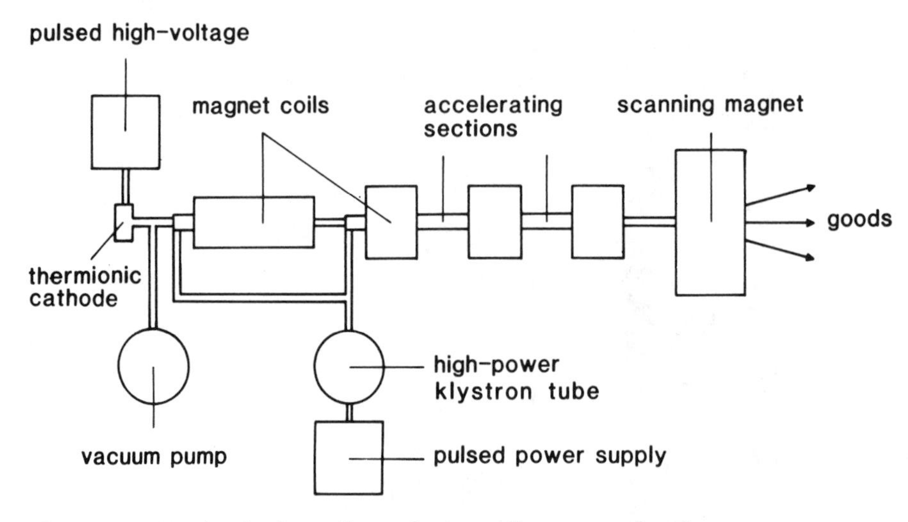

Figure 6 Basic design of an electron linear accelerator.

radiated goods. By scanning the electron beam and simultaneously moving the material to be irradiated perpendicular to the scanning line of the electron beam, an even distribution of radiation energy to the irradiated material can be achieved.

It is very difficult to handle accelerating potentials higher than 5 MV in DC accelerators, whereas linacs can produce electron energies even higher than the 10 MeV allowed for food irradiation. In a linac (Fig. 6), pulses or bunches of electrons produced at the thermionic cathode are acceleratred in an evacuated tube by driving radio frequency (RF) electromagnetic fields along the tube. The electrons ride on a traveling electromagnetic wave, comparable with a piece of wood or a surfer riding the crest of a water wave. The RF generator is designated high-power klystron tube in Figure 6. The electron beam leaving the accelerator tube can be scanned in the same way as for DC accelerators. Linac electrons are also monoenergetic, but the beam is pulsed rather than continuous. An electron pulse of a few microseconds duration may be followed by a dead time of a few milliseconds—different manufacturers use different timings. When the dose rate provided by a pulsed electron beam is described, it is important to indicate whether this is the pulse dose rate or the overall dose rate of pulse plus dead time. Linac designs other than the traveling wave type described here are available [6]. Progress in electron beam irradiation technology has been outlined by several authors [8–9].

A number of linacs have been built specifically for food irradiation studies. Some of these could also be operated in the energy range above 10 MeV for experiments designed to find out at which energy level radioactivity can be induced in the irradiated medium. A linac with a maximum energy of 25 MeV was commissioned for the U.S. Army Natick Research and Development Laboratories in 1963. Its beam power was 6.5 kW at an electron energy of 10 MeV.

Assuming 100% efficiency, a 1 kW beam can irradiate 360 kg of product with a dose of 10 kGy in one hour. The efficiency of electron accelerators is higher than that of gamma sources because the electron beam can be directed at the product, whereas the gamma sources emit radiation in all directions. An efficiency of 50% is a realistic assumption for accelerator facilities. With that and 6.5 kW beam power an accelerator of the type built for the Natick laboratories can process about 1.2 t/hr at 10 kGy.

In Odessa in the Soviet Union, two 20 kW accelerators with an energy of 1.4 MeV installed next to a grain elevator went into operation in 1983. Each of them has the capacity to irradiate 200 t of wheat per hour with a dose of 200 Gy for insect disinfestation. This corresponds to a beam utilization of 56%.

A facility for electron-irradiation of frozen debonded chicken meat commenced operation in Vannes, France, in late 1986. The purpose of irradiation is to improve the hygienic quality of the meat by destroying salmonella and other disease-causing (pathogenic) microorganisms. The electron beam accelerator is a 7 MeV/10 kW Cassitron® built by CGR-MeV [10]. An irradiation facility of this type is shown in Figure 7.

Because of their relatively low depth of penetration, electron beams cannot be used for the irradiation of animal carcasses, large packages, or other thick materials. However, this difficulty can be overcome by converting the electrons to X-rays. As indicated in Figure 8 this can be done by fitting a water-cooled metal plate to the scanner. Whereas in conventional X-ray tubes the conversion of electron energy to X-ray energy occurs only with an efficiency of about 1%, much higher efficiencies can be achieved in electron accelerators. The conversion efficiency depends on the material of the converter plate (target) and on the electron energy. Copper converts 5 MeV electrons with about 7% efficiency, 10 MeV electrons with 12% efficiency. A tungsten target can convert 5 MeV electrons with about 20%, 10 MeV electrons with 30% efficiency. (The exact values depend on target thickness.)

In contrast to the distinct gamma radiation energy emitted from radionuclides and to the monoenergetic electrons produced by accelerators, the energy spectrum of X-rays is continuous from the value equivalent to the energy of the bombarding electrons to zero. The intensity of this spectrum peaks at about one tenth of the maximum energy value. The exact location of the intensity peak depends on the thickness of the converter plate and on some other factors. As indicated in Figure 9, in an experiment with 5 MeV electrons a very

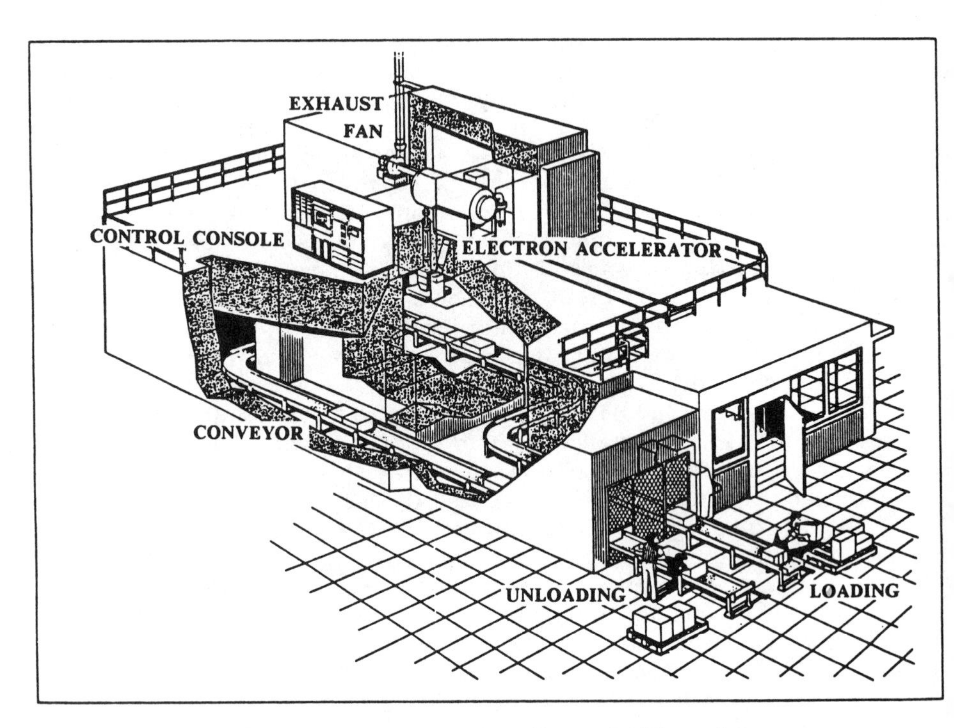

Figure 7 Drawing of an electron irradiator facility with conveyor system. (Courtesy CGR-MeV, F-78530 Buc, France.)

small portion of the X-rays reached an energy level of 5 MeV, whereas the highest intensity was observed near 0.3 MeV [11]. The X-rays are at least as penetrating as cobalt-60 gamma rays so that food products of about 30 cm thickness can be irradiated without difficulty. The possibility of using either X-rays or electrons in the same facility is an attractive feature of this technology [12–14].

IV. COMPARISON OF RADIATION FACILITIES

A comparison of the theoretical throughput of various irradiators is possible on the basis that 67,578 Ci (or about 2.5 TBq) of ^{60}Co or 308,641 Ci (about 11.4 TBq) of ^{137}Cs emit 1 kW of gamma radiation. Furthermore, a 1 kW source, assuming 100% efficiency, can process 360 kg of goods with a dose of 10 kGy in one hour [1]. This immediately allows comparison with electron accelerators, the beam power of which is expressed in kW.

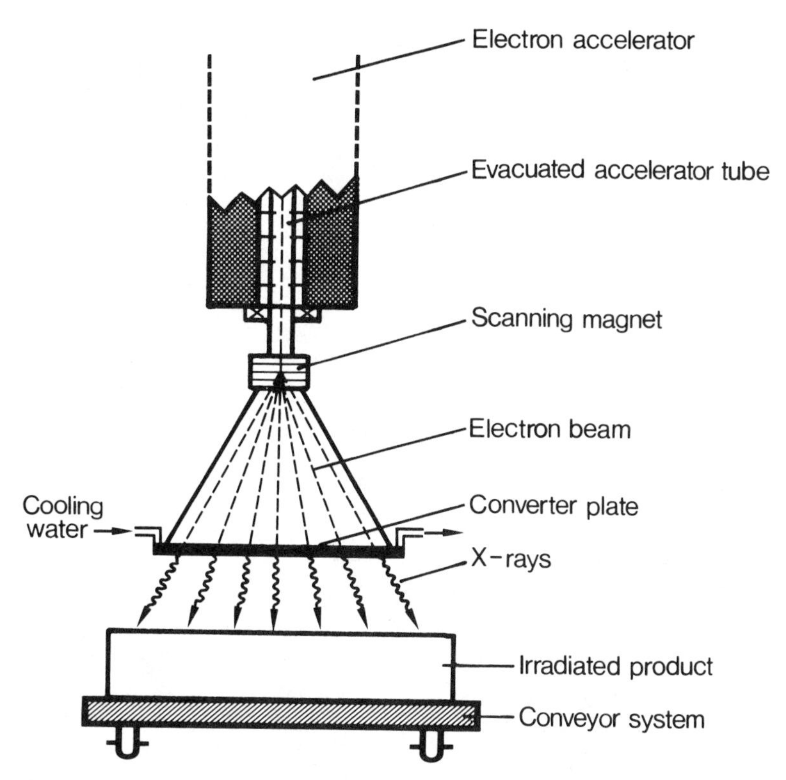

Figure 8 Electron accelerator with X-ray converter.

In well-designed facilities we can assume 50% efficiency for the electron beam, 30% efficiency for a ^{60}Co source and 20% efficiency for a ^{137}Cs source (one third less efficiency for ^{137}Cs than for ^{60}Co because of higher self-absorption). The 1 kW electron beam would thus process 180 kg of food or other materials, the 67,578 Ci ^{60}Co would process 108 kg, and the 308,641 Ci ^{137}Cs would process 72 kg, always with a dose of 10 kGy in one hour. If a 5 MeV electron beam were converted to X-rays with 8% efficiency and the X-ray beam were absorbed by the irradiated food with 50% efficiency, the electron beam power would have to be about 12.5 kW to process 180 kg with a dose of 10 kGy in one hour[13].

If we recalculate this in each case for a throughput of 1.8 t/hr, the beam power or source load indicated in Table 2 is obtained. Many ^{60}Co sources of this size exist all over the world, but not a single ^{137}Cs source approaching 7 MCi exists anywhere. Electron accelerators of the capacity indicated in Table 2 are commercially available.

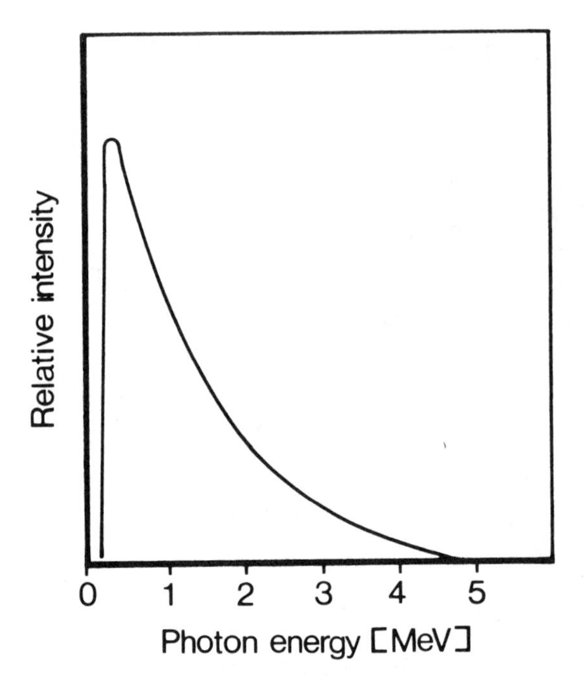

Figure 9 X-ray spectrum of energies produced by 5 MeV electron beam. (Adapted from Ref. 11.)

The possible throughput is only one of the parameters that influence the choice of a particular type of irradiator. Many other technical, economical, and, more and more, sociopolitical aspects must be taken into consideration.

Technical considerations refer primarily to the different penetrating power of various types of radiation and to differences in dose rate. If large crates of potatoes are to be irradiated, it is obvious that a gamma source rather than an electron irradiator must be chosen. When the two types of gamma sources are compared, it must

Table 2 Radiation Facilities Required for Processing 1.8 t of Food per Hour with a Dose of 10 kGy

^{60}Co source	1.1 MCi (42 PBq)
^{137}Cs source	7.7 MCi (290 PBq)
10 MeV electrons	10 kW beam power
5 MeV X-rays	125 kW beam power

be considered that the use of ^{60}Co, due to the higher penetrating power of its gamma radiation compared to that of ^{137}Cs radiation, results in a more uniform dose distribution throughout the irradiated material [13]. If large quantities of a product that can be processed in a thin layer are to be irradiated, e.g., 5,000 to 10,000 t of wheat per day, only electron accelerators are suitable. If maximum flexiblity is desired—products of greatly varying dimensions and densities to be irradiated with varying doses—an electron accelerator with an available X-ray converter plate may be the irradiator of choice.

Economic considerations cannot be presented in this book in any detail. Other authors have analyzed the costs of food irradiation, taking various modifying parameters into consideration [13–19]. ^{137}Cs is much less expensive than ^{60}Co (approximately $0.20 per Ci vs. $1.00 per Ci), but this advantage is largely offset by the much higher source strength required to get the same throughput (see Table 2) and by the lower dose distribution uniformity obtained in a ^{137}Cs irradiator. Theoretical cost calculations show a clear cost benefit of electron beam irradiators as compared to either of the gamma sources [13, 14]. The high penetration power of ^{60}Co gamma rays, the uncomplicated design and proven reliability of ^{60}Co irradiators, the complete service provided by the supplier of ^{60}Co, Atomic Energy of Canada Ltd., and probably a certain hesitancy to handle large inventories of ^{137}Cs have all contributed to the fact that ^{60}Co sources constitute the overwhelming proportion of all existing irradiators. The much improved reliability of a new generation of electron beam accelerators and the flexibility provided by X-ray conversion will probably increase the proportion of machine sources in the near future. If radiation processing continues to grow, the shortage of ^{60}Co, which has caused considerable delays in deliveries in recent years, will become more acute. This also points to an increasingly important role for electron accelerators.

Generalizing conclusions about the relative economics of different types of irradiation may be misleading because the relative costs of different radiation facilities are considerably affected by local conditions such as cost of electricity, labor, transportation, and construction. Economics of operation also depend on the use level of a facility. Where operations can be continued day and night for 12 months a year, a radionuclide source may be more economic, where intermittent operations are more likely, a machine source may be more advantageous.

Sociopolitical considerations relate to the observation that in some countries it is getting more and more difficult to overcome local opposition to the installation of new radioisotope sources. Fears for the safety of the environment in shipping and storing large inventories of ^{60}Co or ^{137}Cs are often cited as the main reason for this opposition. Regardless of whether these fears are justified, planners cannot disregard them. As an example, the National Food Processors Association (NFPA), with support from the U.S. Department of Energy , negotiated in the summer of 1985 for a site in Dublin, California, to build a demonstration and training facility for food irradiation, using 3 million curies of ^{137}Cs. An opposition group, CARD (Citizens Against Radioactive Dublin), was formed which succeeded in producing so much political pressure that the NFPA dropped its plan. Such opposition to electron accelerators has not occurred. It now appears unlikely that NFPA or any other organization interested in food irradiation in the United States is still planning to use ^{137}Cs.

V. PROCESS CONTROL

Irradiated foods should not receive too high a radiation dose because that could affect eating quality. Upper dose limits for permitted irradiation treatments have been established by regulatory authorities in most countries. Avoiding unnecessarily high doses is also of economical importance for the processor, because the cost increases with increasing dose.

On the other hand, it must be assured that all points of an irradiated batch receive at least the minimum dose that governs the effectiveness of the process. Correct measurement of dose and dose distribution in the product helps to provide assurance that the radiation treatment is both effective and legally correct.

Dose and dose distribution are determined by product parameters and by source parameters. Product parameters are primarily the density of the food itself and the density of packing the individual food containers within the tote box or carrier in which irradiation takes place. Source parameters are different for the different types of irradiators. In the case of gamma irradiators the relevant factors are isotope, source strength, and geometry (^{60}Co or ^{137}Cs? how many Ci? how are the ^{60}Co pins or ^{137}Cs capsules arranged in the source rack?); source pass configuration and mechanism (how and at what distance are the carriers or tote boxes stacked around the source and around their own axis?); conveyor speed and dwell time (how fast and how many times is a box or carrier moved around the source?). In the case of machine sources, the relevant factors are type of radiation (electron beam or X-ray); beam energy and beam power (MeV and kW); scan width and scan frequency; pulse repitition rate in the case of pulsed electron beams; beam pass configuration and mechanism; conveyor speed.

Some of these factors are constant for a given irradiator (e.g., type of isotope or radiation, design geometry), others change systematically (the source strength of an isotope source decreases according to the half-life of that isotope), others can be set by the operator according to the requirements of the process (e.g., dwell time). To achieve the desired dose and dose distribution, the operator cannot rely on controlling the operating variables alone. There invariably are product heterogeneities, such as seasonal crop variations, anomalies in bulk density, uneven local density variations, and random packing of agricultural products to be taken into account. Therefore, every time a new batch of food is to be irradiated, the operator must establish the dose and dose distribution by strategically placing *dose meters* into and between the food packages and evaluating the dose meter reading. Once the process is running smoothly, it is usually not necessary to carry out dosimetry on all the product. Monitoring the process parameters and making occasional dosimetric checks is now sufficient [20].

In most countries government regulations require that food irradiation processors maintain records that describe for each food lot the radiation source, source calibration, dosimetry, dose distribution in the product, and certain other process parameters (see Chapter 10).

A. Interaction of Radiation with Matter

The purpose of dose meters is to measure the amount of radiation energy absorbed by the irradiated product. To understand how dose meters function, a short introduction to the interaction of ionizing radiation with matter is appropriate at this point, although the effects of ionizing radiation on food components will be described in more detail in Chapter 3.

The instrument that gives a reading of absorbed dose directly is the *calorimeter*. It measures the total energy dissipated or the rate of energy dissipation in a material in terms of the thermal properties of the absorbing body. This instrument, therefore, may be considered to be an absolute dose meter that can be used for calibrating other dose meters. The principle of radiation calorimetry is implicit in the definition of the radiation dose unit 1 Gy(gray) = 1 J/kg. Ideally the temperature elevation should be measured in the irradiated food product itself, but in practice this is usually not done because the thermal properties of foodstuffs vary widely. A substance with known, reproducible thermal properties is taken instead, which serves as a heat-sensing calorimetric body, included in an adiabatic system (adiabatic = without transmission of heat). Water, graphite, aluminum, or water-equivalent plastics are usually chosen, and the thermal change is determined by small calibrated thermocouples or thermistors embedded in the calorimetric body.

The practice of using radiation calorimetry is not simple, and it is rarely applied in a routine fashion. Because the process of temperature elevation should run under adiabatic or quasi-adiabatic conditions, the dose has to be applied in a very short time. Calorimetry is therefore used successfully as a calibration method for electron accelerator beam doses only. The absorbed dose in the calorimetric body can be converted to that of the material of interest (foodstuff) by taking into consideration the different density and the different energy absorption coefficients of the two materials. The temperature elevation depends on radiation dose and on the specific heat of the material irradiated. A dose of 10 kGy causes temperature elevations as follows:

2.3 K in water (spec. heat 4.2 kJ/kg·K)
6.2 K in dry protein (spec. heat 1.6 kJ/kg·K)
7.1 K in dry carbohydrate (spc. heat 1.4 kJ/kg·K)
12.5 K in glass (spec. heat 0.8 kJ/kg·K)

where K = kelvin. Because of the low temperature elevation in the low dose range, radiation calorimetry is limited in practice to the dose range above 3 kGy.

This small temperature elevation is the gross result of the complex process of radiation interaction with matter. The individual steps of this process depend on the type of radiation used. When high-energy electrons are absorbed by a medium, they lose their kinetic energy by interaction with electrons of the medium. (At very high energy, above that allowed for food irradiation, accelerated electrons can also interact with nuclei of the medium.) The interaction with orbital electrons of the atoms of the medium (the absorber) causes ionizations and excitations. Ionization means that orbital electrons are ejected from atoms of the medium; excitation means that orbital electrons move to an orbit of higher energy. Ejected electrons (secondary electrons), carrying a large portion of the energy of the incident electron, also lose energy through interaction with orbital electrons of the absorber. Electrons at low velocities (subexcitation energy level) can cause molecular vibrations on their way to becoming thermalized. As a result of the collisions with atoms of the absorber material, the incident electrons can change direction. Repeated collisions cause multiple changes of direction. The result is a scattering of electrons in all directions. This is shown schematically in Figure 10a.

When gamma or X-ray photons interact with the absorber, three types of interaction can occur:

the photo electric effect
the Compton effect
pair production (i.e., formation of pairs of electrons and positrons)

Photoelectric absorption occurs largely with photons of energies below 0.1 MeV and pair production primarily with photons of energies

1 Electron beam (primary electrons)

2 Depth of penetration

3 Secondary electrons

4 Irradiated medium

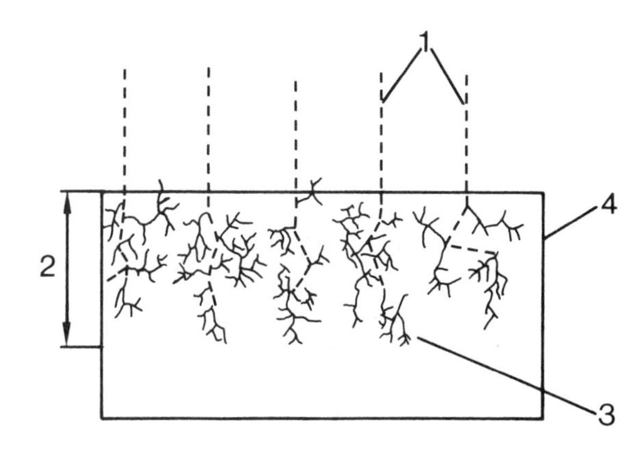

(A)

1 Gamma or X-ray photons

2 Compton electrons

3 Secondary electrons

4 Irradiated medium

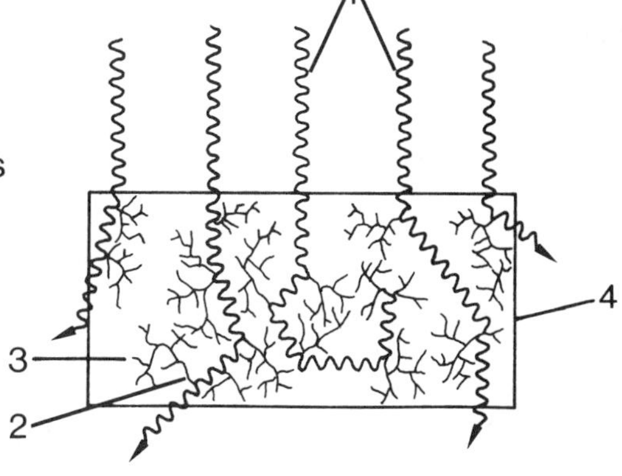

(B)

Figure 10 Interaction of radiation with matter: A) electron radiation; B) gamma or X-radiation.

above 10 MeV. Both are of minor importance in food irradiation, where the *Compton effect* predominates.

As portrayed in Figure 11, in the Compton effect an incident photon interacts with an absorber atom in such a way that an orbital electron is ejected. The incident photon continues after the collision in a changed direction and with less than its original energy. The ejected electron (Compton electron) has enough kinetic energy to cause excitations and ionizations in the absorber atoms. It thus interacts with the absorber in the same way as the incident electrons of an electron accelerator beam (Fig. 10b).

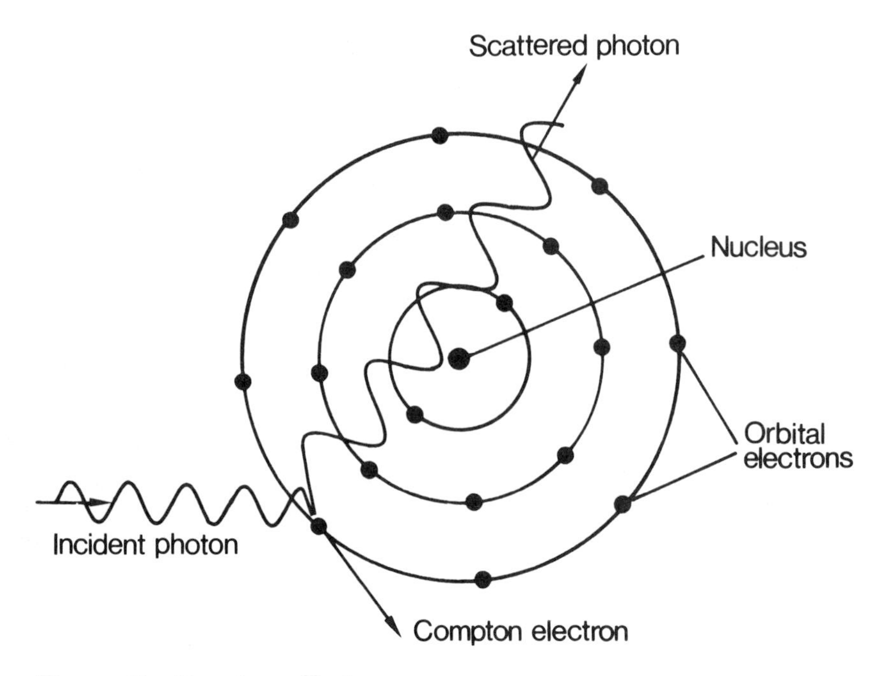

Figure 11 Compton effect.

Because Compton electrons are produced when gamma or X-ray photons interact with a medium, and because the Compton electrons cause ionizations and excitations in the same way as accelerator beam electrons, the radiation-induced chemical changes in the irradiated medium are the same, regardless of the type of radiation used.

B. Chemical Dose Meters

In principal, any chemical change caused by irradiation of a medium can be used to measure the absorbed radiation dose. In practice, only those changes can be evaluated which are stable for a reasonable length time and which can be reliably measured by standard procedures such as titration or spectrophotometry. The chemical change is usually expressed as the *G-value*, which is a measure of the number of atoms, molecules, or ions produced (+G) or destroyed (−G) by 100 eV of absorbed energy. In the new SI system of units the G-value is expressed as per joule instead of per 100 eV.

An important reference dose meter in food irradiation is the ferrous sulfate or Fricke dose meter. It is based on the radiation-induced oxidation of ferrous ions (Fe^{++}) to ferric ions (Fe^{+++}) and consists of measuring the increased optical absorbance of the ferric ions at the absorption peak of 305 nm. For ^{60}Co gamma rays, the G-value for ferric ion yield is 15.6 Fe^{+++} ions per 100 eV, or 9.74 $\times 10^{17}$ ions per joule.

There are many other systems, such as the ethanol-chlorobenzene dose meter, which is based on the formation of hydrochloric acid from chlorobenzene. The hydrochloric acid can be measured by titration or by its effect on the dielectric constant.

PMMA (polymethyl methacrylate) dose meters are very useful for routine dosimetry. Irradiation of PMMA (Perspex®) induces an absorption band in the region of 250–400 nm. PMMA dose meters of 3 mm thickness can be used in the dose range of 3–20 kGy. PMMA containing a red dye develops a radiation-induced absorption band at 640 nm and can be used in the dose range of 1–50 kGy. Such dye film dose meters are widely used for routine dosimetry, process control, and inspection purposes.

In the lower dose range, down to 10 Gy, radiochromic dye dosimetry can be used. When the colorless solution of pararosaniline cyanide in 2-methoxy ethanol and glacial acetic acid is irradiated, an intense red color develops with an absorption maximum at 549 nm. Other dosimetry systems are described in the review of McLaughlin and collaborators [21].

Dosimetry for food irradiation processing has reached a high level of perfection [22]. A standard for this purpose has been issued by the American Society for Testing and Materials [23]. The role of dosimetry in good radiation processing practice is described in the Recommended International Code of Practice for the Operation of Irradiation Facilities Used for the Treatment of Foods [see Appendix II].

C. Dose Distribution

As mentioned earlier, electron beams on the one hand and gamma rays and X-rays on the other hand differ greatly in their ability to penetrate matter. This has important consequences for the dose distribution in the irradiated medium.

Since many foods consist mostly of water, the penetration of radiation in water is shown in Figure 12. When an electron beam penetrates into an aqueous medium, the dose somewhat below the surface is higher than at the surface. This is due to the formation of secondary electrons which, because of their lower energy, are more effectively absorbed than the primary electrons. Also, scattering causes some secondary electrons to escape from the surface in the direction opposite to that of the beam of primary electrons. Thus a 10 MeV electron beam giving a dose of 10 kGy at the surface will deposit about 12.5 kGy at 2 cm below the surface. As more and more primary electrons lose their energy by interacting with water molecules, the absorbed dose decreases with increasing depth, and

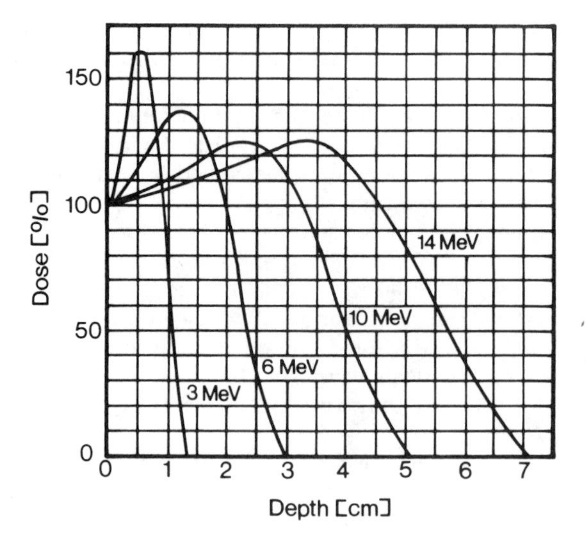

(A)

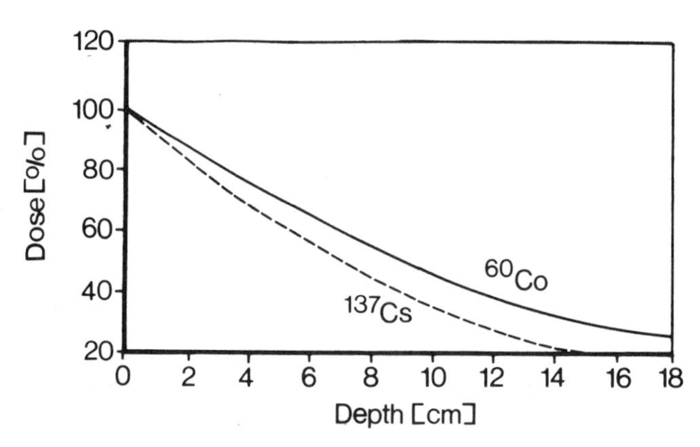

(B)

Figure 12 Dose distribution in water as a function of depth; A) electrons of different energy; B) gamma radiation from ^{60}Co and ^{137}Cs.

at about 5 cm the limit of penetration is reached. In contrast, the dose delivered by gamma rays decreases continuously. The rate of decrease is faster with ^{137}Cs gamma radiation than with ^{60}Co gamma radiation. With X-rays it depends on the energy of the X-ray–producing electrons. For practical purposes the penetration of 5 MeV X-rays is comparable to that of ^{60}Co gamma rays.

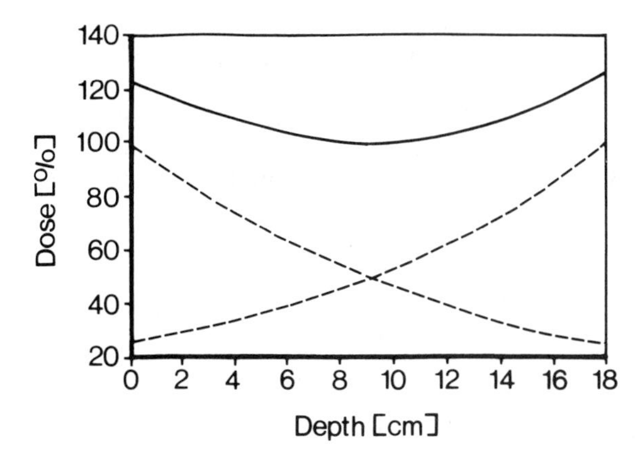

(A)

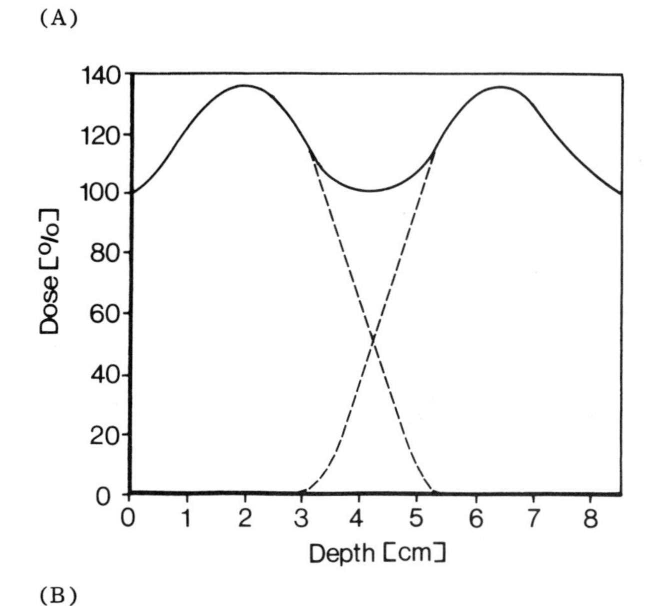

(B)

Figure 13 Depth dose distribution in water irradiated from two opposite sides: A) with ^{60}Co gamma radiation; B) with 10 MeV electrons. (Dashed lines indicate the dose distribution upon one-sided irradiation.)

Two-sided irradiation permits processing of thicker packages with more uniform dose distribution, as indicated in Figure 13. If the density of the irradiated medium is less than that of water, for instance in fatty foods or in dehydrated or porous foods, the depth of penetration is correspondingly greater. The 10 MeV electron beam which barely reaches a depth of 5 cm in water will reach approximately 10 cm at a density of 0.5 g/cm^3.

From Figures 12 and 13 it is clear that an absolutely uniform dose distribution cannot be obtained, even if a material of uniform density is irradiated. If dose meters are distributed throughout the irradiated food product, their evaluation will show where the *highest dose* (D_{max}) and the *lowest dose* (D_{min}) were absorbed. Addition of dose values determined with all dose meters and division by the number of dose meters will give the *average dose* (D_{av}). Many food packages stacked up in a carrier or tote box will receive somewhat different doses D_{max}, D_{min}, and D_{av}, depending on their position in the container [20]. This is illustrated in Figure 14 for a gamma source. Material positioned at the level of the upper or lower end of the source rack (shown as a plaque source in Fig. 14) will receive a lower dose than material positioned at the level of the center of the source rack. Similarly, a scanned electron beam (Fig. 15) will give a higher dose to the material situated in the center of the conveyor and a lower dose to material moving at the outer edges of the conveyor.

The *dose uniformity ratio* D_{max}/D_{min} depends on many factors such as plant design, type, kind of product, and type and energy of radiation. Where this is necessary a very low ratio of D_{max}/D_{min}, i.e., a very high uniformity of dose distribution, can be obtained, for instance, in the situation shown in Figure 14 by placing a food package in the center of the container and surrounding it with low-density plastic material, or in the situation shown in Figure 15 by choosing a narrower dimension of the package, which occupies only the center of the conveyor belt. This would, of course, lower the throughput and efficiency of the irradiation plant. For research purposes a D_{max}/D_{min} of 1.4 or even 1.2 may be considered desirable and can usually be achieved. However, in a commercial situation, where a high throughput is important, a ratio of 2 or 3 is often considered acceptable, especially if material of uneven density is processed [see Appendix II].

The measurement of D_{min} and D_{max} in the product is basic to any consideration of the effectiveness of the process, the legal acceptability of the treatment, and the design and economics of the irradiation plant. It is therefore important that the dose measurements should be as accurate as possible and sufficiently numerous to achieve reliable statistics.

Another important characteristic of a radiation process is the *overall average dose*. It was used by the 1980 FAO/IAEA/WHO Joint Expert Committee on Irradiated Foods when the Committee concluded "that the irradiation of any food commodity up to an overall average dose of 10 kGy presents no toxicological hazard." The overall average dose is the arithmetic mean value of all dose meter readings in a

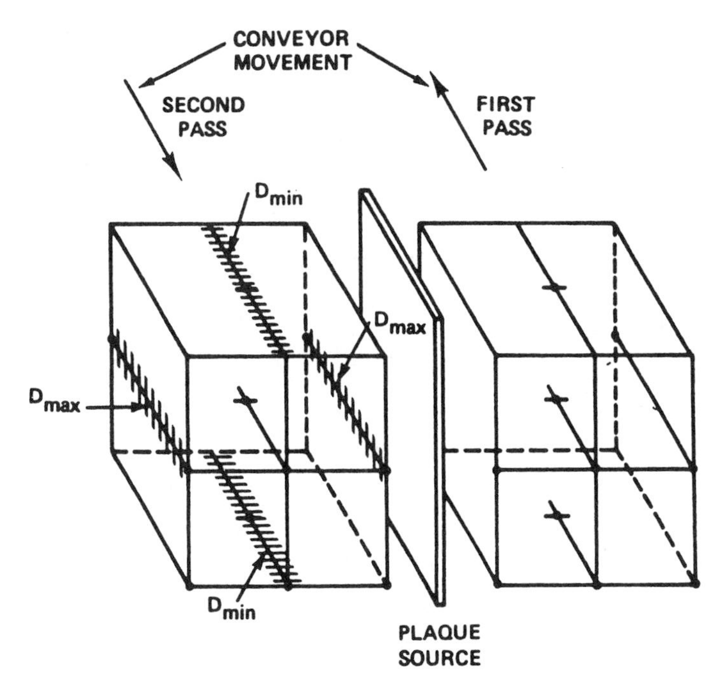

Figure 14 Regions of D_{max} and D_{min} (indicated by hatching) in a rectangular container after two single-direction passes, once on each side of a stationary gamma ray source rack. (From Ref. 20, with permission.)

given irradiation run. To determine this value, an adequate number of dose meters must be randomly distributed in the food as it is exposed to the radiation. The number of dose meters is considered adequate if it permits estimation of the dose distribution in each portion of the food material of different density and if the measurements are representative for all dose and density fluctuations during a usual run. The Joint Expert Committee recognized that "the overall average dose may result in a fraction of the food receiving a maximum absorbed dose up to 50% higher" [24]. The Committee's approval of irradiated foods up to an overall average dose of 10 kGy thus recognized the fact that some parts of the food so irradiated will receive a dose of up to 15 kGy. The International Code of Practice [Appendix II] specifies more precisely that, because of the statistical distribution of the dose, a mass fraction of product of at least 97.5% should receive an absorbed dose of less than 15 kGy when the overall average dose is 10 kGy.

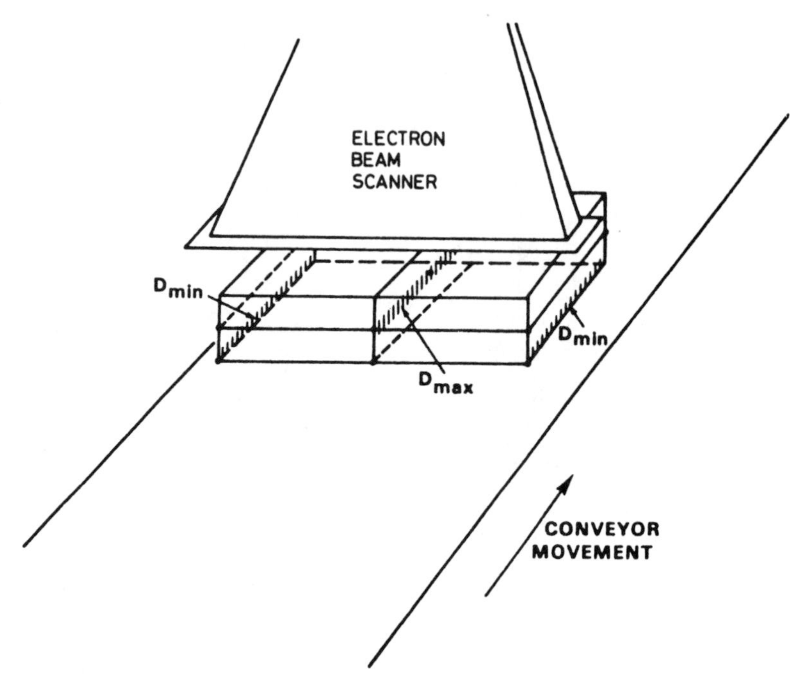

Figure 15 Regions of D_{max} and D_{min} (indicated by hatching) in a rectangular package after single-pass, single-direction irradiation with electrons. (From Ref. 20, with permission.)

REFERENCES

1. Jarett, R. D., Sr., Isotope (gamma) radiation sources, in *Preservation of Food by Ionizing Radiation,* vol. 1, E. S. Josephson and M. S. Peterson, eds. CRC Press, Boca Raton, FL, 1982, p. 137.
2. Brynjolfsson, A., Cobalt-60 irradiator designs, in *Sterilization by Ionization Radiation,* E. R. L. Gaughran and A. J. Gaudie, eds., Multiscience Publication Ltd., Montreal, 1974, p. 145.
3. McMullen, W. H. and D. P. Sloan, Cesium-137 as a radiation source, in *Radiation Disinfestation of Food and Agricultural Products,* J. H. Moy, ed., Hawaii Inst. Trop. Agric. Human Resources, Univ. of Hawaii at Manoa, Honolulu, Hawaii, 1985, p. 245.
4. McKinnon, R. G., and R. D. H. Chu, Pallet irradiators for food processing, *Radiat. Phys. Chem.,* 25:141 (1985).
5. Cuda, J., Optimum plant capacity—technical and economic considerations, *Radiat. Phys. Chem.,* 25:411 (1985).

6. Ramler, W. J., Machine sources, in *Preservation of Food by Ionizing Radiation,* vol. 1, E. S. Josephson and M. S. Peterson, eds., CRC Press, Boca Raton, FL, 1982, p. 165.
7. Cleland, M. R., High-voltage electron beam irradiation facilities, *Rad. Phys. Chem.*, 18:301 (1981).
8. McKeown, J., and N. K. Sherman, Linac based irradiators, *Radiat. Phys. Chem., 25*:103 (1985).
9. Bly, J. H., Electron beam sterilization technology 1985: an update, *J. Indust. Irradiation Tech.*, 3:63 (1985).
10. Gallien, C. L., J. Paquin, C. Ferradini, and T. Sadat, Electron beam processing in food industry—technology and costs, *Radiat. Phys. Chem.*, 25:81 (1985).
11. Matthews, S. M., Food processing with electrically generated photon irradiation, in *Radiation Disinfestation of Food and Agricultural Products,* J. H. Moy, ed., Hawaii Inst. Trop. Agric. Human Resources, Univ. of Hawaii at Manoa, Honolulu, Hawaii, 1985, p. 283.
12. Farrell, J. P., S. M. Seltzer, and J. Silverman, Bremsstrahlung generators for radiation processing, *Radiat. Phys. Chem.*, 22: 469 (1983).
13. Lagunas-Solar, M. C., and S. M. Matthews, Comparative economic factors on the use of radionuclide or electrical sources for food processing with ionizing radiation, *Radiat. Phys. Chem.*, 25:111 (1985).
14. Cleland, M. R., and G. M. Pageau, Electrons versus gamma rays—alternative sources for irradiation processes, in *Food Irradiation Processing,* Proceedings of a Symposium held in Washington, D.C., March 1985, Int. Atomic Energy Agency, Vienna, 1985, p. 397.
15. IAEA, *Factors Influencing the Economical Application of Food Irradiation,* Panel Proceedings Series, International Atomic Energy Agency, Vienna, 1973.
16. Deitch, J., Economics of food irradiation, *Crit. Revs. Food Sci. Nutrit.*, 17:307 (1982).
17. Brown, D. R., Potential value of cesium-137 capsules, *J. Industr. Irradiation Tech.*, 4:21 (1986).
18. Morrison, R. M., and T. Roberts, *Food Irradiation: New Perspectives on a Controversial Technology*, U.S. Dept. of Commerce, National Technical Information Service, Springfield, VA, 1985.
19. Sivinski, J. S., Efficacy testing and market research for the pork industry, *Rad. Phys. Chem.*, 25:263 (1985).
20. IAEA, *Manual of Food Irradiation Dosimetry,* International Atomic Energy Agency, Technical Report Series 178, Vienna, 1977.
21. McLaughlin, W. L., R. D. Jarrett, Sr., and T. A. Olejnik, Dosimetry, in *Preservation of Food by Ionizing Radiation,* vol.

1, E. S. Josephson and M. S. Peterson, eds., CRC Press, Boca Raton, FL, 1982, p. 189.
22. Bögl, K. W., D. F. Regulla, and M. J. Suess, eds., *Health Impact, Identification and Dosimetry of Irradiated Foods*, Report of a WHO Working Group on Health Impact and Control Methods of Irradiated Foods, ISH-Heft 125, Bundesgesundheitsamt, Neuherberg/München, 1988.
23. ASTM, *Standard Practice for Application of Dosimetry in the Characterization and Operation of a Gamma Irradiation Facility for Food Processing*, American Society for Testing and Materials, E 1204-87.
24. WHO, *Wholesomeness of Irradaited Foods*, Technical Reports Series 659, Geneva, 1981.

SUGGESTED READINGS

Boaler, V. J., Electron accelerator facilities for food processing, *J. Food Engineerg.*, 3:285 (1984).
Chadwick, K. H., and W. F. Oosterheert, Dosimetry concepts and measurements in food irradiation processing, *Appl. Radiat. Isot.*, 37:(1) 47 (1986).
IAEA, *Training Manual on Food Irradiation Technology and Techniques*, Technical Reports Series 114, 2nd ed., Vienna, 1982.
Kase, K. R., B. E. Bjarngard, and F. H. Attix, *The Dosimetry of Ionizing Radiation*, Academic Press, Orlando, FL, 1987.
Lundberg, G. D., C. Iverson, and G. Radulescu, Now read this: The SI units are here, *J. Am. Med. Assoc.*, 255:2329 (1986).
McLaughlin, W. L., A. Miller, and R. M. Uribe, Radiation dosimetry for quality control of food preservation and disinfestation, *Rad. Phys. Chem.*, *22*:21 (1983).

3

Chemical Effects of Ionizing Radiation

I. SOME BASIC CONSIDERATIONS

Before describing the very complex events that take place when foodstuffs are irradiated, some concepts and some terminology will be introduced by discussing the radiation chemistry of the simplest organic compound, CH_4 or methane.

When energetic electrons—either coming from an electron-generating machine or produced through Compton scattering—pass through a sample of methane they cause *primary effects*:

$$H-\underset{H}{\overset{H}{C}}-H \rightsquigarrow \begin{cases} \cdot CH_4^+ + e^- & \text{ionization} \\ \cdot CH_3 + \cdot H & \text{dissociation} \\ CH_4^* & \text{excitation} \end{cases}$$

Methane has eight valence electrons, arranged in four electron pairs, indicated on the left of the above scheme as dashes connecting the carbon atom with the hydrogen atoms. A strong interaction of the incident or Compton electrons with the methane molecule may cause *ionization* by removing an electron (e^-), or *dissociation* by splitting off a hydrogen atom. A weak interaction may cause *excitation*, indicated by an asterisk. The excited methane molecule has been merely promoted to a higher internal energy level. The product of the ionization reaction is a cation, characterized by a plus sign. It is also a *free radical*, as indicated by a dot. Free radicals have an unpaired electron and are usually very reactive. The $\cdot CH_4^+$ produced in the ionization reaction is a methane radical cation. The dissociation reaction produces a methyl radical and a hydrogen atom, which is also a free radical. The primary effects are nonspecific; they randomly hit any structure that is in the path of the incident or Compton electrons, without preference for particular atoms or molecules. The electrons removed from atoms or molecules in the primary process, such as the e^- produced in the ionization of methane, may have enough energy to cause further ionizations, dissociations, or excitations. If the irradiated material is a solid, the reactions caused by secondary, tertiary, etc. electrons often occur close to the original ionization, giving rise to clusters, so-called *spurs*, along the trajectory of the incident radiation.

The excited molecules may undergo *deexcitation*, for instance, by giving off energy in the form of light (fluorescence). Or they may receive some additional energy from a further interaction, so that dissociation or ionization can result.

A. Secondary Effects

Because of the high reactivity of the free radicals produced as a result of the primary effect, *secondary effects* will occur. The free radicals may undergo *radical reactions* with each other. In the case of irradiation of methane this may result in *recombination*:

$$\cdot H + \cdot CH_3 \rightarrow CH_4$$

or *dimerization*:

$$\cdot CH_3 + \cdot CH_3 \rightarrow C_2H_6$$

$$\cdot H + \cdot H \rightarrow H_2$$

Another possibility is *electron capture*:

$$\cdot CH_4^+ + e^- \rightarrow CH_4$$

If other substances are present, the free radicals can also react with these. Newly formed compounds, such as C_2H_6 (ethane) in the case of the irradiation of methane, will also interact with radiation. A dissociation reaction could lead to $\cdot C_2H_5$. Two such ethyl radicals could react with each other by dimerization:

$$\cdot C_2H_5 + \cdot C_2H_5 \rightarrow C_4H_{10}$$

to yield butane or by *disproportionation*:

$$\cdot C_2H_5 + \cdot C_2H_5 \rightarrow CH_2{=}CH_2 + C_2H_6$$

to yield ethene and ethane. Disproportionation reactions are those in which one reactant loses a hydrogen atom and the other gains one.

The example of methane irradiation shows that even such a simple compound can yield a rather complex spectrum of stable end products. In irradiated pure methane we could expect to find some hydrogen, ethane, ethene, traces of butane, and a number of other products. Which product or products will predominate depends on various experimental conditions, such as dose, dose rate, and temperature. While primary effects are largely unspecific, secondary effects depend on specific chemical structures. The presence of impurities or additives can have a decisive influence on product formation. A substance which reacts readily with a free radical is known as a *scavenger*, while a substance which produces a more reactive radical is a *sensitizer*.

In a small symmetrical molecule such as methane, the changes of any of the four C–H bonds being attacked and changed are equal. When large molecules are irradiated, the absorbed radiation energy will be unevenly distributed in the excited molecule. The energy is likely to be absorbed in those parts of the molecule having the greatest variation in electron density or where the bonds are weakest. This also happens when molecules absorb thermal energy, and it is therefore not surprising that irradiation and heating cause bond breakages, often at the same locations in the molecules. At the outcome this means that products resulting from irradiation, heating, or from other forms of energy input are often identical or similar.

The overall process that leads to stable end products as a result of irradiation of some medium is called *radiolysis*, and the products of primary and secondary effects are *radiolytic products*. This process occurs within fractions of a microsecond. Because some end products are not completely stable, *postirradiation effects* or *radiation aftereffects* can occur in some systems for days or months after the treatment (this phenomenon will be discussed later).

What has been said about the high reactivity of free radicals does not apply to a situation where dry solids or deep-frozen materials are irradiated. Diffusion is very restricted under these conditions so that the free radicals cannot move about. Because they cannot find reaction partners, these *trapped free radicals* can exist for a long time. If the dry solid is permitted to absorb moisture from the atmosphere or if it is dissolved or if the deep-frozen material is warmed up, movement of free radicals is no longer restricted, and they disappear by reacting together or with other substances.

It should be noted that free radicals occur not only in irradiated materials. Many biochemical reactions, both in plant cells and in the mammalian organism, proceed through radical mechanisms [1]. Grinding of dry powders [2] and heating of protein-rich foods [3] also produces free radicals. The question of whether consumption of foods containing a high concentration of free radicals could be damaging to health has been investigated in animal feeding studies (see Chapter 5).

B. Influence of Dose and Dose Rate

The extent of primary and secondary effects will depend on the radiation energy absorbed by the irradiated substance, i.e., on the the radiation dose. Product formation usually increases linearly with dose: A doubling of the radiation dose will cause a doubling of the amount of radiolytic products. This may not be true in the range of very high doses: If a radiolytic product formed in the low dose range is itself destroyed by radiation, product formation will not continue to increase linearly with dose. It is also possible that a small amount of scavenger is present in an irradiated system. This would suppress product formation initially. When the scavenger is consumed after application of a high dose, product formation would then increase. In food irradiation we can generally expect a linear increase of effects with increasing dose.

Dose rate is the dose absorbed per unit of time. Dose rate from gamma ray sources is usually below 10 Gy/sec, and dose rate effects are not observed in this range. Radiolytic products are the same whether a sample is irradiated in a small or in a large gamma source. Dose rate from electron accelerators is usually between 10^4 and 10^9 Gy/sec. This can mean, particularly at the upper end of this range, that the radiolytic products produced by a given dose are formed within such a short time that they have more chance to react with each other by electron capture and recombination than to react with the molecules of the irradiated substance. The result is less radiolytic change with very high dose rates than with low dose rates. However, practical experience has shown that in food irradiation dose rate effects are usually not very important.

C. Instruments of Research in Radiation Chemistry

The most important instrument for detecting free radicals is the *electron spin resonance* (ESR) spectrometer, also known as the electron paramagnetic resonance (EPR) spectrometer. ESR measurements can detect the paramagnetism (spin) of unpaired electrons even at extremely low concentrations (about 10^{-8} M under suitable conditions). The amplitude of the ESR signal is approximately proportional to the total number of spins (unpaired electrons) in the measured sample. ESR measurements cannot only be used to determine the concentration of free radicals in an irradiated material. The shape of the ESR signal characterizes the kind of free radical present.

Figure 1 demonstrates ESR spectra of some dry foodstuffs at 5 min after electron irradiation and again 6–22 days later. The signal becomes smaller with time, indicating the decrease in free radical concentration. The rate of this decrease depends on water content, as can be seen from Figure 2 in the case of potato starch. When the samples were kept completely dry, a high ESR signal was observed even 12 days after irradiation at a dose of 1 kGy, indicating that the diffusion of free radicals and thus the chances of their reacting together or with starch molecules is severely restricted in the absence of moisture. In contrast, at 15% humidity no ESR signal could be detected after 1 day. Similar observations have been made on spices and spray-dried fruit powder [6].

Besides ESR spectrometry, *pulse radiolysis* is a most important technique in radiation chemistry. It requires a radiation source that produces pulsed electron beams, such as a linear accelerator. During an extremely short time (microseconds or nanoseconds) a pulse of electrons is directed onto a liquid or gaseous sample, usually a substance dissolved in water, and the species produced by primary and secondary effects of radiation are measured with detectors capable of very rapid response. Optical absorption (kinetic spectrophotometry) is mostly used, but kinetic electrical conductivity and ESR have also been very useful for the exact assignment of the observed transient species since these methods determine the charge of the intermediate and the position of the odd electron. The availability of pulsed electron generators has been essential for the exploration of the basics of radiation chemistry. Although much of the information now available on short-lived intermediary species could not have been obtained with gamma sources, there is every reason to believe that the primary and secondary effects identified in electron-irradiated systems also occur during gamma irradiation. At any rate, the end products of radiolysis, which can be determined by various methods of chemical analysis, are usually the same, regardless of which type of ionizing radiation was applied.

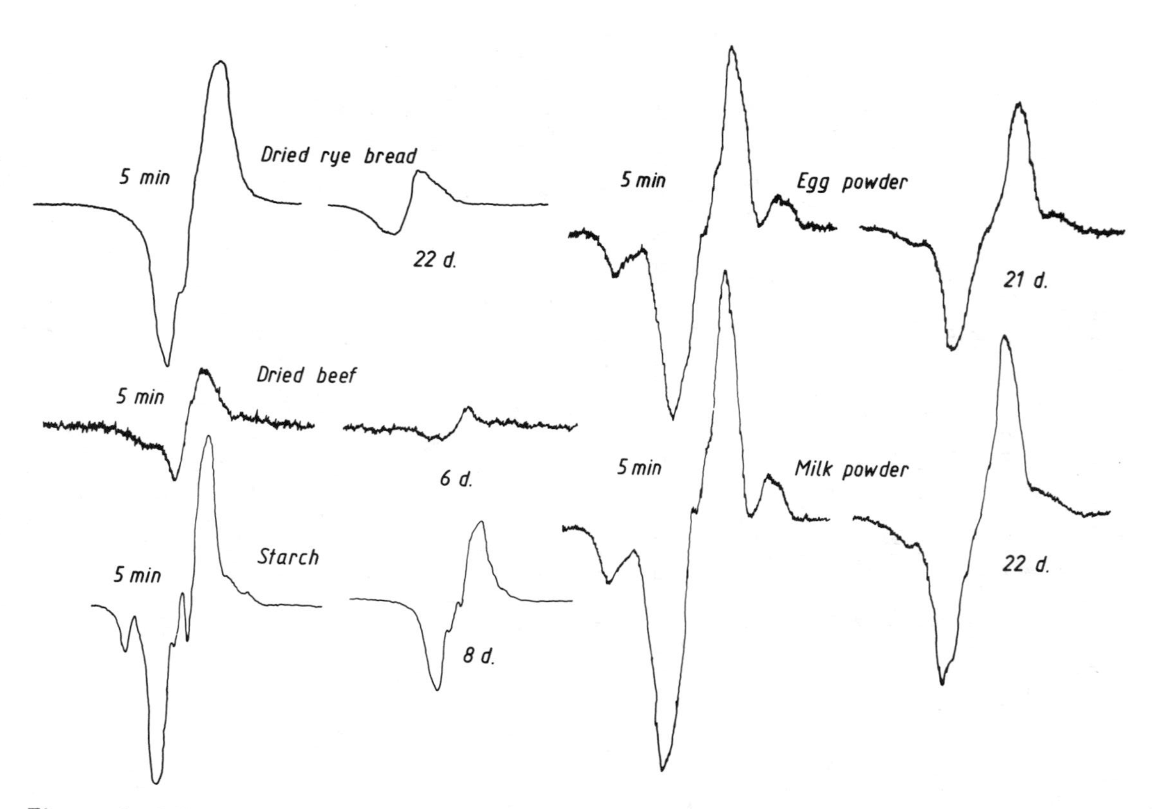

Figure 1 ESR spectra of some dry foodstuffs 5 min and 6 to 22 days after irradiation (10 MeV electrons; 10 kGy; storage without exclusion of air at ambient temperature). (From Ref. 4.)

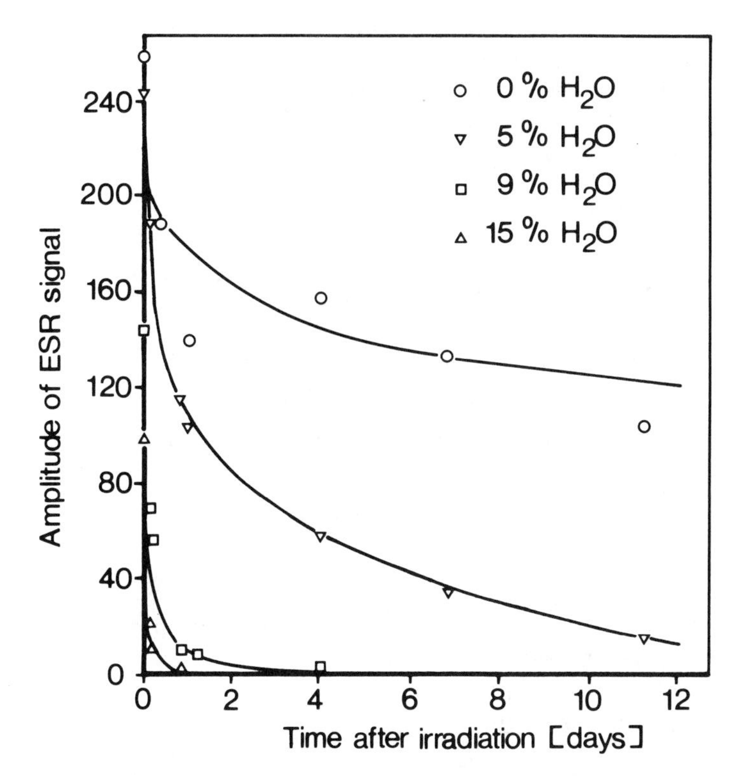

Figure 2 Influence of water content on the decline of the amplitude of ESR signals during 12-day storage of irradiated starch (10 MeV electrons; 10 kGy; storage in sealed quartz ampoules at ambient temperature; signal amplitude in relative units). (From Ref. 5.)

II. EFFECTS OF IONIZING RADIATION ON WATER

Water is present in almost all foods. Its proportion is about 90% in many vegetables, 80% in fruits, 60% in meat, and 40% in bread. Even apparently dry products contain some water: wheat flour 13%, dried vegetables about 10%, nuts 5%. The radiolysis of water is therefore of particular interest in food irradiation. The radiolytic products of water are:

$\cdot OH$	hydroxyl radical
e^-_{aq}	aqueous (or solvated or hydrated) electron
$\cdot H$	hydrogen atom
H_2	hydrogen
H_2O_2	hydrogen peroxide
H_3O^+	solvated (or hydrated) proton

Aqueous or solvated or hydrated means that the electron e^- and the hydrogen ion H^+ cannot exist freely in an aqueous medium. They are closely associated with water molecules.

A. Reactive Intermediates of Water Radiolysis

When water is irradiated, the main primary reactions are:

$$H_2O \rightsquigarrow \cdot H_2O^+ + e^- \quad \text{ionization}$$

$$H_2O \rightsquigarrow H_2O^* \quad \text{excitation}$$

The major decay routes for the excited water molecules are:

$$H_2O^* \rightarrow H_2O \quad \text{deexcitation}$$

$$H_2O^* \rightarrow \cdot H + \cdot OH \quad \text{dissociation}$$

The water cation radical can undergo a *deprotonation* or proton transfer reaction with a water molecule,

$$\cdot H_2O^+ + H_2O \rightarrow H_3O^+ + \cdot OH$$

yielding a solvated proton and a hydroxyl radical.

Some of the reaction products are lost again through recombination,

$$\cdot H + \cdot OH \rightarrow H_2O$$

while more products are formed through combination reactions:

$$e^-_{aq} + \cdot OH \rightarrow OH^-$$

$$e^-_{aq} + H_3O^+ \rightarrow H_2O + \cdot H$$

$$e^-_{aq} + e^-_{aq} + 2H_2O \rightarrow H_2 + 2OH^-$$

The hydroxyl ions are neutralized by reactions with H_3O^+. Dimerization reactions also occur:

$$\cdot OH + \cdot OH \rightarrow H_2O_2$$

$$\cdot H + \cdot H \rightarrow H_2$$

B. End Products of Water Radiolysis

While $\cdot OH$, e^-_{aq}, and $\cdot H$ are very reactive transient species, hydrogen and hydrogen peroxide are the only stable end products of water radiolysis. Because of the reactions

$$H_2O_2 + e^-_{aq} \rightarrow \cdot OH + OH^-$$

$$H_2 + \cdot OH \rightarrow H_2O + \cdot H$$

hydrogen and hydrogen peroxide are largely consumed. They are therefore produced in low yield, even when irradiation doses are quite high. This is why water can be used as a radiation shield in the water pools of gamma sources. As will be explained later, saturation of the water with oxygen can greatly increase production of H_2O_2.

The formation of hydrogen peroxide, known to be an oxidizing agent, might appear to be of great significance in irradiated foods. Actually it is of less significance than the formation of the highly reactive intermediates. The hydroxyl radical is a powerful oxidizing agent, while the hydrated electron is a strong reducing agent. The hydrogen atom is a somewhat less effective reducing agent. Since all foods contain substances that can be oxidized or reduced, we must expect such reactions to occur when water-containg foods are irradiated.

C. Influence of Oxygen

Presence or absence of oxygen during irradiation can have an important influence on the course of radiolysis. Water in equilibrium with the oxygen of air contains a low concentration of oxygen (about 0.27 mM at ambient temperature). Hydrogen atoms can reduce oxygen to the hydroperoxyl radical:

$$\cdot H + O_2 \rightarrow \cdot HO_2$$

which is a mild oxidizing agent. It is in equilibrium with the superoxide anion radical:

$$\cdot HO_2 \rightleftharpoons H^+ + \cdot O_2^-$$

Another pathway to the formation of the superoxide radical is the reaction of solvated electrons with oxygen:

$$e^-_{aq} + O_2 \rightarrow {}^{\bullet}O_2^-$$

Through the removal of the reducing agents e^-_{aq} and $\cdot H$, the importance of the $\cdot OH$ radical and thus the role of oxidizing reactions becomes much greater in oxygenated solutions.

Both the hydroperoxyl radical and the superoxide radical can give rise to hydrogen peroxide:

$$2\ \cdot HO_2 \rightarrow H_2O_2 + O_2$$

$$^{\bullet}O_2^- + {}^{\bullet}HO_2 + H^+ \rightarrow H_2O_2 + O_2$$

Figure 3 demonstrates how much the formation of hydrogen peroxide in an irradiated aqueous system depends on the presence of oxygen. Very little H_2O_2 is produced when oxygen is excluded.

As will be shown later, many other radicals are produced when foods are irradiated. Oxygen can add to some of those radicals, giving rise to peroxy radicals:

$$\cdot R + O_2 \rightarrow \cdot RO_2$$

Through such reactions the small amount of oxygen present in water or in a food can be consumed quickly when that water or food is irradiated. Because diffusion of oxygen from the atmosphere is rather slow, in electron irradiation any dose higher than about 0.6 kGy creates an anoxic (or anaerobic) condition in the irradiated sample. Since gamma sources deliver a radiation dose at a much lower dose rate than electron accelerators, oxygen has time to diffuse into the sample during gamma irradiation, and anoxic conditions are not necessarily created unless the sample is irradiated in vacuum or under nitrogen or other protective gas. Thus, because of the difference in dose rate, the overall effect of electron irradiation on a material can sometimes be different form that of gamma irradiation at the same dose. This apparent dose rate effect is thus actually an oxygen effect.

Not only in aqueous systems will the presence or absence of oxygen have considerable influence on the radiation sensitivity of the irradiated material and on the nature and amount of radiolysis products formed. (An example of this will be shown in another context in Figure 9.)

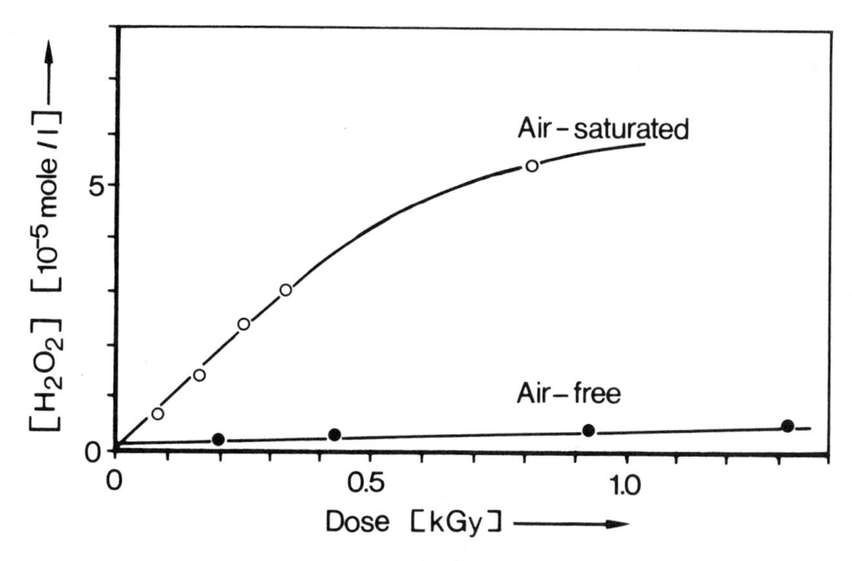

Figure 3 Dose dependence of the formation of hydrogen peroxide in water in absence and presence of air (gamma radiation from a cobalt-60 source). (Adapted from Ref. 7.)

D. Influence of pH

The pH of aqueous systems is another factor that can influence the result of a radiation treatment, because such equilibrium reactions as

$$e^-_{aq} + H^+ \rightleftharpoons \cdot H$$

$$\cdot H + OH^- \rightleftharpoons e^-_{aq}$$

are pH dependent. An acid medium would favor the disappearance of e^-_{aq} by the first reaction, while an alkaline medium would favor the formation of e^-_{aq} by the second reaction.

E. Influence of Temperature

The temperature during irradiation also influences the extent of radiolytic changes. Freezing can have a strong protective effect, as shown in Figure 4 for the example of ascorbic acid. As mentioned earlier, the reactive intermediates of water radiolysis are trapped in deep-frozen materials and are thus kept from reacting with each other or with the substrate. During the warming process they apparently react preferentially with each other, rather than with the substrate. When the material has reached ambient temperature again,

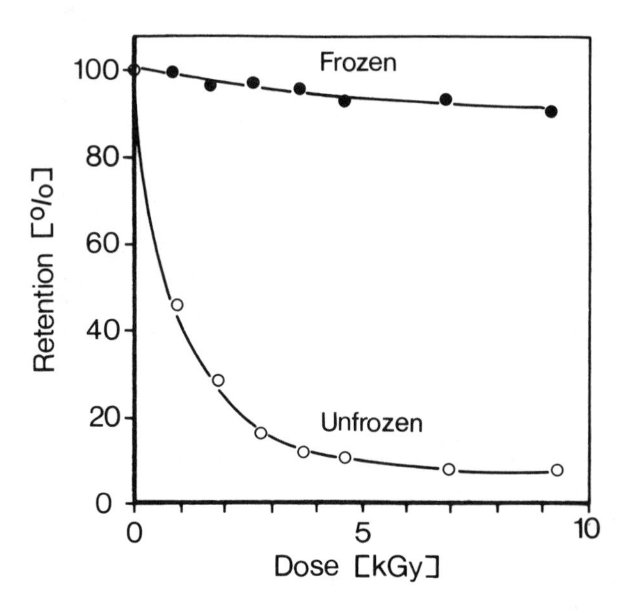

Figure 4 Effect of 3 MeV electron rays on frozen and unfrozen solutions of L-ascorbic acid (8.0 mg/100 ml) in 0.25% oxalic acid solution. (Adapted from Ref. 8.)

the damage to the substrate is much less than when the unfrozen material has been irradiated. This protective effect of freezing is not always as strong as in this example, but it is usually noticeable. It can also play an important role in the irradiation of foods, as demonstrated for the production of volatile products from beef in Figure 5.

The smooth curves in this graph indicate no sharp break between the unfrozen state above 0°C and the frozen state below. Freezing does not mean a complete restriction of diffusion. At −2°C more diffusion of molecules and free radicals is possible than at −10°C, and at −10°C, more than at −80°C.

As discussed in Chapter 9, the sensory quality of meat treated with a high radiation dose deep-frozen is better than that of meat irradiated at ambient temperature.

F. Direct Versus Indirect Effects

When an aqueous solution is irradiated, the molecules of the substrate may be affected directly by the incident electrons or Compton electrons, or they may be affected by reactions with the reactive intermediary species of water radiolysis. The former are *direct ef-*

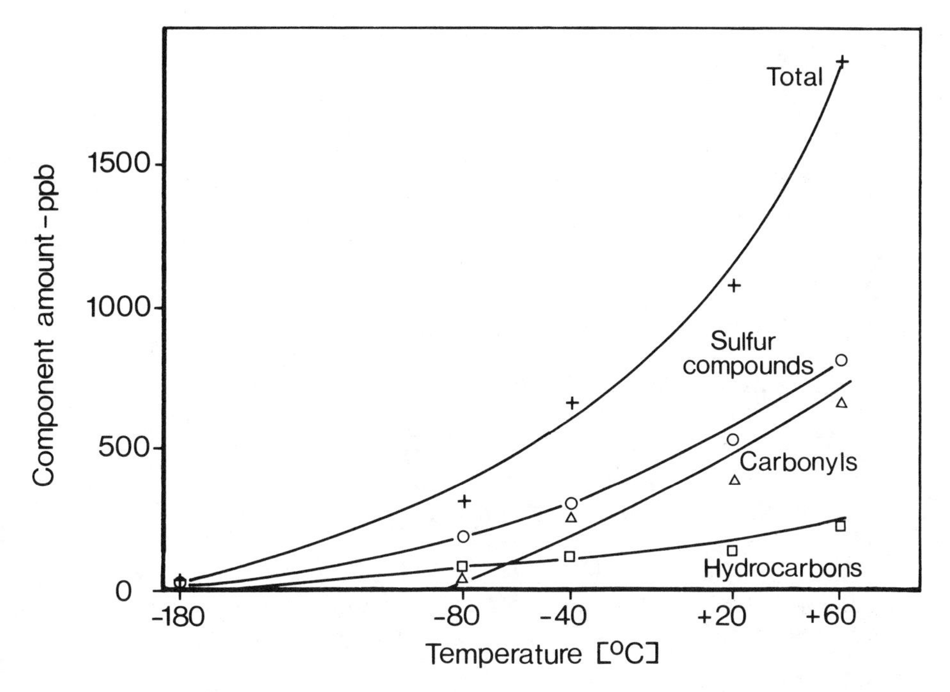

Figure 5 Amounts of volatile components produced in beef irradiated with 45 kGy as a function of temperature. (Reprinted with permission from *J. Agr. Food Chem.* Copyright ©1975 American Chemical Society.)

fects, the latter are *indirect effects* of irradiation. In dilute solutions (less than 0.1 M) indirect effects predominate. Above a concentration of 1 M direct effects become increasingly important.

This terminology is not only applied to aqueous solutions. If a mixture of two substances is irradiated, each substance will be exposed both to direct effects and to the indirect effects caused by reactive radiolytic products of the other substance.

The radiolysis products of direct and indirect action need not be identical. For example, carbon dioxide is evolved when solid benzoic acid or its aqueous solution is irradiated, whereas salicylic acid is formed only in the solution. The formation of salicylic acid depends on hydroxyl radical attack, which can occur only in the presence of water.

The high reactivity of the intermediary radical species which are produced when water is irradiated is responsible for the often reported observation that a given radiation dose will do more damage to a substance dissolved in water than to the pure dry substance, where only direct effects are possible. An example of this is shown in Figure 6. The enzyme pectin esterase had a high radiation resistance

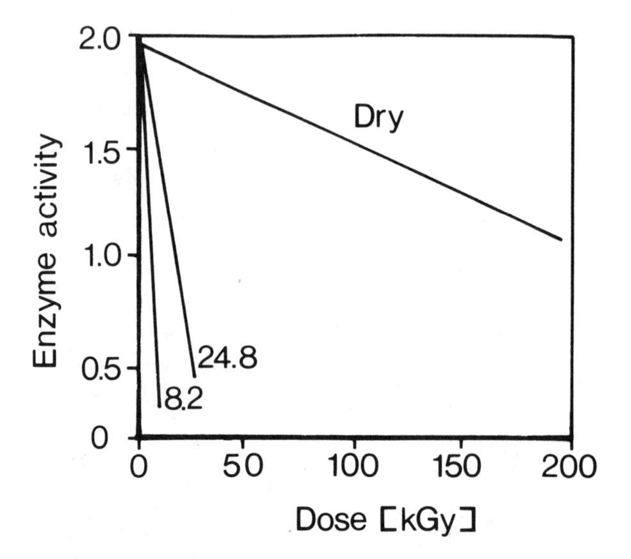

Figure 6 Effect of gamma irradiation on activity (A) of pectin esterase irradiated either in the dry state or in solution (24.8 or 8.2 mg/ml). (Adapted from Ref. 10.)

when irradiated in the dry state but was largely inactivated when irradiated in dilute solution. However, increased radiation sensitivity in the presence of water is not observed in all instances. Radiation-induced decomposition of starch, for instance, decreases with increasing water content [11].

G. The Dilution Effect

When a dilute solution is irradiated the extent of degradation of the solute depends on the number of reactive radicals available for reaction with the solute molecules. The higher the concentration of the solute, the more solute molecules will not find a reaction partner and will remain unchanged. Figure 7 demonstrates this for the example of horseradish peroxidase. A dose of 10 kGy caused 20% loss of enzyme activity in a 1% solution and 60% loss in a 0.5% solution. A dose of less than 1 kGy inactivated the 0.01% solution. This increase of radiation sensitivity with increasing dilution is known as the *dilution effect*.

Instead of reducing the percentage destruction of a solute by irradiating it at a higher concentration, one can also reduce it by adding other solutes which will also compete for the free radical intermediates of the solvent. An example may again demonstrate this (Fig. 8): When the enzyme pepsin was irradiated in dilute solution

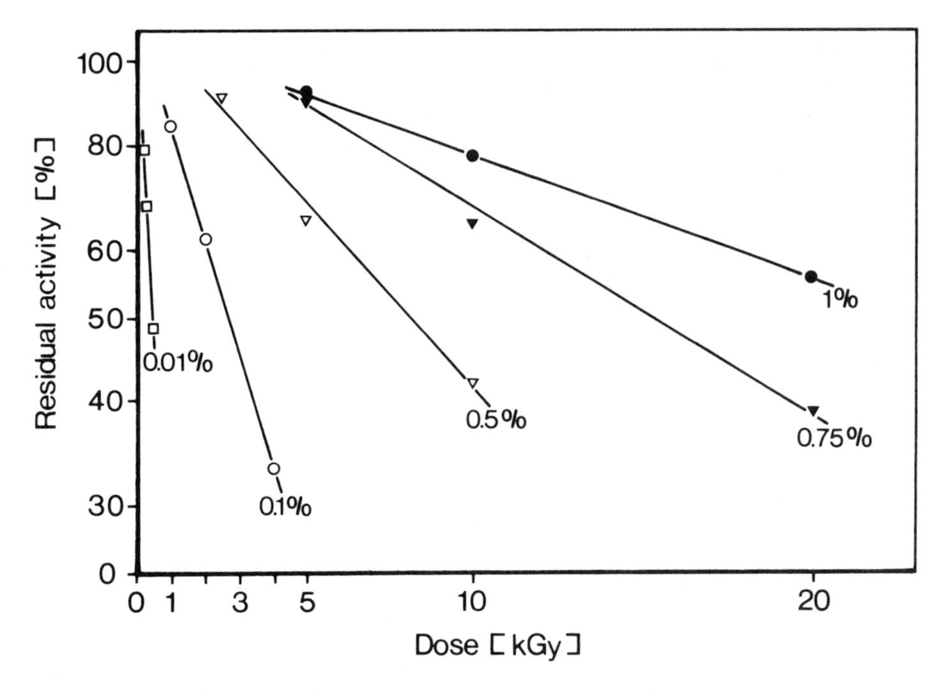

Figure 7 Effect of dilution on the inactivation of the pI 7.1 isoenzyme of horseradish peroxidase by cobalt-60 gamma rays. (From Ref. 12.)

(0.1%), a dose of about 5 kGy caused 80% destruction. Increasing concentrations of D-isoascorbic acid provided increasing protection. The presence of 0.1% isoascorbic acid reduced the destruction caused by 5 kGy to 20%. Isoascorbic acid (like ascorbic acid) is particularly effective in this regard because it acts as a scavenger of electrons and free radicals. (Isoascorbic acid, also called D-araboascorbic or erythorbic acid, is used as an antioxidant but does not have the scurvy-preventive activity of ascorbic acid.) Most other compounds that can react with the intermediary radiolysis products of water would also exert some protection on other solutes present (sensitizers are the exception). This principle applies not only to aqueous systems. As can be seen from Figure 9, alpha-tocopherol (vitamin E) dissolved in isooctane was completely destroyed by a radiation dose of 50 kGy. When trilinolein was simultaneously present, increasing protection of alpha-tocopherol was provided by increasing concentrations of the triglyceride. With 15 mM trilinolein present, no loss of alpha-tocopherol was observed in a nitrogen atmosphere, a loss of about 50% in air. (The differential destruction

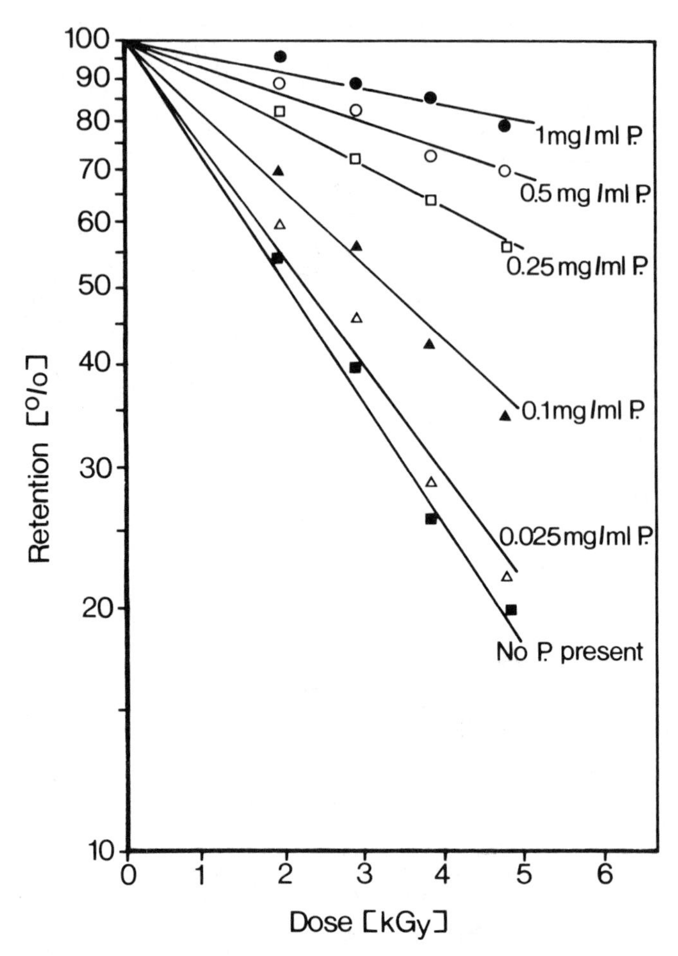

Figure 8 Protective effect of sodium D-isoascorbate (P = protector) on pepsin (1 mg/ml acetate buffer, pH 4.3) irradiated with 3-MeV cathode rays. (Adapted from Ref. 13.)

of alpha-tocopherol in air and in nitrogen may serve as a reminder of what was said earlier about the role of oxygen.)

H. Single-Component Vs. Multicomponent Systems

Because of the mutual protection exerted when different substances are irradiated together, and because most foodstuffs consist of a great number of compounds, food irradiation will generally not cause much chemical change in any one of these compounds. The radiation

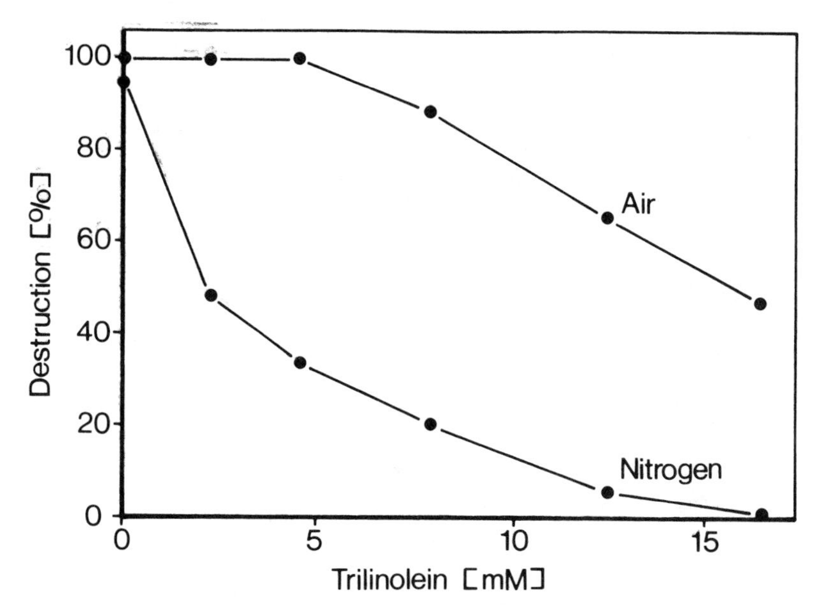

Figure 9 Effect of increasing concentrations of trilinolein on radiolysis of alpha-tocopherol (0.5 mM in isooctane) in air or nitrogen (10 MeV electrons, 50 kGy). (From Ref. 14.)

damage will rather be distributed to all components, though not evenly. A scavenger, like ascorbic acid, will show a greater percentage loss than a less reactive component.

The principle of mutual protection of substances irradiated in a mixture is of great practical significance. Regrettably, it is often overlooked—both by some who oppose radiation processing and by some who advocate it. Some authors have made dire predictions about the extent of radiation-induced changes in food, basing their arguments on experiments carried out with pure solutions of vitamins or other food components. Others propose to use radiation processing as a method for destroying aflatoxin or other contaminants in foods. They also base their proposals on experiments done on pure solutions, which have shown substantial radiation-induced losses of such contaminants. Experiments done on actual foods have usually shown minimal destruction in the acceptable dose range.

III. EFFECTS OF IONIZING RADIATION ON OTHER FOOD COMPONENTS

Besides water, the main components of foodstuffs are carbohydrates, lipids, and proteins. The radiation chemistry of these groups of substances will be the subject of this section. Radiation effects on some minor constituents, particularly the vitamins, will be discussed in Chapter 7. Minerals and trace elements require no comments in this context because they cannot be affected under the process conditions of food irradiation.

A. Carbohydrates

In the presence of water, carbohydrates are attacked mainly by •OH radicals. Solvated electrons and hydrogen atoms play only a minor role. The •OH radicals abstract predominantly the hydrogen of C—H bonds, forming water:

$$\cdot OH + H{-}\overset{|}{\underset{|}{C}}{-}OH \rightarrow \cdot\overset{|}{\underset{|}{C}}{-}OH + H_2O \quad \text{hydrogen abstraction}$$

The resulting radicals react further by various mechanisms:

$$\cdot\overset{|}{\underset{|}{C}}{-}OH + \cdot\overset{|}{\underset{|}{C}}{-}OH \rightarrow \overset{|}{\underset{|}{C}}{=}O + H\overset{|}{\underset{|}{C}}OH \quad \text{disproportionation}$$

$$\cdot\overset{|}{\underset{|}{C}}{-}OH + \cdot\overset{|}{\underset{|}{C}}{-}OH \rightarrow HO{-}\overset{|}{\underset{|}{C}}{-}\overset{|}{\underset{|}{C}}{-}OH \quad \text{dimerization}$$

$$\begin{matrix} | \\ \cdot C{-}OH \\ | \\ H{-}C{-}OH \\ | \end{matrix} \rightarrow \begin{matrix} | \\ C{=}O \\ | \\ \cdot C{-}H \\ | \end{matrix} + H_2O \quad \text{dehydration}$$

Depending on the molecular position of the $\overset{|}{\underset{|}{C}}{=}O$ formed by disproportination or dehydration, the resulting product can be an acid, a ketone, or an aldehyde. Thus, hydrogen abstraction at the C—1 of glucose can lead to gluconic acid:

GLUCOSE → (−H) → GLUCOSE RADICAL → (−H) → GLUCONIC ACID

Through loss of CO the 6-carbon sugar glucose can also be converted to the 5-carbon sugar arabinose:

GLUCOSE RADICAL → (−CO) → 5-CARBON RADICAL → (−H) → ARABINOSE

Ring opening can lead to 5-deoxygluconic acid:

GLUCOSE RADICAL → 6-CARBON RADICAL → (+H) → 5-DEOXY-GLUCONIC ACID

Two further products of hydrogen abstraction at the C—1 position of glucose are 2-deoxyribose and 2-deoxygluconic acid. Since •OH radicals can abstract hydrogen from all six carbon atoms of glucose, a great variety of reaction products can be formed. Von Sonntag's review listed 34 radiolytic products of glucose [15].

The formation of deoxy compounds is suppressed when glucose is irradiated in the presence of oxygen, whereas yields of sugar acids and keto sugars increase. The formation of acids leads to a decrease in the pH of irradiated sugar solutions. In glucose solution a dose of 25 kGy results in a pH decrease of 3 units. As pointed out by Schubert, this has sometimes been overlooked in experiments where effects of irradiated sugar solutions on bacteria and on cell cultures were studied [16].

Many investigations have been carried out with monosaccharides other than glucose, and similar reaction mechanisms were postulated. When disaccharides or polysaccharides are irradiated, the reactions observed with monosaccharides can also occur. Additionally, the

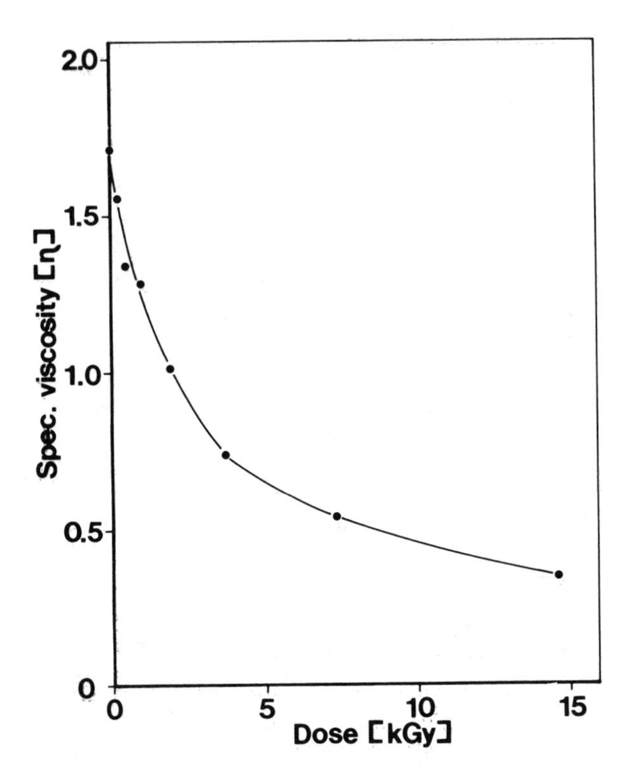

Figure 10 Effect of radiation dose on spec. viscosity of potato starch. Cobalt-60 irradiation of starch containing 15.7% water, viscosity of 0.4% solution in N-KOH at 25°C. (Adapted from Ref. 17.)

glycosidic bonds which connect the monosaccharide units can be broken. Irradiation of starch produces dextrins, maltose, and glucose. This reduces the degree of polymerization in polysaccharides and leads to reduced viscosity of polysaccharides in solution. In Figure 10 this is demonstrated for the example of potato starch. The solubility of starch in water increases with increasing radiation dose.

The main radiolytic products identified in irradiated maize starch are listed in Table 1. The nature and quantity of these products is essentially the same when other varieties of starch (wheat, potato etc.) are irradiated [19]. Hydrogen and carbon monoxide must surely be added to this list. Presumably the analytical method used by the authors on whose data Table 1 is based did not permit detection of these gases.

Table 1 Main radiolytic Products of Maize Starch Containing 12–13% Water, Gamma Irradiated Without Exclusion of Oxygen

Radiolytic product	mg/kg per 10 kGy
Formic acid	100
Acetaldehyde	40 (up to 8 kGy)
Formaldehyde	20
Maltose	9.8
Glycolaldehyde	9
Hydrogen peroxide	6.6 (1–4 kGy)
Glucose	5.8
Glyceraldehyde and/or dihydroxyacetone	4.5
Glyoxal	3.5
Methanol	2.8
Acetone	2.1 (above 20 kGy)
Malondialdehyde	2
Erythrose	1.2
Hydroxymethylfurfural and others at still lower concentrations	1

Source: Ref. 18.

Winchester [20] has reported that the radiation-induced formation of malondialdehyde permits identification of irradiated starch even when a low radiation dose was applied. However, the formation of malondialdehyde and its rate of disappearance after irradiation depend on the humidity of the starch. This should make it rather difficult to use the analytical determination of this substance as an indicator of irradiation. The extent to which factors unrelated to radiation, such as production and storage conditions, influence levels of malondialdehyde must also be taken into account.

Very different results can be obtained when crystalline sugars are irradiated rather than aqueous sugar solutions. This could hardly be relevant in food irradiation because there would be no good reasons for irradiating crystalline sugars. However, the pharmaceutical industry is using crystalline sugars as carriers for medications in making tablets. If radiation processing were to be used for sterilizing such tablets, attention would have to be paid to the particular radiation response of some of the crystalline sugars. For instance, when crystalline D-fructose is irradiated, predominantly one product is obtained: 6-deoxy-D-threo-2,5-hexodiulose. This is caused by a chain reaction, which can be explained on the basis of the crystal structure of D-fructose [21]. Crystalline lactose also undergoes chain reaction with high one-product yield, but glucose does not.

The influence of water in the complete range from dry ($a_w = 0$) to wet ($a_w = 0.97$) on product formation in an irradiated polysaccharide was studied, using dextran as a model (Fig. 11). With increasing a_w the yield of 5-deoxy-xylohexodialdose decreased linearly, while threo-tetrodialdose and threonic acid lactone increased. Formation of glucose remained constant up to $a_w = 0.76$ and increased slightly in 30% aqueous solution ($a_w = 0.97$). When solid dextran was irradiated, degradation reactions predominated; when aqueous solutions were irradiated, cross-linking and branching reactions were more pronounced [22].

The relatively high yield of radiolytic products in irradiated pure sugars or polysaccharides, according to Table 1, for instance, 100 mg of formic acid per kg of starch irradiated with a dose of 10 kGy, recalls what was said in Sec. II.H about the high radiation sensitivity of single-component systems. When carbohydrates are irradiated as components of a food, they are much less radiation-sensitive than in pure form. For example, when radiolysis products of pure starch and of wheat flour, where the starch is protected by the presence of proteins, were compared, product formation from starch irradiated with a dose of 5 kGy was about as much as from flour irradiated at 50 kGy (Fig. 12). (The gas chromatography detector was run at a higher sensitivity for the wheat flour samples, as indicated by the stronger peak of the internal standard.)

The protection of a carbohydrate by proteins is illustrated in Figure 13. The disaccharide trehalose was irradiated in aqueous solution in the presence of three different proteins: bovine serum albumin, sperm whale myoglobin, or papain. The radiolytic destruction of trehalose was much less pronounced in the presence of the proteins. The different effectiveness of different proteins is partly due to their different amino acid compositions. Some amino acids, like cysteine, methionine, or phenylalanine, are better scavengers than other amino acids, like glycine, and therefore exert better protection (Fig. 14).

B. Proteins

Proteins consist of chains of amino acids connected by peptide bonds. A discussion of the effects of radiation on proteins must therefore be

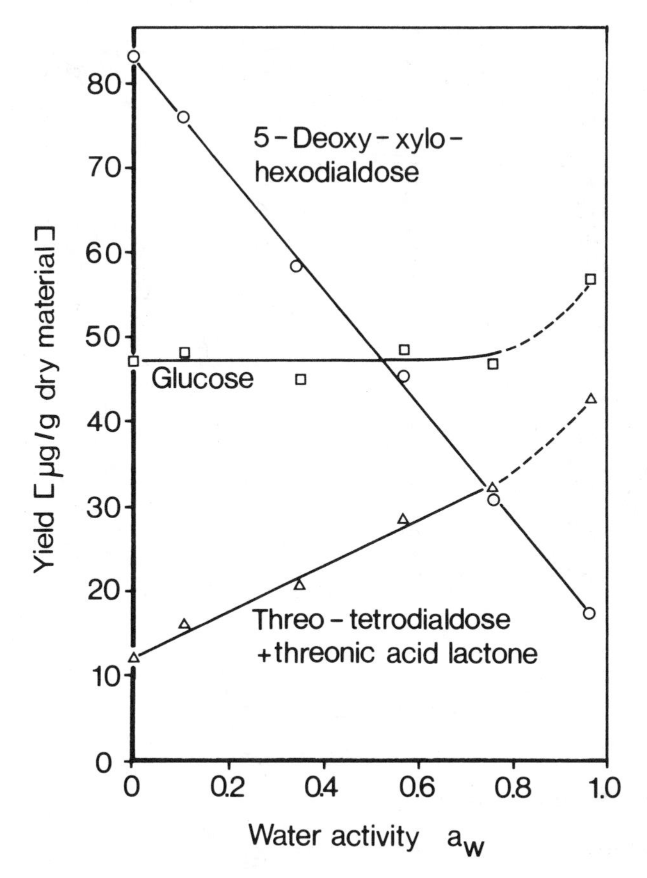

Figure 11 Influence of water activity on the yields of monomeric products from electron-irradiated dextran (10 MeV electrons, 50 kGy). (From Ref. 22.)

based on an understanding of the radiation chemistry of amino acids. Taking the aliphatic amino acid alanine as an example, the following reactions describe the major radiolytic events resulting from irradiation in aqueous systems:

$$\cdot OH + H_3\overset{+}{N}-\underset{\underset{CH_3}{|}}{\overset{H}{C}}-COO^- \rightarrow H_3\overset{+}{N}-\underset{\underset{CH_3}{|}}{\overset{\cdot}{C}}-COO^- + H_2O \quad \text{abstraction of H}$$

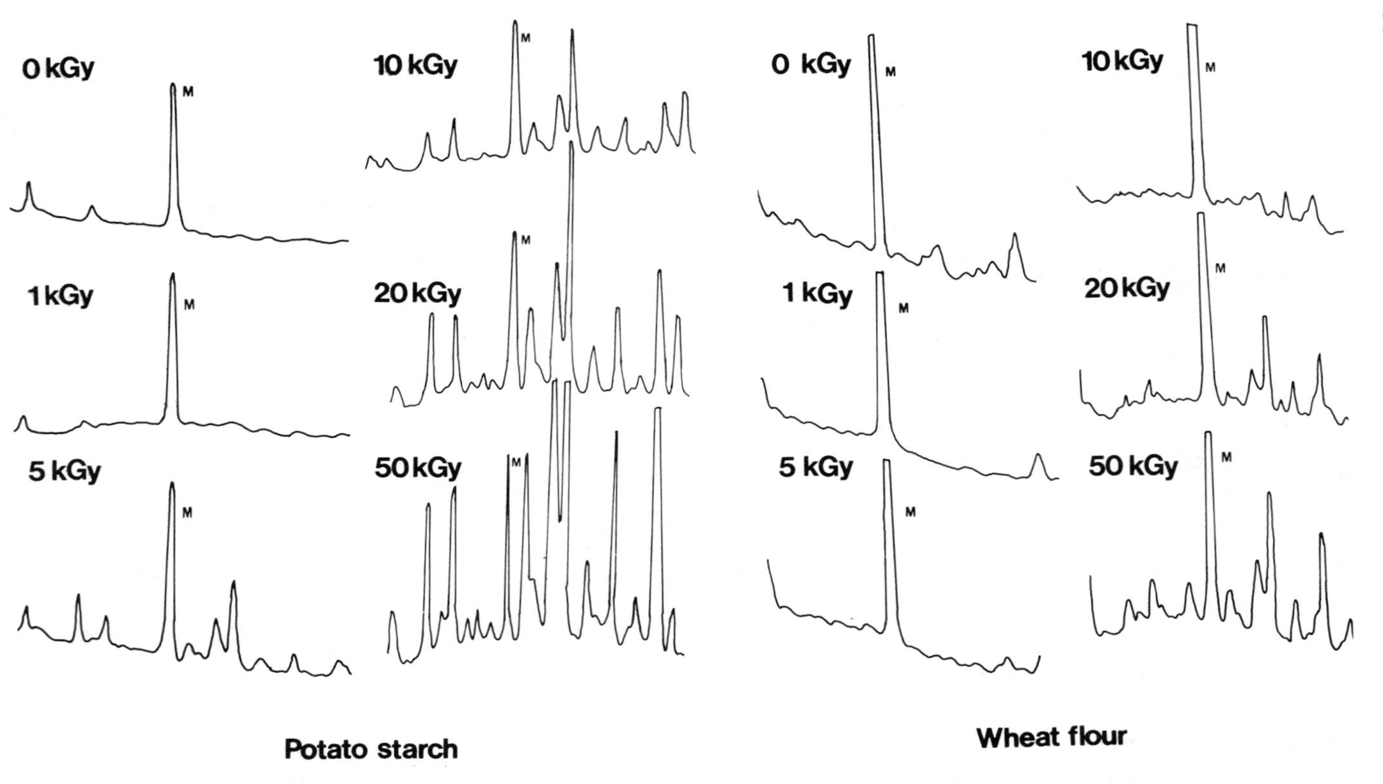

Figure 12 Gas chromatograms of extracts of unirradiated and irradiated starch and wheat flour (10 MeV electrons; samples extracted with ethylacetate-acetone-water 4:5:1; reduction with KBH_4 and trimethylsilylation. M = mesoerythritol, added as an internal standard). (From Ref. 23.)

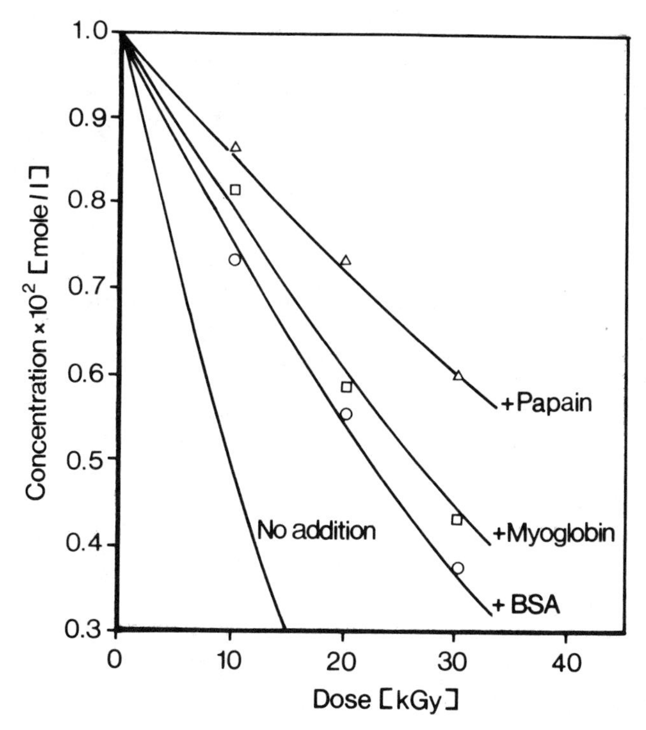

Figure 13 Dose dependence of the decomposition of 0.36% aqueous trehalose solution (10^{-2} M) upon gamma irradiation with and without addition of 0.36% proteins: BSA, bovine serum albumin; sperm whale myoglobin; papain. (From Ref. 24.)

$$\cdot H + H_3\overset{+}{N}-\underset{CH_3}{\overset{H}{\underset{|}{C}}}-COO^- \rightarrow H_3\overset{+}{N}-\underset{CH_3}{\underset{|}{\dot{C}}}-COO^- + H_2$$ abstraction H of H

$$e^-_{aq} + H_3\overset{+}{N}-\underset{CH_3}{\overset{H}{\underset{|}{C}}}-COO^- \rightarrow NH_3 + \cdot\underset{CH_3}{\overset{H}{\underset{|}{C}}}-COO^-$$ reductive deamination

The produced radicals will react further, for instance, by disproportionation:

$$H_3\overset{+}{N}-\overset{\cdot}{\underset{CH_3}{C}}-COO^- + H_3\overset{+}{N}-\overset{\cdot}{\underset{CH_3}{C}}-COO^- \longrightarrow H_2\overset{+}{N}=\underset{CH_3}{C}-COO^- + \text{ALANINE}$$

The unstable imino acid is immediately hydrolyzed:

$$H_2\overset{+}{N}=\underset{CH_3}{C}-COO^- \xrightarrow{+\,H_2O} \overset{+}{N}H_4 + O=\underset{CH_3}{C}-COO^-$$

PYRUVIC ACID

$$H_3\overset{+}{N}-\overset{\cdot}{\underset{CH_3}{C}}-COO^- + \cdot\overset{H}{\underset{CH_3}{C}}-COO^- \longrightarrow H_2\overset{+}{N}=\underset{CH_3}{C}-COO^- + H\overset{H}{\underset{CH_3}{C}}-COO^-$$

PROPIONIC ACID

(The imino acid $H_2\overset{+}{N}=C(CH_3)-COO^-$ reacts $+\,H_2O$ to give $\overset{+}{N}H_4 + O=C(CH_3)-COO^-$, pyruvic acid.)

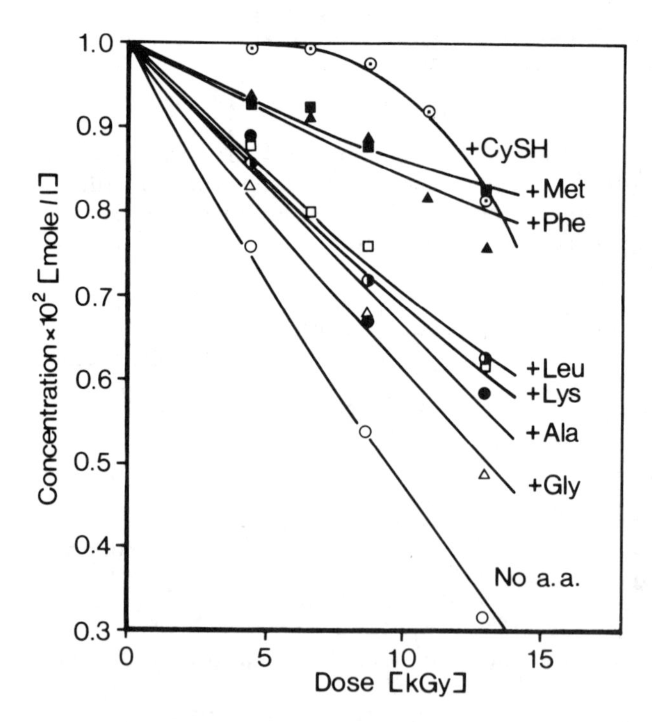

Figure 14 Dose dependence of the decomposition of trehalose upon gamma-irradiation in air-saturated aqueous solution (10^{-2} M) with and without added amino acids (10^{-2} M). (From Ref. 24.)

Radical transfer from the carboxyl group to the alpha-carbon can give rise to another pathway:

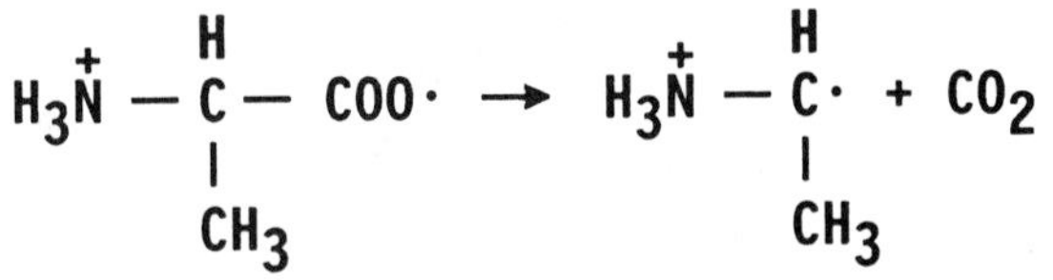

$$H_3\overset{+}{N}-\underset{\underset{CH_3}{|}}{\overset{H}{C}}-COO\cdot \rightarrow H_3\overset{+}{N}-\underset{\underset{CH_3}{|}}{\overset{H}{C}}\cdot + CO_2 \quad \text{decarboxylation}$$

The resulting radical can react with another one by disproportionation:

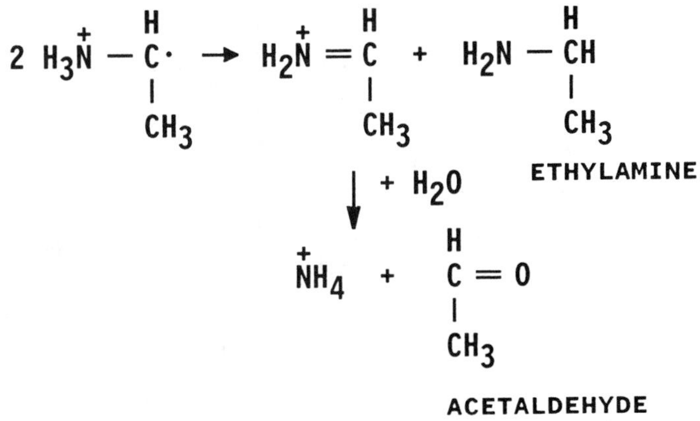

$$2\ H_3\overset{+}{N}-\underset{\underset{CH_3}{|}}{\overset{H}{C}}\cdot \rightarrow H_2\overset{+}{N}=\underset{\underset{CH_3}{|}}{\overset{H}{C}} + H_2N-\underset{\underset{CH_3}{|}}{\overset{H}{CH}}$$

ETHYLAMINE

$$H_2\overset{+}{N}=\underset{\underset{CH_3}{|}}{\overset{H}{C}} \xrightarrow{+H_2O} \overset{+}{N}H_4 + \underset{\underset{CH_3}{|}}{\overset{H}{C}}=O$$

ACETALDEHYDE

These reactions do not all have the same importance, as can be seen from the product yields listed in Table 2.

The prevalence of ammonia and pyruvic acid indicates that deamination plays a greater role than decarboxylation. Other studies have shown that, in contrast to the radiolysis of carbohydrates, reactions initiated by the solvated electron play a major role in the radiolysis of amino acids and proteins.

Presence of oxygen during irradiation does not radically alter the spectrum of radiolytic products, although it does influence product yields. Oxidative deamination occurs instead of reductive deamination:

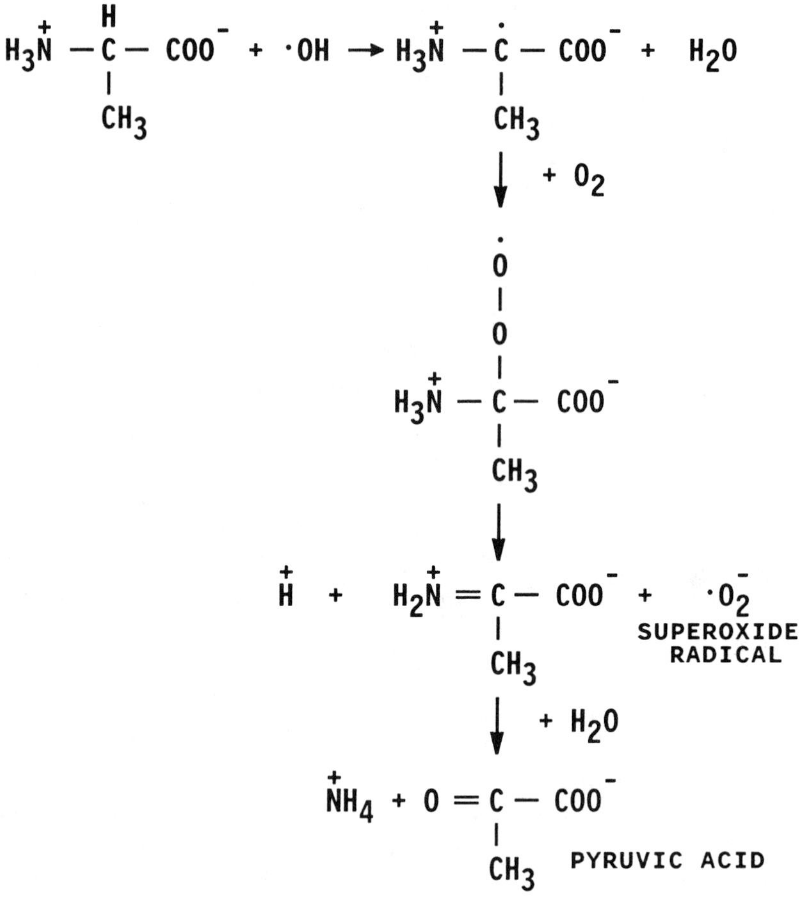

$$H_3\overset{+}{N}-\underset{CH_3}{\overset{H}{C}}-COO^- + \cdot OH \rightarrow H_3\overset{+}{N}-\underset{CH_3}{\dot{C}}-COO^- + H_2O$$

$$\downarrow + O_2$$

$$H_3\overset{+}{N}-\underset{CH_3}{\overset{\dot{O}-O}{C}}-COO^-$$

$$\downarrow$$

$$\overset{+}{H} + H_2\overset{+}{N}=\underset{CH_3}{C}-COO^- + \cdot O_2^- \quad \text{SUPEROXIDE RADICAL}$$

$$\downarrow + H_2O$$

$$\overset{+}{N}H_4 + O=\underset{CH_3}{C}-COO^- \quad \text{PYRUVIC ACID}$$

As mentioned earlier, cysteine, cystine, and methionine act as scavengers. They react more readily with free radicals than the aliphatic amino acids. With cysteine the following reactions take place preferentially:

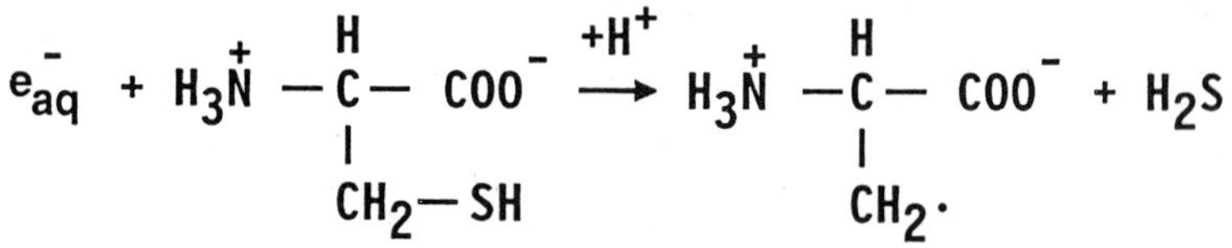

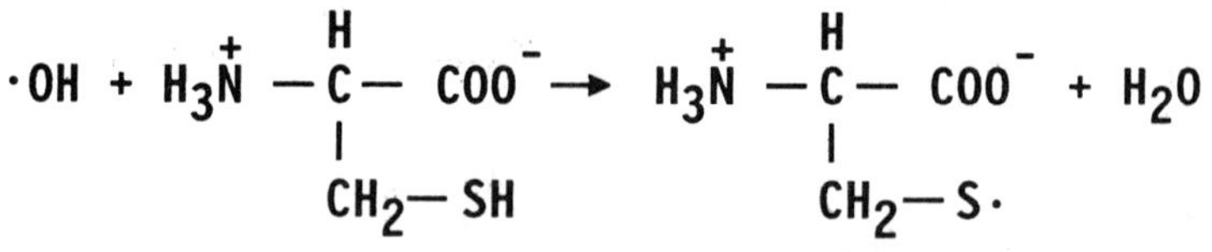

$$e^-_{aq} + H_3\overset{+}{N}-\underset{CH_2-SH}{\overset{H}{C}}-COO^- \xrightarrow{+H^+} H_3\overset{+}{N}-\underset{CH_2\cdot}{\overset{H}{C}}-COO^- + H_2S$$

$$\cdot OH + H_3\overset{+}{N}-\underset{CH_2-SH}{\overset{H}{C}}-COO^- \rightarrow H_3\overset{+}{N}-\underset{CH_2-S\cdot}{\overset{H}{C}}-COO^- + H_2O$$

Table 2 Main Radiolytic Products from Gamma Irradiated Aqueous Oxygen-Free Solution of Alanine (1 M) at pH 5.9

Radiolytic products	mg/kg per 10 kGy
Pyruvic acid	176
Propionic acid	80
Ammonia	79
Carbon dioxide	27
Acetaldehyde	27
Ethylamine	8
Hydrogen	2.3

Source: Ref. 25.

$$\cdot H + H_3\overset{+}{N}-\underset{\underset{CH_2-SH}{|}}{\overset{H}{C}}-COO^- \rightarrow H_3\overset{+}{N}-\underset{\underset{CH_2-S\cdot}{|}}{\overset{H}{C}}-COO^- + H_2$$

The alanine radical or cysteine radical formed by these reactions can undergo various subsequent reactions, such as:

$$H_3\overset{+}{N}-\underset{\underset{CH_2\cdot}{|}}{\overset{H}{C}}-COO^- + H_3\overset{+}{N}-\underset{\underset{CH_2-SH}{|}}{\overset{H}{C}}-COO^- \rightarrow \text{ALANINE} + H_3\overset{+}{N}-\underset{\underset{CH_2-S\cdot}{|}}{\overset{H}{C}}-COO^-$$

$$2\ H_3\overset{+}{N}-\underset{\underset{CH_2-S\cdot}{|}}{\overset{H}{C}}-COO^- \rightarrow H_3\overset{+}{N}-\underset{\underset{CH_2}{|}}{\overset{H}{C}}-COO^- \quad H_3\overset{+}{N}-\underset{\underset{CH_2}{|}}{\overset{H}{C}}-COO^- \quad (CH_2-S-S-CH_2)$$

Stable end products of cysteine hydrolysis are mainly hydrogen, hydrogen sulfide, alanine, and cystine.

Minor products result from reactions of the type described with alanine, namely, decarboxylation and deamination. However, the much higher reactivity of the sulfhydryl group largely prevents these other pathways.

When cystine is irradiated, splitting of the disulfide bridge is the most important reaction, while methionine undergoes reactions of the type:

$$e^-_{aq} + R{-}S{-}CH_3 \rightarrow RS^- + \cdot CH_3$$

or

$$e^-_{aq} + R{-}S{-}CH_3 \rightarrow R\cdot + {}^-SCH_3$$

The aromatic amino acids phenylalanine and tyrosine also react readily with the transient species of water radiolysis, hydroxylation of the aromatic ring being the principal reaction. In the case of phenylalanine o-, m-, and p-tyrosine are formed. Oxidation converts these to various isomers of dihydroxyphenylalanine (dopa). The main product formed with tyrosine is 3,4-dopa. Subsequent oxidation of 3,4-dopa and polymerization can produce melanine-type pigments.

Histidine also belongs to the more radiation-sensitive amino acids. Deamination of histidine occurs to a greater extent than with any other amino acid. Both the side chain and the imidazole ring contribute ammonia.

When proteins are irradiated in the presence of water, all the reactions that are possible with amino acids are also possible with a protein containing these amino acids. Irradiation in the absence of water has also been studied but is of much less interest in the context of food irradiation.

With some 20 amino acids as constituents of the proteins, and with three reactive species of water radiolysis, very complex interactions are possible. Additional effects are exerted by the spatial configuration of the protein chains, as determined by hydrogen bonds, disulfide bridges, hydrophobic bonds, and ionic bonds. Amino acids which are sensitive to radical attack when irradiated by themselves are often much less sensitive when they are deeply buried in a protein structure and thus more or less inaccessible to radical reactions. Another factor probably also contributes to the much higher radiation resistance of proteins as compared to isolated amino acids: Due to the more or less rigid spatial structure of the protein molecule, radicals formed as a result of irradiation are held in position and have a high chance of recombination.

Metal ions of Fe, Cu, and Zn, particularly when bound to a porphyrin ring, modify mainly the reactions of secondary radicals formed on the protein. They can introduce new pathways for reaction, presumably involving intramolecular electron transfer.

Only a few of the many possible pathways of radiolysis in proteins can be shown:

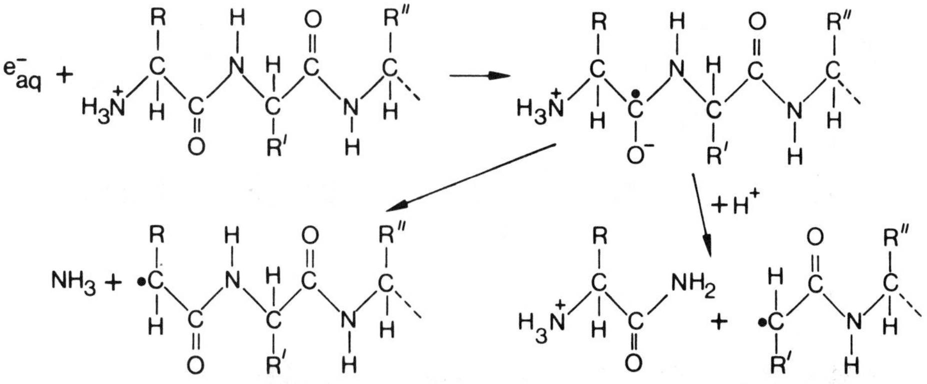

Besides scission of C—N bonds in the backbone of the polypeptide chain, other reactions, such as the splitting of disulfide bridges (described earlier in the case of cystine), can cause degradation to smaller proteins. Such degradation reactions, indicated by a decrease of viscosity of the protein solution, have been observed with fibrous proteins. On the other hand, globular proteins irradiated in dilute solutions undergo aggregation reactions, resulting in increased viscosity. The nature of bonds responsible for the aggregate formation is not yet known, but it is conceivable that radicals produced in the backbone or in the side chains, as shown in this hypothetical example,

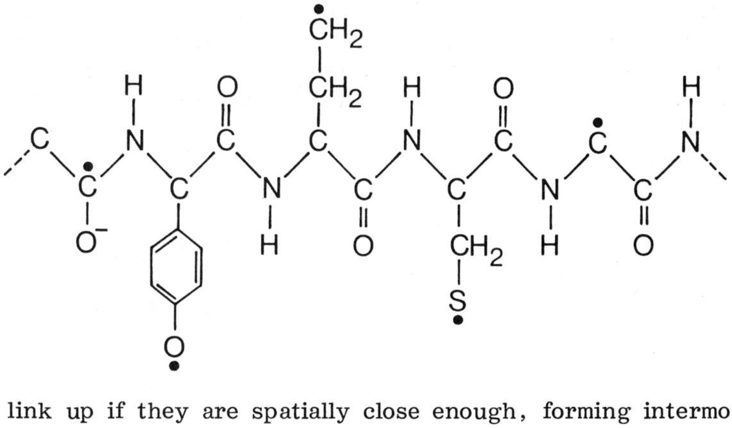

will link up if they are spatially close enough, forming intermolecular bonds:

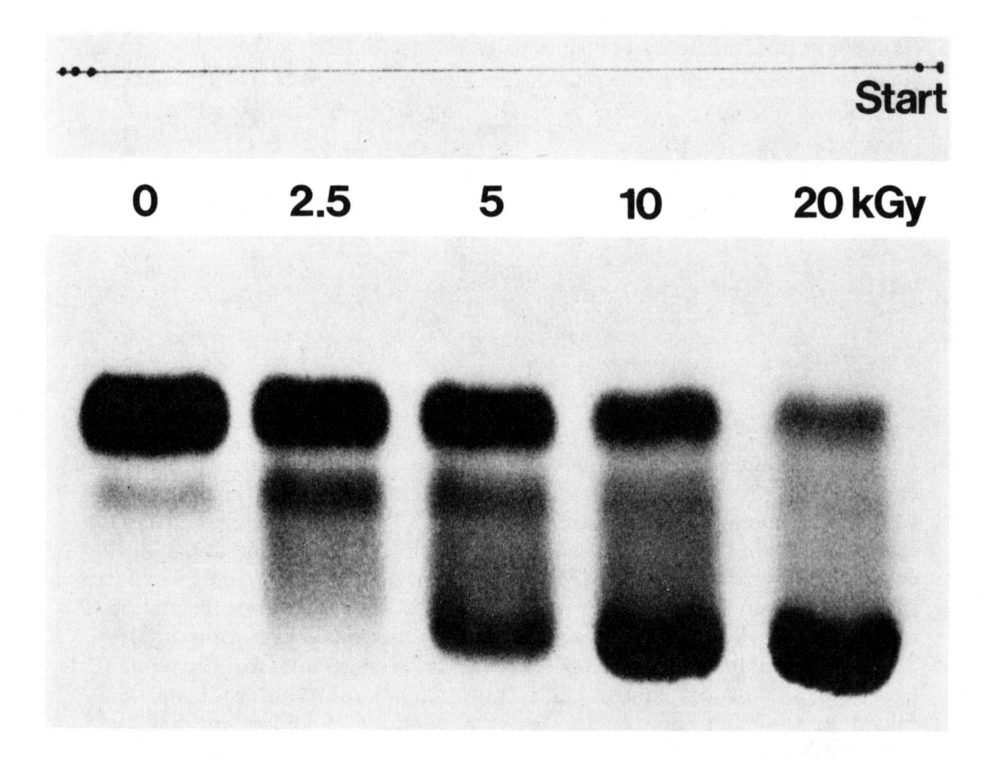

Figure 15 Radiation-induced aggregation of ribonuclease. Thin-layer gel chromatography on Sephadex G-75 Superfine, of RNase gamma irradiated in 1% solution. Starting line at the top. Proteins stained with Amidoblack 10 B. The more a protein has traveled away from the starting line, the higher is its molecular weight. (From Ref. 26.)

A good method to demonstrate changes in molecular size is gel filtration. In the thin-layer technique, a glass plate is coated with the molecular sieving gel, and after migration of the proteins through a certain distance of the gel, a paper print is taken and stained. In

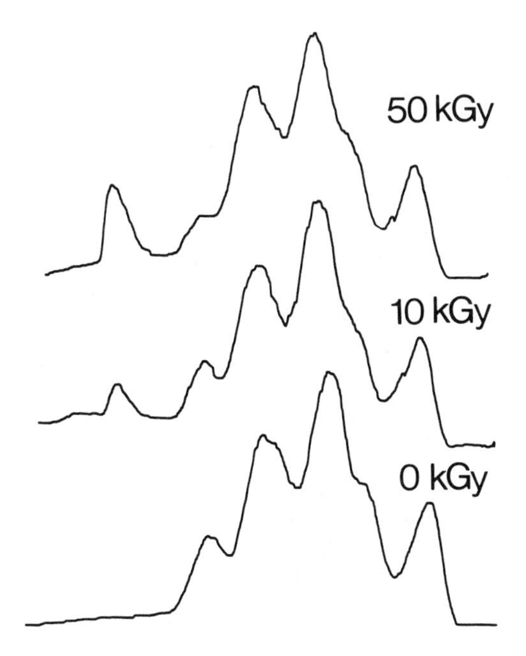

Figure 16 Densitograms of sarcoplasmatic proteins of beef, unirradiated or irradiated at 10 or 50 kGy (10 MeV electrons; irradiation temperature 0°C); thin layer gel chromatography on Sephadex® G-200, stained with Coomassie Blue® G-250). Starting point on the right-hand side. (From Ref. 27.)

Figure 15 the effect of irradiation on the molecular size of the enzyme ribonuclease is exhibited. While the unirradiated sample contained only a trace of dimeric material, a dose of 2.5 kGy resulted in the formation of more dimers and some polymers. Higher doses decreased the proportion of monomer and increased the proportion of higher polymers. Such aggregation reactions also take place when meat is irradiated, as demonstrated in Figure 16. Sarcoplasmatic proteins of beef were separated by the thin-layer gel filtration technique, and the optical density of the stained protein bands was measured. The four peaks visible in the densitogram of the unirradiated sample correspond to protein fractions with molecular weights of approximately (from right to left): 18,000; 68,000; 157,000; and 340,000. The fifth peak that appears in the 10 kGy sample and more strongly in the 50 kGy sample is attributable to proteins having molecular weights of >465,000. It should be noted that in the experiment underlying Figure 16, equal protein concentrations of extracts from irradiated and control meat samples were applied to the thin-layer plate. Radiation-induced aggregation leads to decreased solubility of proteins. Unless an adjustment for this lower protein solubility is made, lower peaks are obtained in the densitogram of irradiated meat. A 30% decrease in total soluble proteins of beef irradiated at 10 kGy has been reported [28]. Radiation induced aggregation of enzymes is not necessarily associated with loss of enzymatic activity [26].

A large proportion of the radiation energy deposited in an irradiated protein apparently goes into protein denaturation, i.e., changes in secondary and tertiary structure, rather than into destruction of constituent amino acids. Even this denaturation is much less extensive than that caused by heating. This is why radiation-sterilized foods destined for long-term storage must receive a heat treatment in addition to the radiation treatment. Due to the resistance of endogenous enzymes even to the high radiation doses used for sterilization, such foods would undergo enzymatic spoilage, unless they were enzyme-inactivated by blanching.

Residual enzyme activity in lamb's liver irradiated with doses of up to 400 kGy (!) is depicted in Figure 17. While phospholipase and total lipase required doses of about 100 kGy to affect enzyme activity at all, protease activity declined by about 20% after a dose of 10 kGy.

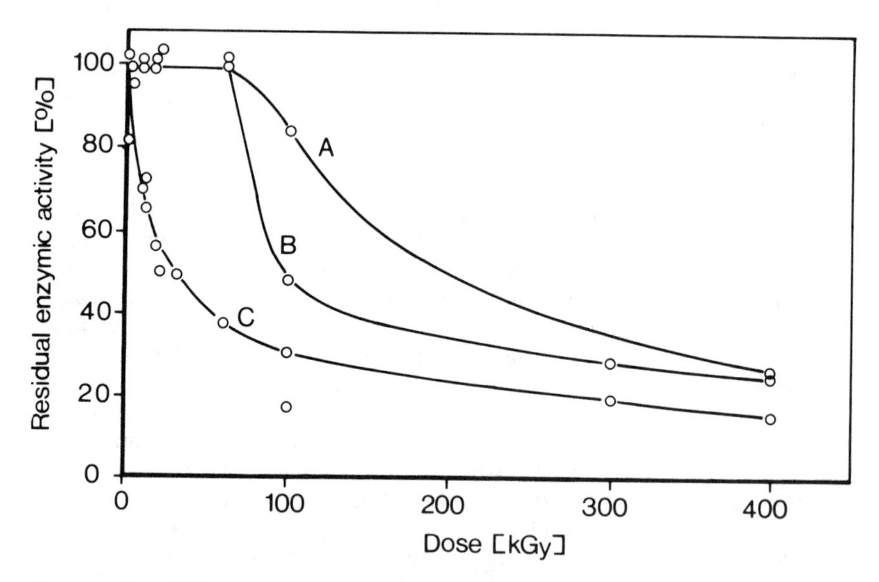

Figure 17 Effect of electron irradiation (2 MeV) on proteolytic and lipolytic activity in lamb's liver: A, total lipase; B, phospholipase; C, protease. (Adapted from Ref. 29.)

Other evidence that radiation damage to constituent amino acids of proteins in irradiated foods is very limited comes from chromatographic analysis of the amino acids in protein hydrolyzates. In most experiments no significant change in amino acid composition was found in the dose range below 50 kGy. As an example, the amino acid composition of proteins of cod fish is shown in Table 3. If there was any effect of irradiation, it was within the range of biological variability and analytical error.

More extensive accounts of the radiation chemistry of amino acids and proteins are available [31–33] and may be consulted for details.

C. Lipids

The lipid or fat portion of foods consists predominantly of triglycerides. Milk fat, for instance, contains 94%, soybean oil contains 88% triglycerides. To simplify matters, the following discussion of the radiation chemistry of lipids will be limited to this major fraction.

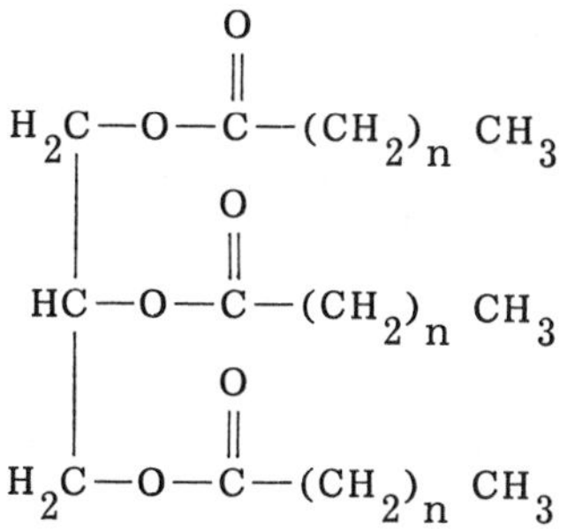

$$
\begin{array}{l}
H_2C-O-C(=O)-(CH_2)_n\,CH_3 \\
\quad | \\
HC-O-C(=O)-(CH_2)_n\,CH_3 \\
\quad | \\
H_2C-O-C(=O)-(CH_2)_n\,CH_3
\end{array}
\qquad \text{triglyceride}
$$

When the abbreviation $RCH_2-O-CO-(CH_2)_nCH_3$ is used in the following, R stands for the diglyceride portion of the molecule.

In contrast to the radiation chemistry of food proteins and food carbohydrates, where indirect radiation effects mediated through water play a major role, reactions of lipids with a reactive species of water radiolysis play only a minor role in most situations—quantitatively at least. Qualitatively they may be important in the lipid membranes bordering on aqueous cell fluids.

Upon irradiation of fats, the primary effect of incident electrons or Compton electrons leads to cation radicals and excited molecules:

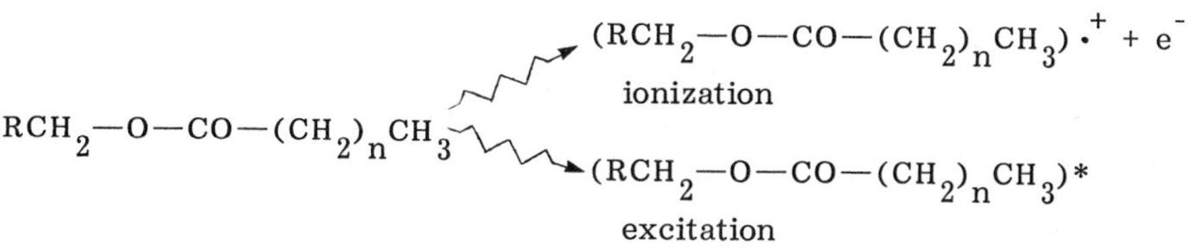

$$
RCH_2-O-CO-(CH_2)_nCH_3 \rightsquigarrow
\begin{cases}
(RCH_2-O-CO-(CH_2)_nCH_3)^{\cdot +} + e^- & \text{ionization} \\
(RCH_2-O-CO-(CH_2)_nCH_3)^{*} & \text{excitation}
\end{cases}
$$

Table 3 Amino-Acid Composition (g/16 g N) of Nonirradiated Cod, and Cod Irradiated at Different Dose Levels

Amino acid	Nonirradiated	Irradiation dose (kGy) 1	3	6	10	25	45
Lysine	11.35	10.68	10.33	11.73	9.68	10.95	10.93
Methionine	3.19	3.10	2.78	3.21	3.06	3.54	3.56
Cystine	1.66	0.98	0.83	1.08	1.03	1.14	1.18
Aspartic acid	10.95	9.97	10.04	11.43	10.02	11.40	11.83
Threonine	4.44	4.43	4.17	4.68	4.07	4.57	4.92
Serine	4.26	4.51	4.16	4.27	4.14	4.50	4.65
Glutamic acid	16.11	15.82	14.69	17.35	15.14	16.20	16.71
Glycine	4.68	4.64	4.63	5.26	4.83	4.81	5.03
Alanine	5.52	5.60	5.36	6.54	5.80	6.14	6.34
Valine	4.82	4.44	4.32	4.82	4.23	4.51	5.06
Isoleucine	4.33	3.82	3.74	4.24	3.75	3.96	4.47
Leucine	7.75	7.95	7.13	7.86	7.16	7.71	8.30
Tyrosine	3.45	3.23	3.09	3.49	3.05	3.17	3.32
Phenylalanine	4.00	3.86	3.72	4.28	3.63	3.80	4.12
Arginine	6.39	6.51	6.01	6.91	5.49	5.74	5.91

Averages of 10 parallel samples.
Source: Ref. 30.

The cation radical is shown generally, with the localization of the charge unspecified. Regardless of the localization, the main reaction of the cation radical is deprotonation:

$$(RCH_2{-}O{-}CO{-}(CH_2)_nCH_3)^{\dot{+}} \longrightarrow RCH_2{-}O{-}CO{-}\dot{C}H(CH_2)_{n-1}CH_3 + H^+$$

$$\searrow RCH_2{-}O{-}CO{-}(CH_2)_n\dot{C}H_2 + H^+$$

followed by dimerization or disproportionation:

$$2\ RCH_2{-}O{-}CO{-}\dot{C}H(CH_2)_{n-1}CH_3 \nearrow \begin{matrix} RCH_2{-}O{-}CO{-}CH(CH_2)_{n-1}CH_3 \\ | \\ RCH_2{-}O{-}CO{-}CH(CH_2)_{n-1}CH_3 \end{matrix}$$

$$2\ RCH_2{-}O{-}CO{-}\dot{C}H(CH_2)_{n-1}CH_3 \searrow \begin{matrix} RCH_2{-}O{-}CO{-}CH{=}CH(CH_2)_{n-2}CH_3 \\ + \\ RCH_2{-}O{-}CO{-}(CH_2)_nCH_3 \end{matrix}$$

Similar reactions take place with other C-radicals, such as:

$$RCH_2{-}O{-}CO{-}(CH_2)_n\dot{C}H_2$$

Another primary effect is electron attachment:

$$e^- + RCH_2{-}O{-}CO{-}(CH_2)_nCH_3 \rightarrow RCH_2{-}O{-}\underset{\displaystyle |\atop O^-}{\dot{C}}{-}(CH_2)_nCH_3$$

which may be followed by dissociation and decarbonylation or dimerization:

$$RCH_2{-}O{-}\underset{\displaystyle |\atop O^-}{\dot{C}}{-}(CH_2)_nCH_3 \rightarrow RCH_2O^- + O\dot{C}(CH_2)_n\ CH_3 \quad \text{dissociation}$$

$$O\dot{C}(CH_2)_nCH_3 \rightarrow \begin{matrix} OC(CH_2)_nCH_3 \\ | \\ OC(CH_2)_nCH_3 \end{matrix} \quad \text{dimerization}$$

$$O\dot{C}(CH_2)_nCH_3 \rightarrow CO + \dot{C}H_2(CH_2)_{n-1}CH_3 \quad \text{decarbonylation}$$

$$\dot{C}H_2(CH_2)_{n-1}CH_3 \rightarrow CH_3(CH_2)_{n-1}CH_2CH_2(CH_2)_{n-1}CH \quad \text{dimerization}$$

$$\dot{C}H_2(CH_2)_{n-1}CH_3 \rightarrow \begin{matrix} CH_3(CH_2)_{n-2}CH{=}CH_2 \\ + \\ CH_3(CH_2)_{n-1}CH_3 \end{matrix} \quad \text{disproportionation}$$

Additional reactions can initiate from the excited triglyceride molecules. The radiolysis of triglycerides of unsaturated fatty acids proceeds similarly, but the presence of double bonds, particularly if conjugated, will modify the spectrum of products.

Obviously a multitude of products are possible. If the cleavage sites in the triglycerides are indicated by letters *a* to *f* in the following way:

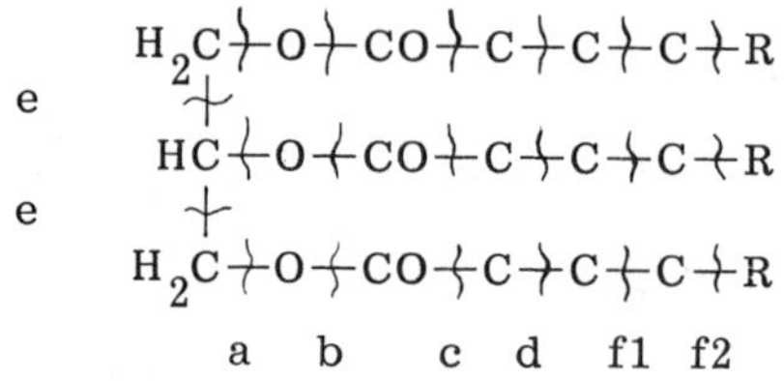

```
       H2C┼O┼CO┼C┼C┼C┼R
e        ┼
        HC┼O┼CO┼C┼C┼C┼R
e        ┼
       H2C┼O┼CO┼C┼C┼C┼R
           a  b   c  d  f1 f2
```

the possible radiolytic products of triglycerides can be summarized as listed in Table 4, taken from Delincée's review of the radiation chemistry of lipids [34].

Very detailed studies of the radiolysis products of various food lipids have been reported, particularly from Nawar's laboratory at the University of Massachusetts, Amherst, and by Merritt and co-workers at the U.S. Army Natick Research and Development Center. Table 5, taken from Nawar's work, lists the volatile radiolysis products obtained when various vegetable oils were irradiated with a dose of 60 kGy.

As is conventionally done in lipid chemistry, the products are characterized by a system of figures indicating the length of the carbon chain and the number of double bonds. Hexadecadiene, for instance, in 16:2. Nawar's reviews [35—38] make reference to many investigations carried out on other irradiated lipids. Similarly, recent publications of Merritt and co-workers [39, 40] give access to their earlier studies.

What has been said up until now about radiolysis of lipids refers to irradiation in the absence of air, and most investigations of radiation effects on lipids have been carried out under anoxic conditions. It is generally assumed that irradiation in the presence of oxygen leads to accelerated autoxidation, and that the pathways are the same as in light-induced or metal-catalyzed autoxidation. However, there have been few systematic investigations of this. In a collaborative study of the Amherst and Natick groups, tripalmitin was irradiated either in the presence or in the absence of air (Table 6). There were no drastic differences in product formation, but the presence of air increased the formation of tridecane, tetradecane, and gamma-palmitolactone, and decreased formation of pentadecane. The authors explained this as a result of the formation of oxy free radicals produced at the beta and gamma positions, leading to greater formation of the n-2 and n-3 alkanes [41].

Table 4 Possible Radiolytic Products of Triglycerides

Site of cleavage	Primary products	Recombination products
a	C_n fatty acid	C_n fatty acid esters
	Propanediol diesters	Alkanediol diesters 2-Alkyl-1,3-propanediol diesters Butanetriol triester
	Propenediol diesters	
b	C_n aldehyde	Ketones Diketones Oxoalkylesters
	Diglycerides	
	Oxo-propanediol diesters	Glyceryl ether diesters Glyceryl ether tetraesters
	2-Alkylcyclobutanones (C_n)	
c	C_{n-1} alkane	Longer hydrocarbons
	C_{n-1} 1-alkene	
	Formyl diglycerides	Triglycerides with shorter fatty acids
d	C_{n-2} alkane	
	C_{n-2} 1-alkene	Hydrocarbons
	Acetyl diglycerides	Triglycerides with shorter or longer fatty acids
e	C_n fatty acid methyl ester	C_n fatty acid esters
	Ethanediol diester	Alkanediol diesters Erythritol tetraester
f_i	C_{n-x} hydrocarbons	Hydrocarbons
	Triglycerides with shorter fatty acids	Triglycerides with longer fatty acids

i = 1, 2, . . . , n − 3.
x = any carbon number from 3 up to n − 1.
Source: Ref. 34.

Table 5 Quantitative Analysis of Products Formed in Vegetable Oils Irradiated at 60 kGy and 25°C (μg/100 g fat)

Carbon number	Safflower oil	Soybean oil	Coconut oil	Corn oil	Olive oil
Hydrocarbons					
8:0	[a]	0	620		[a]
8:1	[a]	Trace	710	200	[a]
9:0	650	Trace	1150	300	0
9:1	Trace	Trace	210	200	0
10:0	15	Trace	240	382	200
10:1	12	Trace	5810	163	440
10:2	220	160		256	15
11:0	20	37	9010	48	108
11:1	50	160	806	56	54
11:2	[a]	[a]		56	54
12:0	25	120	165	Trace	150
12:1	18	80	2750	48	104
12:2	20	100		26	Trace
13:0	37	120	3200	12	54
13:1	15	80	570	80	270
13:2	220	400		Trace	180
14:0	20	70	57	75	120
14:1[c]		34		100	250
14:1	1395	2000	1325	2260	2220
14:2	180	200	0	114	270
15:0	2150	2830	2040	2500	2220
15:1[c]	215	150	90	100	176
15:1	170	150	714	150	232
15:2	365	150	0	140	196
16:0	180	Trace	140	160	Trace
16:1[c]		810		48	207
16:1	308	Trace	260	300	332

Table 5 (Continued)

Carbon number	Safflower oil	Soybean oil	Coconut oil	Corn oil	Olive oil
Hydrocarbons (continued)					
16:2	3200	3500	735	3824	13400
16:3	18100	16000	0	6430	1200
16:4		a			
17:0	Trace	a			
17:1[c]	2000		a	820	
17:1	0	2150	a	a	9500
17:2	19300	20000	230	6420	3820
17:3	2980	4100	a	1500	320
Aldehydes					
10:0			400		
12:0			a		
12:1			4460		
14:0			2360		
16:0	1400	3560	677	1600	3700
16:1	140	600		70	274
18:0	180	286		70	170
18:1	1160	6500	2500	8250	
18:2	11900	12000		4800	9850
18:3		2000		2700	
Methyl esters					
10:0			Trace		
12:0			a		
14:0			Trace		
16:0	0	a	Trace	a	
18:0	0	a		a 0	

Table 5 (Continued)

Carbon number	Safflower oil	Soybean oil	Coconut oil	Corn oil	Olive oil
Ethyl esters					
8:0			0		
10:0			Trace		
12:0			Trace		
14:0			Trace		
16:0	0	Trace	0	0	
18:0	0	0		0	

[a]Present but could not be quantitatively determined due to overlap.
[b]Trace = less than 10 μg/100 g fat.
[c]Unsaturation is internal and not terminal.
Source: Ref. 35.

Table 6 Radiolysis Products from Tripalmitin Irradiated with 250 kGy in the Presence and Absence of Air at 25°C

Compound	Vacuum (mg/g)	Air (mg/g)
Tridecane	0	0.26 ± 0.07
Tetradecane	0.09 ± 0.01	0.81 ± 0.1
Pentadecane	3.10 ± 0.1	0.81 ± 0.1
Hexadecanal	0.28 ± 0.02	0.16 ± 0.03
Methyl palmitate	0.21 ± 0.04	0.45 ± 0.09
Ethyl palmitate	1.10 ± 0.1	1.20 ± 0.1
Palmitic acid	10.40 ± 1.2	13.40 ± 1.7
Gamma-palmitolactone	0.26 ± 0.02	1.50 ± 0.1
1,2-Propanediol dipalmitate	3.40 ± 0.01	3.20 ± 0.07
1,3-Propanediol dipalmitate	1.80 ± 0.01	1.50 ± 0.1

Source: Ref. 41.

Detailed knowledge of the kind and amount of radiolytic products formed in foods and food components is very useful in the evaluation of the health safety of irradiated foods, as will be discussed in Chapter 5. Such studies have made it clear, for instance, that radiation does not cause

formation of aromatic rings
condensation of aromatic rings
formation of heterocyclic rings

reactions known to take place at higher temperatures of cooking. Some of these ring structures are considered to be carcinogens, and they are therefore of particular interest to the toxicologist.

Much information is available on increased autoxidation of irradiated fats. If air is not excluded, peroxide values can reach high levels, as indicated for mackerel oil in Figure 18. This is not necessarily the case when foodstuffs are irradiated in which lipids constitute only a portion of the total. Evidence from several studies on meat irradiation indicates that proteins or, possibly, protein-carbohydrate interaction products exert an antioxidant effect that increases with radiation dose, thus protecting the lipids against oxidative changes.

Using TBA number as an indicator of oxidation, Green and Watts [43] found less oxidation in irradiated than in unirradiated ground beef samples (Fig. 19). Other examples of increased oxidative stability in irradiated meat are mentioned in the author's review of radiolytic changes in irradiated foods [44].

As indicated in Figure 18, there was no difference in peroxide values of the different samples immediately after irradiation. Only during storage (and only in the presence of air) did the result of irradiation become apparent. Other such *postirradiation effects* are occasionally observed, not only in lipid systems. As mentioned previously, irradiation of aqueous systems may produce hydrogen peroxide, particularly in the presence of oxygen. During postirradiation storage, hydrogen peroxide will gradually disappear, while some other constituents of the system are being oxidized. Obviously, some oxidized compounds not present or present in lower concentration immediately after irradiation will be present in higher concentration after hours or days. It is important to be aware of this possibility. All too often, research publications do not indicate at what time after irradiation certain measurements or analyses were carried out. The aforementioned review [44] documents various examples of postirradiation effects. Such post processing effects are not limited to irradiation. Many substances or foodstuffs undergo different chemical changes during storage depending on whether they have been cooked, frozen, dried, or left untreated.

In this context it is interesting to compare the extent of chemical changes caused by irradiation with that caused by other processes,

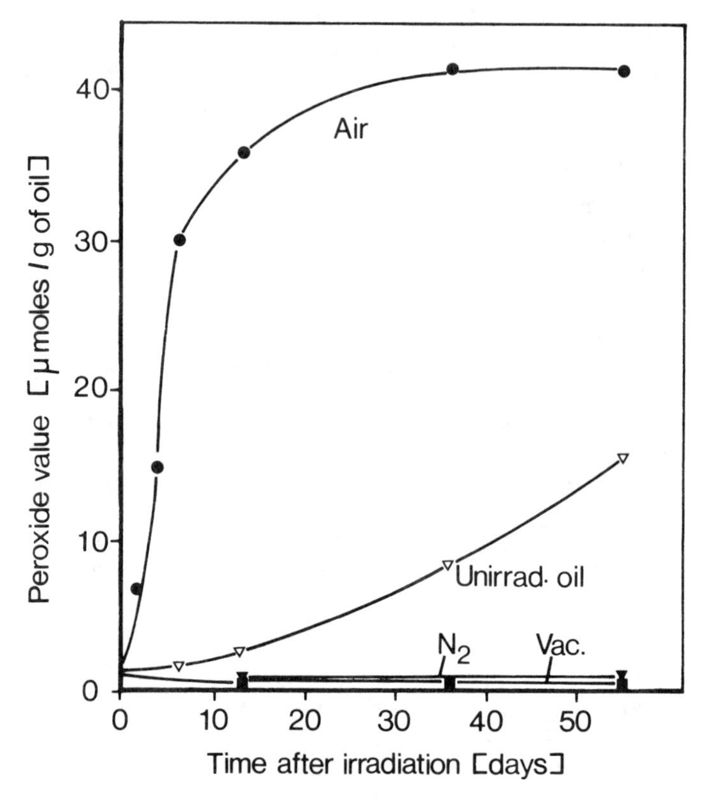

Figure 18 Development of peroxide, during storage at 25°C, in mackerel oil irradiated (14 kGy; 3 MeV electrons) in air, nitrogen, or vacuum at 20°C. Storage in closed containers with limited head space. (From Ref. 42.)

especially heating. Nawar found mostly the same volatile reaction products in various fats that had been either heated or irradiated. Yields were different, some higher after irradiation (250 kGy), some higher after heating (1 hr at 180°C) [45]. The overall yield of reaction products was higher after the heat treatment. Irradiation of pure amino acids and peptides at 60 kGy produced far fewer volatiles than heating the same components at 170°C for 1 hr [38]. Schubert has estimated that consumption of heat-processed foods results in a daily intake of thermal decomposition products that is 50 to 500 times higher than the intake of radiolytic products from an irradiated diet [46]. Merritt concluded in 1972 that no volatile compounds produced in foods by irradiation had been found that were

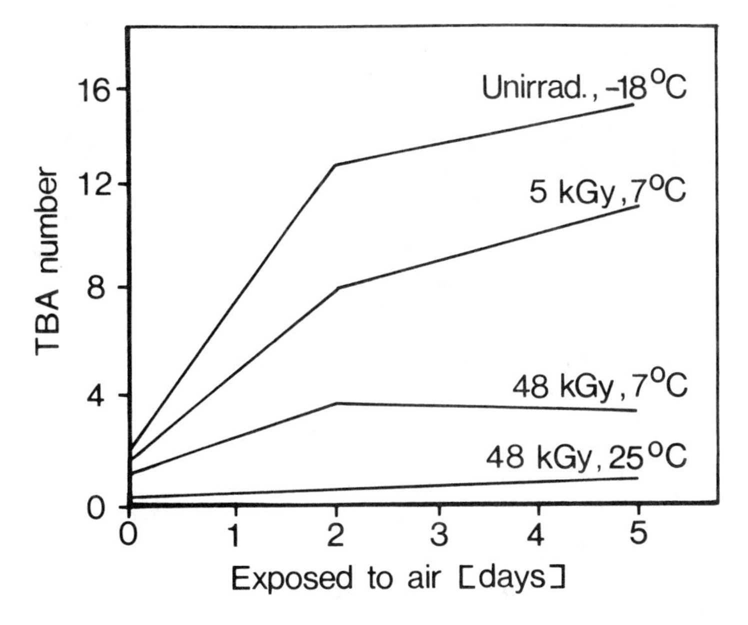

Figure 19 Effect of radiation dose (hard X-rays) and storage temperature on TBA numbers of irradiated ground beef stored for 83 days in sealed containers and subsequently exposed to air in the refrigerator or at room temperature. (Adapted from Ref. 43.)

not found qualitatively and quantitatively in other products resulting naturally [47]. This has been confirmed by numerous studies since then. There have been many speculations about "unique radiolytic products" (URPs), but their existence in foods has not yet been ascertained.

D. Some Conclusions from Chemical Studies

When irradiation processing was new, it had the fearful connotations of the unknown, the unexplored. It was something that could be full of surprises—and to those who do not know about the wealth of data available, it still has this connotation. It seemed therefore important enough in this book to convey some idea of the extent of the knowledge now available. Readers who want to know more about current state-of-the-art radiation chemistry may wish to turn to the excellent text edited by Farhataziz and Rodgers [48].

As will be described in more detail in Chapter 5, those responsible for deciding on the health safety and legal acceptability of irradiated foods in the United States and in other countries initially demanded long-term animal feeding studies for each food under consideration. The uncertainty over the nature and extent of chemical changes in irradiated foods was so great that evidence considered satisfactory for permitting irradiation of wheat was not considered sufficient for permitting irradiation of rye; acceptance of irradiated beef did not mean acceptance of irradiated pork; after long-term animal feeding studies on irradiated cod had been completed successfully, others were demanded on irradiated haddock—and so on, ad infinitum. German authorities questioned the acceptability of animal feeding studies with fish irradiated with 28 or 56 kGy as evidence for the safety of fish irradiated with a dose of 1 kGy [49]. Some authorities would not accept safety evidence obtained on gamma-irradiated food for permitting electron-irradiated food. This impass could only be overcome on the basis of radiation chemical studies.

The recognition of commonality and predictability of radiolytic changes in irradiated foods [39, 50] has finally shown the absurdity of demanding ever more animal feeding studies with irradiated foods.

IV. EFFECTS OF IONIZING RADIATION ON PACKAGING MATERIALS

With the exception of such applications as sprout inhibition in potatoes or onions, insect disinfestation in bulk grain, or delay of postharvest ripening of fruits, irradiation of foodstuffs is usually carried out with packaged food items.

There may be different reasons for this: prevention of microbial reinfection or insect exposure, prevention of water loss, exclusion of oxygen, prevention of mechanical damage during transport, or simply improved handling and marketing.

The packaging material used must not release significant amounts of radiation-induced reaction products or additives to the food, nor should it lose its functional qualities such as mechanical strength, seal stability, or impermeability to water.

Radiation resistance of various packaging materials has been extensively tested. Some are more radiation resistant than others, but in the dose range of up to 10 kGy, almost all materials commonly used for food packaging are suitable because radiation effects are minor in this range.

Plastic films for flexible packaging are of greatest interest in this context. They all consist of polymeric materials of which some suffer main-chain scission and a loss in mechanical properties when exposed to radiation, while others undergo cross-linking between adjacent polymer chains, leading to improved mechanical properties [51, 52]. The first group are called *degrading polymers*, and polyisobutylene, polymethacrylate, polytetrafluoroethylene, and some others belong to this group. Among the *cross-linking polymers* are polyethylene, polypropylene, polystyrene, polysiloxane, polyvinylchloride, and others. The fact that their functional properties are improved by irradiation has led to the extensive use of ionizing radiation in the plastics and cable manufacturing industries [53]. Electrical cables exposed to high mechanical or thermal stress, for instance, those used in certain household appliances, in space capsules, or in nuclear installations, are now routinely produced by radiation curing of the insulating material surrounding the metal core.

When packaged food is irradiated, not only the packaging material itself must be suitable for this purpose but also any adhesive used for sealing. If the package carries a printed message, the printing colors must be tested for radiation resistance. Many years of experience with the irradiation of packaged medical items such as syringes, surgeon's gloves, catheters, and suture thread (catgut) have shown which plastics, adhesives, and colors are suitable.

In the United States packaging materials used in the irradiation of prepackaged foods must be approved for this purpose by the Food and Drug Administration. A list of FDA-approved materials will be presented in Chapter 10.

When glass is irradiated it acquires a brown tint, and glass containers are therefore usually unsuitable for food irradiation. After a low-dose treatment this is barely visible, but the color intensity increases with increasing dose until a saturation level is reached, high above the doses used in food irradiation. The brown color is stable at room temperature but disappears when the glass is heated. Some plastic materials also darken upon irradiation in a dose-dependent way. This can be used for dosimetry purposes (see Chapter 2).

Paper and cardboard materials consist of cellulose, and cellulose belongs to the group of degrading polymers. With increasing radiation dose paper becomes increasingly brittle. FDA allows glassine paper and wax-coated cardboard only to a maximum dose of 10 kGy, while some of the cross-linking polymers like polyethylene are allowed up to a maximum dose of 60 kGy.

For long-term storage of radiation-sterilized meat at room temperature, certain flexible pouches consisting of multilayered laminated plastic films and aluminum foil [54] and tinplate cans have been tested [55].

V. IDENTIFICATION OF IRRADIATED FOODS

In countries where food irradiation is not allowed, the authorities would greatly appreciate any method permitting reliable identification of irradiated foods. In other countries permissions for irradiation are limited to specific commodities or groups of commodities and to specific dose levels. Ideally, the authorities would like to have methods that allow not only detection of irradiated foods but also an estimation of the applied radiation dose.

The availability of such methods would not only be in the interest of government agencies but also in the interest of radiation processors. In countries where the marketing of irradiated commodities is permitted, these products usually achieve a somewhat higher price than their nonirradiated counterparts, because the seller can claim desirable qualitites for his product, such as absence of salmonella, longer shelf life, no sprouting, etc. Dishonest merchants may be tempted to sell some of their products as "irradiated" even if they are untreated. In this case a reliable method for clearly separating irradiated and nonirradiated foodstuffs would help to discover fraud.

As presented in previous sections of this chapter, the chemical and physical changes caused in foods by irradiation are small and unspecific. When first attempts were made to develop suitable detection methods in the 1960s, the task seemed almost hopeless, especially for foods irradiated with low doses of radiation. The Health Directorate of the Commission of the European Communities initiated and financed a multinational research program in this field in 1965. Work performed under this program was reviewed at a final meeting in 1973, and the proceedings of this conference give a good overview of various approaches taken [56]. Some of the participating laboratories used chemical analysis to determine radiolytic products such as malondialdehyde, others used physical methods, such as electron spin resonance or electrical conductivity, or biological methods based, for instance, on the inability of irradiated potatoes to sprout.

With many of the proposed methods it is relatively easy to find a difference between two samples of the same material, one of the samples being irradiated, the other not. However, identification of an irradiated sample is much more difficult when an unirradiated control sample is not available. Foods are biological products and as such they have widely varying properties and characteristics, even within the same type of food. In plant foods, factors such as cultivar, climate, soil, ripeness at time of harvest, postharvest storage conditions, processing, and packaging can be very influential. In foods of animal origin, breed, feeding conditions, age of the animal, processing, storage, and packaging can play a similarly important role. The influence of these factors on product characteristics is usually much greater than the influence of irradiation, particularly in the low dose range. Claims that a new method for the detection of irradiated foods has been developed have been published year after year—but when these methods were tested round robin style between several laboratories and with blind samples, they have usually not fulfilled the expectations.

Many attempts were made to find *unique radiolytic products* (URPs) in radiation-processed foods, and none were successful. At one time o-tyrosine, produced from the amino acid phenylalanine by attack of radiation-induced •OH radicals, was thought to be a URP in irradiated meat [57]. More recent work at the National Bureau of Standards has shown that o-tyrosine also occurs in unirradiated meat [58].

Some progress has been made in identifying irradiated spices by measuring chemiluminescence and thermoluminescence [59]. However, with increasing storage time after irradiation, these methods of detection become more and more unreliable, especially if the samples are not kept completely dry. Spice samples of different origin vary greatly in their luminescence reaction, even if they have not been irradiated. Factors such as particle size have a profound effect on the results of such measurements [60].

Whereas luminescence methods require expensive equipment, another recently proposed method for the identification of spices and other dehydrated carbohydrate-rich foods can be carried out with relatively simple viscosimeters. As described in Sec. III.A, the viscosity of gels prepared from irradiated starch (and other polysaccharides) decreases with increasing radiation dose, and this is the basis of a proposal to detect irradiated spices by the lowered viscosity of a 10% suspension of the spice, heated to 80°C and subsequently cooled [61]. Extensive studies with 13 different spices led to the conclusion that 8 of these showed effects of irradiation on viscosity. However, because of natural variations, low viscosity could not be taken as proof of irradiation. In the case of ginger, pepper, and cinnamon, radiation treatment could be reliably excluded if the viscosities were high [62].

If it is difficult to differentiate irradiated from nonirradiated foodstuffs, it is much more difficult to determine the radiation dose applied to a food sample by studying the properties of that sample. Electron spin resonance spectroscopy of bones has been suggested for postirradiation dosimetry of meat [63], and the dose-dependence of ESR signals in irradiated spices and fruit powders has been determined. How reliable this method will be under practical conditions remains to be seen.

In addition to methods based on chemical, physical, or biological changes in the food itself, methods based on radiation-induced changes in the microbial flora are also under investigation [64].

Studies based on a variety of methods are progressing in laboratories all over the world, and the continued improvement of analytical techniques will probably provide reliable methods for the identification of certain "first-generation" irradiated foods in the near future. The present state of these efforts is described in the report of a WHO Working Group [65]. From the results obtained to date it appears unlikely that "second-generation" foods, i.e., foods containing irradiated ingredients, can be detected by any of these methods.

REFERENCES

1. Pryor, W.A., ed., *Free Radicals in Biology*, 4 vols., Academic Press, New York, NY 1976/1980.
2. Hansen, P. M. T., W. J. Harper, and K. K. Sharma, Formation of free radicals in dry milk proteins, *J. Food Sci.*, 35:598 (1970).
3. Uchiyama, S., and M. Uchiyama, Free radical production in protein rich food, *J. Food Sci.*, 44:1217 (1979).
4. Diehl, J. F., and S. Hofmann, Electron spin resonance studies on radiation-preserved foods. 1. Influence of radiation dose on spin concnetration (in German), *Lebensm.-Wiss. Technol.*, 1:19 (1968).
5. Diehl, J. F., Electron spin resonance studies on radiation-preserved foods. 2. The influence of water content on spin concentration (in German), *Lebensm.-Wiss. Technol.*, 5:51 (1972).
6. Yang, G. C., M. M. Mossoba, U. Merin, and I. Rosenthal, An EPR study of free radicals generated by gamma-radiation of dried spices and spray-dried fruit powders, *J. Food Quality*, 10:287 (1987).
7. Henglein, A., W. Schnabel, and J. Wendenburg, *Introduction to Radiation Chemistry* (in German), Verlag Chemie, Weinheim, 1969, p. 378.
8. Proctor, B. E., and J. P. O'Meara, Effect of high-voltage cathode rays on ascorbic acid, *Ind. Eng. Chem.*, 43:718 (1951).
9. Merritt, C., Jr., P. Angelini, E. Wierbicki, and G. W. Shults, Chemical changes associated with flavor in irradiated meat, *J. Agr. Food Chem.*, 23:1037 (1975).
10. Vas, K., Preliminary studies on the nature of radiation inactivation of pectin methyl esterase and cellulase preparations, in *Enzymological Aspects of Food Irradiation*, Int. Atomic Energy Agency, Panel Proceedings Series Vienna, 1969, p. 37.
11. Ehrenberg, L., M. Jaarma, and E. C. Zimmer, The influence of water content on the action of ionizing radiation on starch, *Acta Chem. Scand.*, 11:950 (1957).
12. Delincée, H., and B. Radola, Effect of gamma-irradiation on the charge and size properties of horseradish peroxidase: individual isoenzymes, *Radiat. Res.*, 59:572 (1974).
13. Proctor, B. E., S. A. Goldblith, C. J. Bates, and O. A. Hammerle, Biochemical prevention of flavor and chemical changes in foods and tissues sterilized by ionizing radiation, *Food Technol.* (Chicago), 6:237 (1952).
14. Diehl, J. F., Influence of irradiation conditions and of storage on radiation-induced vitamin E losses in foods (in German), *Chem. Microbiol. Technol. Lebensm.*, 6:65 (1979).
15. von Sonntag, C., Free-radical reactions of carbohydrates as studied by radiation techniques, *Adv. Carbohydr. Chem. Biochem.*, 37:7 (1980).

16. Schubert, J., Inadequacies in radiobiological experiments on medium effects, *Int. J. Radiat. Biol.*, 13:297 (1967).
17. Kertesz, Z. I., E. R. Schulz, G. Fox, and M. Gibson, Effects of ionizing radiation on plant tissues. 4. Some effects of gamma radiation on starch and starch fractions, *Food Res.*, 24:609 (1959).
18. Dauphin, J., and L. R. Saint-Lèbe, Radiation chemistry of carbohydrates, in *Radiation Chemistry of Major Food Components*, P. S. Elias and A. J. Cohen, eds., Elsevier, Amsterdam, 1977, p. 131.
19. Raffi, J. J., J.-P. L. Agnel, C. J. Thiery, C. M. Fréjaville, and L. R. Saint-Lèbe, Study of gamma-irradiated starches derived from different foodstuffs: a way for extrapolating wholesomeness data, *J. Agr. Food Chem.*, 29:1227 (1981).
20. Winchester, R. V., Detection fo corn starch irradiated with low doses of gamma rays. 2. Disappearance of malonaldehyde from starches of various moisture contents, *Starch/Stärke*, 26:278 (1974).
21. Dizdaroglu, M., J. Leitich, and C. von Sonntag, Conversion of D-fructose into 6-deoxy-D-threo-2,5-hexodiulose by gamma-irradiation: a chain reaction in the crystalline state, *Carbohydr. Res.*, 47:15 (1976).
22. Adam, S., R. Blankenhorn, and J. F. Diehl, Effect of water activity upon electron-radiolysis of dextran, *Starch/Stärke*, 31:423 (1979).
23. Scherz, H., Radiolysis of starch (in German), *Starch/Stärke*, 31:423 (1979).
24. Diehl, J. F., S. Adam, H. Delincée, and V. Jakubick, Radiolysis of carbohydrates and of carbohydrate-containing foods, *J. Agr. Food Chem.*, 26:15 (1978).
25. Liebster, J., and J. Kopoldova, The radiation chemistry of amino acids, *Adv. Rad. Biol.*, 1:157 (1964).
26. Delincée, H., and B. J. Radola, Structural damage of gamma-irradiated ribonuclease revealed by thin-layer isoelectric focusing, *Int. J. Radiat. Biol.*, 28:565 (1975).
27. Radola, B. J., Identification of irradiated meat by thinlayer gel chromatography and thinlayer isoelectric focusing (in German), in *Identification of Irradiated Foodstuffs*, Commission of the European Communities, EUR 5126, Luxembourg, 1974, p. 27.
28. Uzunov, G., C. Tsoluva, and N. Nestorov, Changes in the soluble muscle proteins and isoenzymes of lactate dehydrogenase in irradiated beef meat, *Int. J. Radiat. Biol.*, 22:437 (1972).
29. Rhodes, D. N., and C. Meegungwan, Treatment of meats with ionizing radiations. 9. Inactivation of liver autolytic enzymes, *J. Sci. Food Agric.*, 13:13 (1962).
30. Underdal, B., J. Nordal, G. Lunde, and B. Eggum, The effect of ionizing radiation on the nutritional value of fish (cod) protein, *Lebensm. Wiss. Technol.*, 6:279 (1973).

31. Urbain, W. M., Radiation chemistry of proteins, in *Radiation Chemistry of Major Food Components*, P. S. Elias and A. J. Cohen, eds., Elsevier, Amsterdam, 1977, p. 63.
32. Delincée, H., Recent advances in the radiation chemistry of proteins, in *Recent Advances in Food Irradiation*, P. S. Elias and A. J. Cohen, eds., Elsevier, Amsterdam, 1983, p. 129.
33. Simic, M. G., Radiation chemistry of water soluble food components and Taub, I. A., Reaction mechanisms, irradiation parameters, and product formation, in *Preservation of Food by Ionizing Radiation*, vol. 2, E. S. Josephson and M. S. Peterson, eds., CRC Press, Boca Raton, FL, 1983, p. 1 and 125.
34. Delincée, H., Recent advances in the radiation chemistry of lipids, in *Recent Advances in Food Irradiation*, P. S. Elias and A. J. Cohen, eds., Elsevier, Amsterdam, 1983, p. 89.
35. Kavalam, J. P., and W. W. Nawar, Effects of ionizing radiation on some vegetable fats, *J. Am. Oil Chem. Soc.*, 46:387 (1969).
36. Nawar, W. W., Radiation chemistry of lipids, in *Radiation Chemistry of Major Food Components*, P. S. Elias and A. J. Cohen, eds., Elsevier, Amsterdam, 1977, p. 21.
37. Nawar, W. W., Radiolysis of nonaqueous components of foods, in *Preservation of Food by Ionizing Radiation*, vol. 2, E. S. Josephson and M. S. Peterson, eds., CRC Press, Boca Raton, FL, 1983, p. 76.
38. Nawar, W. W., Volatiles from food irradiation, *Food Revs. Internat.*, 2:45 (1986).
39. Merritt, Ch., Jr., and I. A. Taub, Commonality and predictability of radiolytic products in irradiated meats, in *Recent Advances in Food Irradiation*, P. S. Elias and A. J. Cohen, eds., Elsevier, Amsterdam, 1983, p. 27.
40. Vajdi, M., and C. Merritt, Jr., Identification of adduct radiolysis products from pork fat, *J. Am. Oil Chem. Soc.*, 62:1252 (1985).
41. Vajdi, M., W. W. Nawar, and C. Merritt, Jr., Effects of various parameters on the formation of radiolysis products in model systems, *J. Am. Oil Chem. Soc.*, 59:38 (1982).
42. Astrack, A., O. Sorbye, A. Brasch, and W. Huber, Effects of high intensity electron bursts upon various vegetable and fish oils, *Food Res.*, 17:570 (1952).
43. Green, B. E., and B. M. Watts, Lipid oxidation in irradiated cooked beef, *Food Technol.* (Chicago), 20:(8) 111 (1966).
44. Diehl, J. F., Radiolytic effects in foods, in *Preservation of Foods by Ionizing Radiation*, vol. 1, E. S. Josephson and M. S. Peterson, eds., CRC-Press, Boca Raton, FL, 1982, p. 279.
45. Nawar, W. W., Comparison of chemical consequences of heat and irradiation treatment of lipids, in *Recent Advances in Food Irradiation*, P. S. Elias and A. J. Cohen, eds., Elsevier, Amsterdam, 1983, p. 115.

46. Schubert, J., Toxicological studies on irradiated food and food constituents, in *Food Preservation by Irradiation*, vol. 2, Proceedings of a Symposium held in Wageningen, Nov. 1977, Int. Atomic Energy Agency, Vienna, 1978, p. 3.
47. Merritt, C., Jr., Qualitative and quantitative aspects of trace volatile components in irradiated foods and food substances, *Rad. Res. Rev.*, 3:353 (1972).
48. Farhataziz, and M. A. J. Rodgers, eds., *Radiation Chemistry, Principles and Applications*, VCH Publishers, New York, 1987.
49. Diehl, J. F., Irradiated food, *Science*, 180:214 (1973).
50. Basson, R. A., M. Beyers, D. A. E. Ehlermann, and H. J. van der Linde, Chemiclearance approach to evaluation of safety of irradiated fruits, in *Recent Advances in Food Irradiation*, P. S. Elias and J. A. Cohen, eds., Elsevier, Amsterdam, 1983, p. 59.
51. Killoran, J. J., Packaging irradiated foods, in *Preservation of Food by Ionizing Radiation*, vol. 2, E. S. Josephson and M. S. Peterson, eds., CRC Press, Boca Raton, FL, 1983, p. 317.
52. Charlesby, A., Radiation chemistry of polymers, in *Radiation Chemistry, Principles and Applications*, Farhataziz and M. A. J. Rodgers, eds., VCH Publishers, New York, 1987, p. 451.
53. Bly, J. H., Radiation curing of elastomers, *J. Indust. Irradiation Tech.*, 1:51 (1983).
54. Killoran, J. J., J. S. Cohen, and E. Wierbicki, Reliability of flexible packaging of radappertized beef under production conditions, *J. Fd. Process. Preserv.*, 3:25 (1979).
55. Killoran, J. J., J. J. Howker, and E. Wierbicki, Reliability of the tinplate can for packaging of radappertized beef under production conditions, *J. Fd. Process. Preserv.*, 3:11 (1979).
56. CEC, *The Identification of Irradiated Foodstuffs*, Proceedings of an International Colloquim held in Karlsruhe, 24/25 October 1973, EUR 5126, Commission of the European Communities, Luxembourg, 1974.
57. Simic, M. G., M. Dizdaroglu, and E. DeGraff, Radiation chemistry—Extravaganza or an integral component of irradiation processing of food? *Radiat. Phys. Chem.*, 22:233 (1983).
58. Karam, L. R., and M. G. Simic, Ortho-tyrosine as a marker in postirradiation dosimetry (PID) of chicken, in *Health Impact, Identification, and Dosimetry of Irradiated Foods*, Report of a WHO Working Group on Health Impact and Control Methods of Irradiated Foods, K. W. Bögl, D. F. Regulla and M. J. Suess, eds., ISH-Heft 125, Bundesgesundheitsamt, Neuherberg/Munich, 1988, p. 297.
59. Bögl, W., and L. Heide, Chemiluminescence measurements as an identification method for gamma-irradiated foodstuffs, *Radiat. Phys. Chem.*, 25:25 (1985).
60. Sattar, A., H. Delincée, and J. F. Diehl, Detection of gamma

60. Sattar, A., H. Delincée, and J. F. Diehl, Detection of gamma irradiated pepper and papain by chemiluminescence, *Radiat. Phys. Chem.*, 29:215 (1987).
61. Mohr, E., and G. Wichman, Decreased viscosity as an indicator for cobalt irradiation of spices (in German), *Gordian*, 85:(5) 96 (1985).
62. Heide, L., E. Mohr, G. Wichmann, S. Albrich, and K. W. Bögl, Viscosity measurements—A method for the identification of irradiated spices? (in German), ISH-Heft 120, Bundesgesundheitsamt, Neuherberg/Munich, 1988.
63. Desrosiers, M. F., and M. G. Simic, Postirradiation dosimetry of meat by electron spin resonance spectroscopy of bones, *J. Agric. Food Chem.*, 36:601 (1988).
64. Betts, R. P., L. Farr, P. Bankes, and M. F. Stringer, The detection of irradiated foods using the direct Epifluorescent Filter Technique, *J. Appl. Bacteriol.*, 64:329 (1988).
65. Bögl, K. W., D. F. Regulla, and M. J. Suess, eds., *Health Impact, Identification, and Dosimetry of Irradiated Foods*, Report of a WHO Working Group on Health Impact and Control Methods of Irradiated Foods, ISH-Heft 125, Bundesgesundheitsmat, Neherberg/Munich, 1988.

4

Biological Effects of Ionizing Radiation

I. GENERAL CONSIDERATIONS

Many of the applications of food irradiation aim at the destruction of certain microorganisms—those causing food spoilage and/or those causing disease in man. Fundamental research in radiation biology and applied research aimed at improved hygiene and reduction of food losses have contributed to the present knowledge of how radiation affects living organisms.

How can radiation kill a cell? What sort of cell damage is responsible for the lethal effect of radiation? Many hypotheses have been proposed and tested. Some scientists, especially in the Soviet Union, thought "radiotoxins" were responsible, toxic substances produced in the irradiated cell. Others proposed that radiation was directly damaging to cellular membranes, so that the functioning of subcellular structures was impaired. Radiation effects on enzymes or on the energy metabolism were postulated. It is now universally accepted that the deoxyribonucleic acids (DNA) in the chromosomes represent the most critical target of ionizing radiation. Effects on the cytoplasmic membrane appear to play an additional role in some circumstances [1].

When ionizing radiation is absorbed in biological material, there is a possibility that it will act directly on the critical targets in the cell. The nucleic acid molecules may be ionized or excited, so initiating the chain of events that leads to biological change and to cell death if the change is serious enough. This is the so-called *direct effect* of radiation, which is the dominant process when dry spores of spore-forming microorganisms are irradiated. Alternatively, the radiation may interact with other atoms or molecules in the cell, particularly water, to produce free radicals which can diffuse far enough to reach and damage the DNA. This *indirect effect* of radiation is important in vegetative cells, the cytoplasm of which contains about 80% water. The free radical chemistry of nucleic acids and their base constituents has been elucidated in considerable detail [2] and the chemical basis of radiation biology is well established [3].

In the previous section it was shown that radiation-induced chemical changes of food constituents, even if they are large when pure solutions are irradiated, are usually small when foods of complex composition are irradiated. If this is so, and if radiation damage to cells is the result of chemical change in DNA, how can a radiation dose that causes little chemical change in food cause lethal change in a bacterial cell? The answer is that DNA has particular properties which set it apart from all other constituents of the cell.

DNA carries the genetic information. The sequence of purines and pyrimidines in the DNA chain serves as a template for assembling a DNA copy in the process of cell division. Through RNA (ribonucleic acid) templates this sequence determines the synthesis of proteins, including the enzymes which regulate the cell's metabolism.

DNA molecules are enormously large in comparison with other molecules in the cell and thus provide a large target.

The function of DNA as the carrier of genetic information depends on the intactness of this large molecule.

Only one copy (or very few copies) of a DNA molecule is (are) present in a cell.

Considering that the radiation sensitivity of macromolecules is roughly proportional to their molecular weight, Pollard [4] has estimated that a dose of 0.1 kGy will damage 2.8% of the DNA in bacterial cells, while the same dose will damage 0.14% of the enzymes and only 0.005% of the amino acids. The 2.8% DNA damage will be lethal to a large fraction of the irradiated cells, and this will have consequences easily recognized with the bare eye: Fewer colonies develop upon inoculation of a culture medium. In contrast, the 0.14% of damaged enzyme molecules will be difficult to detect, even by sophisticated analytical methods, and a 0.005% change of the amino acids in a biological system cannot be detected at all.

With other figures and other words, Brynjolfsson [5] has come to similar conclusions: The probability (X) that a molecule with the moelcular weight M will change upon irradiation is given by

$$X = 10^{-7} \cdot G \cdot M \cdot D$$

where

G = the number of changes per 100 eV of absorbed energy
D = the dose in kGy

The value for G is usually not above 4 when aqueous systems are irradiated. If water (M = 18) is irradiated with a dose D = 10 kGy, then

$$X = 10^{-7} \cdot 4 \cdot 18 \cdot 10 = 7.3 \cdot 10^{-5}$$

that is, 7.3 out of every 100,000 water molecules have changed. If glucose (M = 180) is irradiated, then

$$X = 10^{-7} \cdot 4 \cdot 180 \cdot 10 = 7.3 \cdot 10^{-4}$$

that is, 7.3 our of every 10,000 glucose molecules have changed. If DNA with a moelcular weight of $M = 10^9$ is irradiated, we get X = 4,000, which means that each DNA molecule has been changed 4,000 times or at 4,000 locations. Most of these changes may not be lethal, but double-strand breaks usually are. The G-value for double-strand breaks is about 0.07. This DNA would then have received

$$X = 10^{-7} \cdot 0.007 \cdot 10^9 \cdot 10 = 7$$

double-strand breaks.

These considerations explain why a given dose may have a lethal effect on microorganisms in an irradiated food sample without causing much change in the chemical composition of that food. They also explain some of the factors that influence radiation sensitivity. One of them is the molecular weight of DNA. The DNA in the nuclei of the cells of insects represents a target much larger than the genome of bacteria. Not surprisingly, bacteria are less radiation sensitive than insects. The cells of mammalian organisms, containing DNA molecules which must provide much more genetic information than those of insects, are correspondingly larger and more sensitive to radiation.

Another factor influencing radiation effects is the structural arrangement of the DNA in the cell. During cell division the normally double-stranded DNA of bacteria separates, and by means of a polymerase new chains of DNA are assembled along the template. This process is associated with changes in radiation sensitivity. In the double-stranded form DNA is much less sensitive to radiation than in the single-stranded form.

If a growth medium is inoculated with bacteria, there will be no increase in the number of cells for some time (lag phase); then the

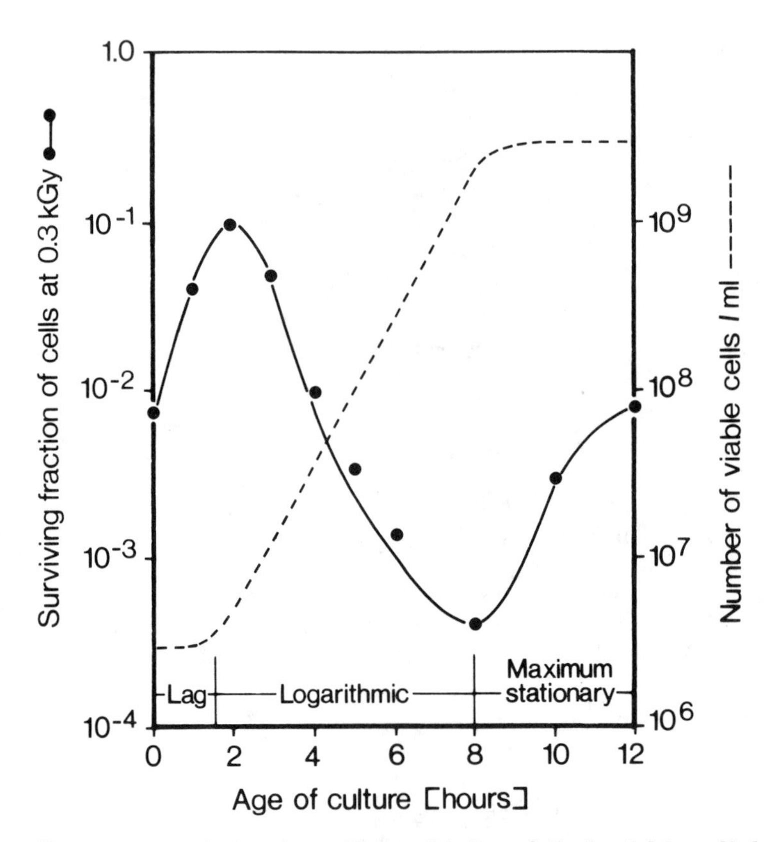

Figure 1 Variation in radiosensitivity of *Escherichia coli* during the growth cycle. (Adapted from Ref. 6.)

number of cells will increase logarithmically, until competition for nutrients and space inhibits further growth (stationary phase). As demonstrated for *Escherichia coli* in Figure 1, radiation resistance is high during the lag phase, decreases during the phase of logarithmic growth, and increases again in the stationary phase. However, even under laboratory conditions the influence of growth cycle on radiation sensitivity is not always as pronounced as it is in this example, and under the practical conditions of food irradiation one would rarely have such a clear separation of different growth cycles.

Factors other than size and arrangement of DNA also have much influence on radiation sensitivity. The DNA in the cell is associated with basic peptides, nucleoproteins, RNA, lipids, lipoproteins, and metal cations. In different species of microorganisms these substances may modify indirect effects of radiation differently. Most importantly, however, the DNA in different species is aligned with replication and repair enzymes of greatly differing effectiveness. These enzymes can carry out DNA degradation or DNA synthesis. They can thus excise a damaged sequence of nucleotides and resynthesize the missing sequence. These repair systems may be induced not only by irradiation but by a variety of natural or laboratory stresses causing DNA damage, such as heat, freezing, and chemical reagents. The detection and characterization of these repair processes has been one of the aims of molecular biology research since DNA repair was first discovered in the mid-1960s.

A. Inactivation Dose

When a population of microorganisms is irradiated with a low dose, only a few of the cells will be damaged or killed. With increasing radiation dose the number of surviving organisms decreases exponentially (as it does with increasing heat treatment). Different species and different strains of the same species require different doses to reach the same degree of inactivation. In order to characterize organisims by their radiation sensitivity, the *mean lethal dose* (MLD) is sometimes used. It is the dose required to kill 63% of a population, letting 37% survive (D_{37}).

A more commonly used measure of radiation sensitivity is the *D_{10} dose*, which is required to kill 90% of a population. A typical inactivation curve is shown in Figure 2. In this case the slope of the curve indicates a D_{10} value of 550 Gy or 0.55 kGy for the inactivation of *Salmonella typhimurium* in ground beef.

If N_0 is the initial number of the organisms present, N the number after irradiation with a dose D, and D_{10} the decimal reduction dose, then

$$\log \frac{N}{N_0} = -\frac{1}{D_{10}} D$$

or

$$D_{10} = \frac{D}{\log N_0 - \log N}$$

The slope of the regression line in Figure 2 is $-1/D_{10}$. The value of D_{10} can be determined graphically, as indicated in the figure, or by inserting values for N_0, N, and D in the above equation. Interconversion of D_{37} and D_{10} values is possible by using the equation $D_{10} = 2.303 \cdot D_{37}$. Inactivation curves do not always show a straight line as in Figure 2. Sometimes a "shoulder" appears in the low dose range before the linear slope begins. (An example of this will be shown in Figure 4.) In that case, to estimate D_{10}, a least-square

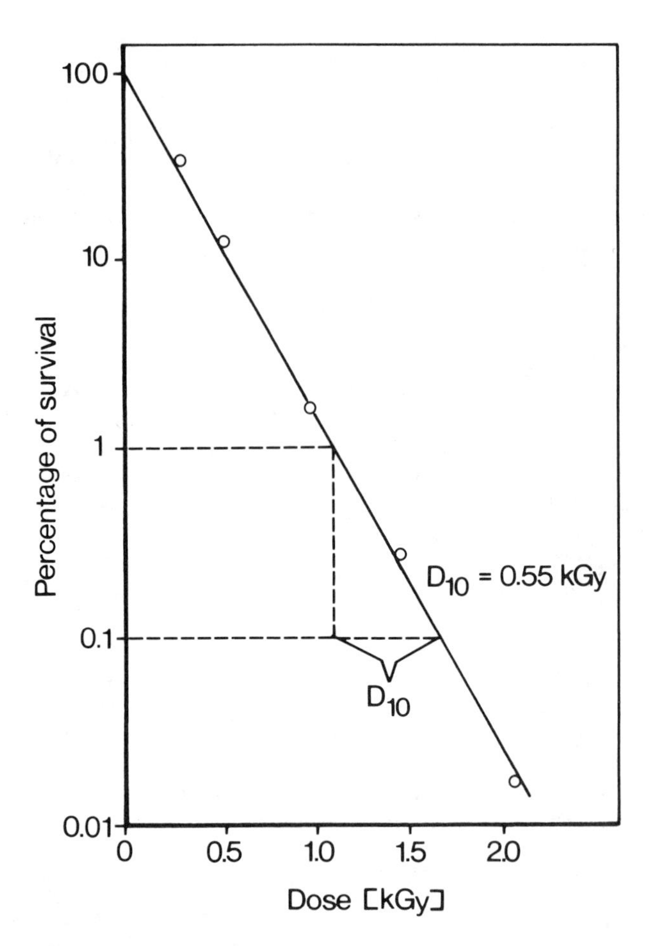

Figure 2 Effect of irradiation on *S. typhimurium* in ground beef. (Adapted from Ref. 7.)

regression line is fitted through the data points in such a way that the 0-dose (or N_0) value is excluded.

The specification of the medium (ground beef in Figure 2) is important, because D_{10} values can differ considerably in different media. The data presented in Table 1 demonstrate this for a strain of *Salmonella senftenberg*. Many other studies have also shown that a complex medium, such as liquid egg, provides a certain radiation protection to microorganisms, as compared to irradiation in a buffer solution. This parallels the higher resistance to radiation-induced chemical changes in multicomponent versus single component systems, which was described in the previous chapter. Other studies have

Table 1 Influence of Different Media on D_{10} Values of *S. senftenberg* N.C.T.C. 9959

Medium	D_{10} value (Gy)
Liquid whole egg	504
Desiccated coconut	1340
Bone meal	557
Phosphate buffer	
aerated	130
anoxic	389
frozen	299

Source: Ref. 8.

shown that D_{10} values even depend on whether the organisms were irradiated in low fat or in high fat ground beef. Maxcy and Tiwari [9] reported, for example, a D_{10} of 580 Gy and of 280 Gy when *Staphylococcus aureus* was irradiated in low fat and high fat ground beef, respectively. Low fat means high protein. Possibly the radical-scavenging properties of the proteins provide higher protection to the organisms in the high protein matrix.

A dry medium, such as desiccated coconut, provides better protection than an aqueous medium, such a liquid egg. D_{10} values in a frozen medium are usually higher than in a medium that is at room temperature during irradiation. The protective effect of drying and freezing can be explained as a result of the suppression of the indirect effects caused by the reactive intermediates (primarily •OH radicals of water radiolysis).

This effect of water is not only noticeable in the comparison dry vs. wet or frozen vs. unfrozen, but also when bacteria are irradiated in solutions of variable water activity a_w ($a_w = p/p_o$, where p is the aqueous vapor pressure of the given solution and p_o is that of water at the same temperature). This is shown in Figure 3 for *Salmonella thompson*, an organism of low radiation resistance, and in Figure 4 for the much more radiation-resistant spores of *Bacillus stearothermophilus*. While the data for these figures were obtained from organisms suspended in solutions, Figure 5 is based on results obtained with *S. senftenberg* in egg powder adjusted to different water activities. Variations in a_w below $a_w = 0.5$ had not much effect on D_{10}, whereas the increase to $a_w = 0.8$ caused a considerable decrease in D_{10}.

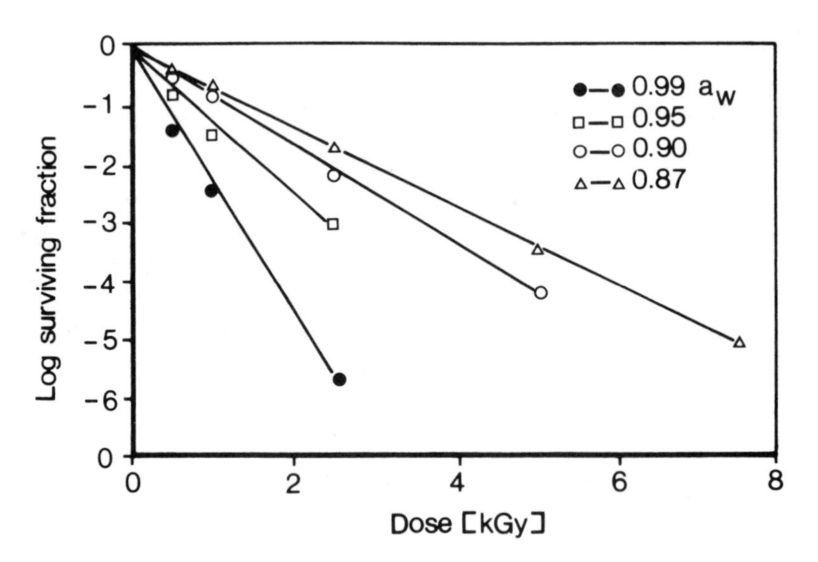

Figure 3 Survival curves of *S. thompson* in sucrose solutions of different concentration after irradiation with 10 MeV electrons at 0°C. (From Ref. 10.)

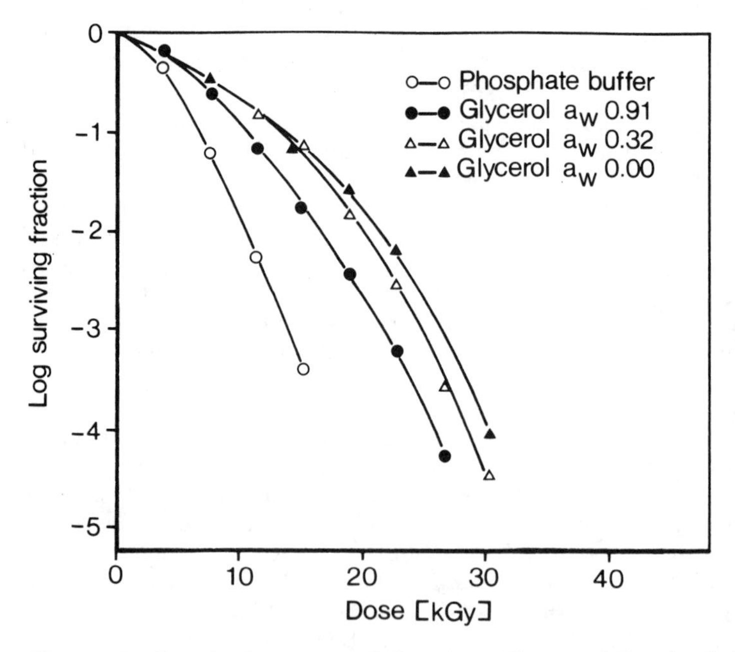

Figure 4 Survival curves of *B. stearothermophilus* in 0.1 M phosphate buffer and in glycerol solutions of different concentration after irradiation with gamma rays at 0°C. (Adapted from Ref. 11.)

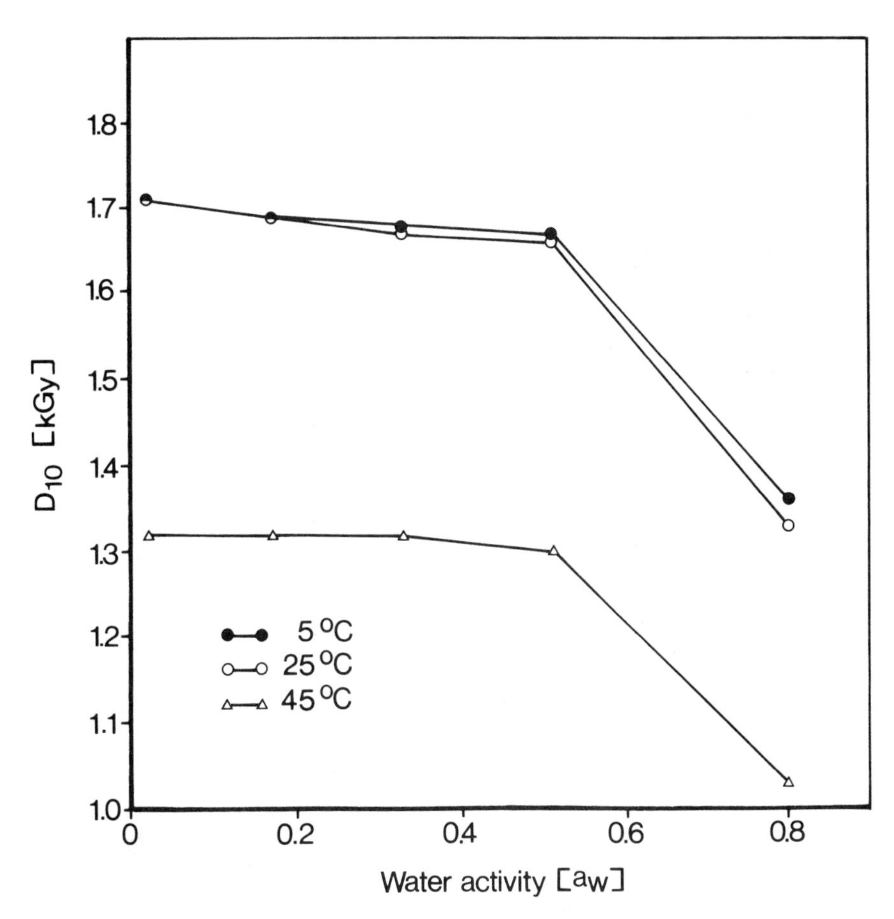

Figure 5 Influence of water activity on radiation resistance of *S. senftenberg* in egg powder irradiated at different temperatures with 10 MeV electrons. (From Ref. 10.)

The higher radiation sensitivity at 45°C as compared to 25°C is also noticeable. Other studies have shown a precipitous drop of D_{10} values when irradiation was carried out at still higher temperatures. This is obviously due to the combined effect of radiation and heat damage.

D_{10} values are also influenced by postirradiation environmental conditions. Unless one can incubate in the food itself, it is necessary to transfer samples to artificial media in order to find the cells able to multiply after irradiation. The proportion of radiation-damaged cells able to recover may be influenced by the composition of the medium, by the atmosphere, and by the temperature during

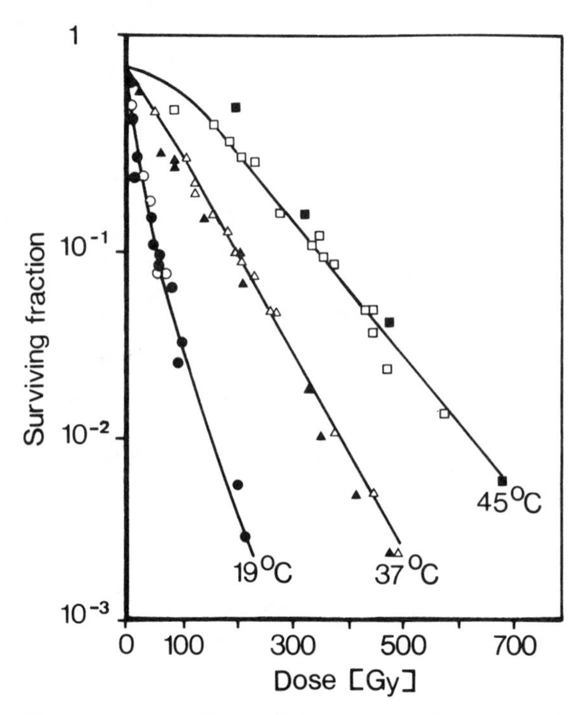

Figure 6 Survival of *E. coli* strain B irradiated in the absence of oxygen. Incubated for about 2 h, immediately after irradiation and plating, at 19°, 37°, or 45°C. Open symbols: several runs, 250 Kvp x-rays. Filled symbols: several runs, 8 MeV electrons. (Adapted from Ref. 12.)

incubation. The effect of postirradiation incubation temperature on survival of *E. coli* is demonstrated in Figure 6. Surprisingly, an incubation temperature of 45°C allowed better recovery than 37°C. The efficiency of repair systems is apparently improved at the higher temperature. Incidentally, the experimental points obtained after X-ray or electron irradiation indicate that more or less identical results are obtained with different types of ionizing radiation [12].

The observations concerning the effects of medium, water activity, and temperature on the radiation sensitivity of microorganisms are of considerable practical interest. They show that D_{10} values taken from the literature must be used critically. The environmental factors which may alter radiation resistance must be carefully considered. The facts presented should also make clear that there is no simple answer to the question, "What radiation dose is needed to kill the bacteria in foods?" The effectiveness of a given dose depends on the medium, i.e., on the food, on the temperature during irradiation, on the initial extent of contamination, and if a nonsterilizing dose of radiation was used, on postirradiation storage conditions.

The higher D_{10} values obtained under anoxic conditions as compared to irradiation in an aerated system (Table 1) are more of academic than of practical interest. Bacterial sensitivity increases 2- to 5-fold if oxygen is present during treatment. However, this *oxygen effect* is not always so clearly observed because—even if no special efforts are made to exclude oxygen—irradiation itself causes more or less anoxic conditions in a sample, especially when electron radiation is used. The reasons for this have been explained previously. The full extent of the oxygen effect is only seen if oxygen is continuously repleted during irradiation, for instance, by bubbling air or oxygen through an irradiated suspension of bacteria. This is not feasible in food irradiation. At any rate, the presence of oxygen also increases radiation damage to the food components. In a practical situation one would therefore not try to increase oxygen levels in food samples during irradiation, even if that would lower the D_{10} values.

B. Radappertization, Radicidation, Radurization

In the beginning of food irradiation research, treatments aimed at inactivation of microorganisms were categorized into two groups:

Radiation sterilization, with the objective of producing an indefinitely stable article, by destroying most or, ideally, all of the microorganisms in the food

Radiation pasteurization, with the objective of destroying only a part of the microbial population, either to delay the onset of spoilage or to eliminate a particular group of organisms important for public health

For a number of reasons this terminology was considered unsatisfactory. Radiation sterilization could not be equated with heat sterilization, because the former does not inactivate viruses to the same extent as the latter (this will be discussed in some detail below). The use of the word pasteurization for two treatments with very different objectives was confusing. An international group of experts [13] therefore suggested three new terms in 1964:

Radappertization: The application to foods of a dose of ionizing radiation sufficient to reduce the number and/or activity of viable microorganisms to such an extent that very few, if any, are detectable in the treated food by any recognized method (viruses being excepted). No microbial spoilage or toxicity should become detectable in a food so treated, no matter how long or under what conditions it is stored, provided the package remains undamaged. The required dose is usually in the range of 25–45 kGy.

Radicidation: The application to foods of a dose of ionizing radiation sufficient to reduce the number of viable specific non–spore-forming pathogenic bacteria to such a level that none are detectable when the treated food is examined by any recognized method. The required dose is in the range of 2–8 kGy. The term may also be applied to the destruction of parasites like tapeworm and trichina in meat, in which case the required dose is in the range of 0.1–1 kGy. When the process is used specifically for destroying enteropathogenic and enterotoxinogenic organisms belonging to the genus *Salmonella*, it is referred to as "Salmonella radicidation."

Radurization: The application to foods of a dose of ionizing radiation sufficient to enhance its keeping quality by causing a substantial decrease in numbers of viable specific spoilage microorganisms. The required dose is in the range of 0.4–10 kGy.

These terms are still not often used in spoken communication, probably because the first two are considered to be tongue twisters, and they are not consistently used in the research literature.

The importance of assuring the absence of the highly radiation-resistant spore-forming pathogens in radappertized foods and precautions required to prevent safety risks from surviving spore-formers in radicidized and radurized foods will be discussed in the following.

II. BACTERIA

As can be seen from Tables 2 and 3, the radiation sensitivity of various bacterial species differs greatly. In the microflora of food it ranges from the extremely sensitive pseudomonads and vibrio species to the highly resistant spore-formers. Some non-spore forming bacteria (certain micrococci and *Moraxella-Acinetobacter*) are extremely resistant, but they are not of public health significance [31]. Interestingly, high heat resistance does not necessarily imply high radiation resistance. The highly radiation-resistant *Moraxella-Acinetobacter* are easily inactivated by heat, while the highly heat-resistant *B. stearothermophilus* is much less radiation resistant.

If radappertization is the purpose of an irradiation treatment, the need to inactivate reliably the most radiation-resistant pathogenic bacteria determines the required dose. In thermal sterilization (canning) it is customary to demand that the process be capable of reducing the viable spores of the most heat-resistant strain of *Clostridium botulinum* by a factor of 10^{-12}. This 12 *D concept* is based on the arbitrary assumption that the initial population of spores is 10^{11} and that "no-survival" means 10^{-1}. In other words, the 12 D concept guarantees, at least theoretically, that the probability of survival of a dangerous microorganism is no greater than 1×10^{-12}.

At the first international meeting on the microbiology of irradiated foods in 1960, C. F. Schmidt of Continential Can Company recommended that radiation sterilization processes be based on the same principle as thermal sterilization [32]. The 12 D concept applied to an irradiated food means that the dose has to be high enough to reduce viable spores of the most radiation resistant strain of *C. botulinum* by a factor of 10^{-12}.

The *minimum required dose* (MRD) for radiation sterilization is equal to 12 times the D_{10} value of the most radiation-resistant strain of *C. botulinum*. Taking the value for type A-33 (Table 3), this amounts to $12 \times 3.3 = 39.6$ kGy. Schmidt's suggestion was adopted by the FAO/IAEA/WHO Joint Expert Committee on Irradiated Food in 1964 [33] and has become standard practice in radappertization.

Because the determination of D_{10} values in the laboratory may not exactly duplicate the practical situation, D values for the actual process are usually determined by inoculated-pack studies. In a typical study of this kind, Anellis and co-workers [34], using 10 different strains of *C. botulinum*, inoculated cans of ham with approximately 10^6 spores per strain, 10^7 spores per can, and irradiated them at 5 kGy intervals from 5 to 45 kGy in lots of 1000 cans per dose. Following irradiation the cans were incubated at 30°C for 6 months. Cans in lots which had received 35 kGy or more were unswollen, nontoxic, and sterile. A few cans (5 out of 2000) which had received doses of 25 or 30 kGy contained inert but recoverable spores. The highest 12 D dose for any strain tested was 29 kGy, corresponding to a D_{10} value of 2400 Gy. Significantly increased rates of spoilage occurred over that of the controls in cans that received 5, 10, and 15 kGy doses. The following possible reasons were suggested for this enhanced susceptibility to swelling and toxin production with lower radiation doses: (a) decreased competition from the less radiation resistant indigenous microbial flora; (b) decreased nitrate-nitrite concentration due to radiation-induced losses; (c) radiation activation of the spores. Evidence for all three factors exists, but (a) is probably of the greatest importance.

Instead of the costly inoculated-pack studies, a miniaturized system for the determiantion of D values has been developed in Pflug's laboratory at the University of Minnesota [35]. It requires very small samples (1 cm × 1.5 cm slices) of inoculated meat which are sealed and irradiated inside polypropylene cryotubes. The D values obtained from the analysis of the number of surviving *C. botulinum* spores in this study were in good agreement with those determined by the traditional inoculated-pack studies.

Table 2 D_{10} Values of Selected Nonsporogenic Bacteria

Bacterium	Medium	Irradiation temperature	D_{10} (Gy)	Reference
**Vibrio parahaemolyticus*	Homog. fish	Ambient	30–60	14
Pseudomonas fluorescens	Low fat ground beef	Ambient	120	9
Leuconostoc mesenteroides	Water on filter paper	Ambient	120–140	15
**Campylobacter jejuni*	Ground beef	Ambient	140–160	16
**Aeromonas hydrophila*	Ground beef	2°C	140–190	17
**Proteus vulgaris*	Homog. oysters	5°C	200	18
**Yersinia enterocolytica*	Ground beef	Ambient	100–210	16
**Shigella dysenteriae*	Homog. shrimp	Frozen	220	19
**Shigella flexneri*	Homog. shrimp	Frozen	410	19
**Brucella abortus*	Ground beef	Ambient	340	9
**Escherichia coli*	Low fat ground beef	Ambient	430	9
**Salmonella anatum*	Ground beef	Ambient	670	16
**Salmonella enteritides*	Low fat ground beef	Ambient	700	9
**Salmonella newport*	Liquid whole egg	0°C	320	20

*Salmonella oranienburg	Liquid whole egg	0°C	320	20
**Salmonella panama*	Ground beef	Ambient	660	16
**Salmonella paratyphi A*	Homog. oysters	5°C	750	18
**Salmonella paratyphi B*	Homog. oysters	5°C	850	18
**Salmonella stanley*	Ground beef	Ambient	780	16
**Salmonella typhimurium*	Ground beef	Ambient	550	16
**Salmonella typhosa*	Homog. oysters	5°C	750	18
**Staphylococcus aureus*	Low fat ground beef	Ambient	580	9
Lactobacillus species	Ground beef	Ambient	300–880	21
**Streptococcus faecalis*	Homog. shrimp	5°C	750	18
**Streptococcus faecium*	Buffer solution	5°C	900	22
Micrococcus radiodurans	Nutrient broth	Ambient	3500	23
**Moraxella-Acinetobacter* from marine fish	Nutrient agar	Ambient	950–1900	24
**Moraxella-Acinetobacter* from beef	Nutrient broth	Ambient	4700	23

*Potential pathogen.

Table 3 D_{10} Values of Selected Spore-Forming Bacteria

Bacterium	Medium	Irradiation temperature	D_{10} (Gy)	Reference
**Clostridium botulinum* A-33	Buffer solution	Ambient	3300	25
**Clostridium botulinum* A-62	Buffer solution	Ambient	2200	25
**Clostridium botulinum* B-53	Buffer solution	Ambient	3300	25
**Clostridium botulinum* B-51	Buffer solution	Ambient	1300	25
**Clostridium botulinum* E-Alaska	Beef stew	Ambient	1400	26
**Clostridium botulinum* E-Iwanai	Beef stew	Ambient	1250	26
**Clostridium perfringens* C	Water	Ambient	2100	27
**Clostridium perfringens* E	Water	Ambient	1200	27
**Bacillus cereus*	Nutrient broth	Ambient	3200	23
Bacillus pumilus E-601	Dry	Ambient	3000	28
Bacillus subtilis 6051	Saline	Ambient	640	29
Bacillus stearothermophilus	Buffer solution	0°C	1000	30

*Potential pathogen.

Table 4 Minimum Required Doses (MRD) for Radappertization of Various Food Items

Food	MRD (kGy)
Bacon	25.2
Beef	41.2
Chicken	42.7
Corned beef	26.9
Ham	31.4
Pork	43.7
Pork sausage	25.5
Codfish cakes	31.7

Temperature during irradiation −30° ± 10°C, except bacon, which was irradiated at ambient temperature.
Source: Ref. 36.

On the basis of inoculated-pack studies, 12 D values for various radappertized products are listed in Table 4.

These MRD values apply only to the specific process developed at Natick Laboratories, which involves an enzyme-inactivating heat treatment (65–70°C), vacuum-packaging, and an irradiation temperature of −30°C in most cases. Variations in salt and nitrate/nitrite level in the cured products would also influence the MRD. It is apparent from Table 4 that uncured products (beef, chicken, pork) require the highest dose.

Experience at Natick has shown that the MRDs based on 12 D inactivation of *C. botulinum* also guarantee the absence of other pathogenic or spoilage organisms. More radiation-resistant nonpathogenic organisms, such as *Micrococcus radiodurans* or certain strains of *Moraxella-Acinetobacter*, do not grow well in meat and have low heat resistance. The heat treatment required for enzyme inactivation in the Natick process is apparently sufficient to inactivate them. Thus, with high-dose irradiation aimed at achieving commercial sterility in food, no public health problems can be foreseen [38]. A more recent review has again confirmed that "highly acceptable microbiologically safe products" can be produced by this process [39].

In contrast to radappertization, the lower radiation doses used for radicidation or radurization permit viable microorganisms to survive. As in other processes short of sterilization (such as chemical preservation, smoking, heat pasteurization), certain precautions are required to prevent spoilage and possible hygienic risks. Such problems exist particularly with perishable foods.

Perishable foods are those in which dangerous microorganisms can multiply and which are therefore liable to cause food poisoning. These include meat, fish, poultry, eggs, and their readily perishable products.

Nonperishable foods are those with a good public health record, in which pathogenic bacteria cannot develop. Such foods are, for example, fruits which are protected by their acidity, sufficiently salted or sugared foods, or dry foods, which are protected by their low water activity. A pH below 4.5, a salt content above 10%, or an equilibrium relative humidity below 0.92 suffices to put a food into this category.

Foods, even those that are themselves not perishable or that are kept from spoiling by such methods as freezing, can be carriers of pathogenic microorganisms. Several factors have worked together in recent years to increase the risks of foodborne disease. Mass tourism, worldwide trade with foodstuffs and feedstuffs, mass production of food animals and mass slaughtering, mass catering, a growing preference for ready-prepared foods which require very short cooking times—all of these developments have probably contributed to the worldwide rise in the frequency of outbreaks of food poisoning.

Mossel and Stegeman [19] list four epidemiological groups of disease-causing foodborne organisms:

The "big four": *Salmonella* species, *Campylobacter* species, *Staphylococcus aureus*, and *Bacillus cereus*

The "minor culprits" *Shigella, Yersinia enterocolitica; Vibrio parahaemolyticus*, various enterophathogenic and enterotoxinogenic types of *Escherichia coli, Clostridium perfringens, Aeromonas hydrophila, Edwardsiella tarda*

The very aggressive, but fortunately less frequently involved organisms, including *C. botulinum*

Organisms whose etiological role in food-transmitted disease has only recently or not definitely been established, such as *Cryptosporidium parvum* or *Vibrio vulnificus*

Fortunately, the most common and most troublesome bacteria are most sensitive to radiation [31] and can be reliably eliminated by doses below 10 kGy. The more resistant ones, like *C. botulinum* or *B. cereus*, will at least be reduced in numbers, and the surviving flora is generally less resistant to heat and to other environmental factors such as pH-changes, salt concentration, and antibiotics [39]. Numerous proposals have been made to use irradiation, in the interest of public health, in the form of radicidation processing [7, 19, 31, 40–42].

As mentioned earlier, the study of Anellis and co-workers [34] has shown that canned ham inoculated with 10^7 spores of different strains of *C. botulinum* incubated at 30°C spoiled faster when irradiated with doses of 5, 10, or 15 kGy, as compared to unirradiated control cans. Decreased competition from the less radiation-resistant microbial flora was thought to be the most likely explanation of this phenomenon. How general is this observation? Does irradiation of perishable foods with nonsterilizing doses generally imply an increased risk of botulism?

Many authors have studied this question and it appears that the use of such a massive inoculum and high incubation temperature, while appropriate for determination of the 12 D dose, is not a suitable basis for judging the risk of low dose applications. Safe meat products can be produced with nonsterilizing doses of radiation if the radiation effect is combined with the effect of curing salts and/or low storage temperature.

Nitrite for the curing of bacon is commercially used at a level of 120 μg/g. Rowley and co-workers [43] showed that 40 μg/g is sufficient to prevent toxin formation in bacon inoculated with 2 spores of *C. botulinum* (10 strains of type A and B) per g and irradiated with a dose of 15 kGy, then stored for 60 days at 27°C. The incidence of *C. botulinum* in bacon normally ranges from 10^{-3} to 6.4×10^{-5} cells/g. These authors therefore considered that a spore level of 2/g, as used in their experiments, provided a sufficient safety margin.

A study on turkey frankfurters inoculated with *C. botulinum* A and B has shown that the use of sodium chloride together with irradiation (5 and 10 kGy) was more effective at inhibiting botulinal toxin production than the combination of irradiation with potassium or magnesium salts [44]. With a sodium chloride concentration of 2.5% an inoculum of 450–500 spores/g and an incubation temperature of 27°C, unirradiated samples were toxic after 4 days, while samples irradiated at 5 kGy were toxic after 40 days. Samples irradiated at a dose of 10 kGy showed no toxin even after the maximum incubation time of 50 days.

Microbiological safety can be achieved without curing salts and without irradiation when meat is stored at 10°C or below. In studies with uncured chicken skin [45] inoculated with *C. botulinum* types A and B and incubated at 30°C, growth and toxin production was delayed by a radiation dose of 3 kGy.

The situation is different with fishery products. They may be contaminated with *C. botulinum* type E, which can grow and produce toxin at temperatures down to 3.3°C. In experiments described by Eklund [46], fillets from different species of fish were inoculated with *C. botulinum* E (100 spores/g) and stored under different conditions of temperature, packaging, and irradiation. Samples were periodically checked for toxin production and for spoilage as judged

Table 5 Maximum Storage Life and Earliest Toxin Production by *C. botulinium* E (100 spores/g) in Irradiated and Unirradiated Petrale Sole Fillets in Oxygen-Impermeable Packages

Storage temperature (°C)	Irradiation dose (kGy)	No. of days	
		Maximum storage	Earliest toxicity
3.3	0	18	>40
	1	32	>70
	2	43	>70
5.6	0	12	>42
	1	20	32
	2	30	28

Source: Ref. 46.

by odor evaluation. Data obtained with petrale sole fillets are presented in Table 5. At a storage temperatue of 3.3°C there was a large margin of safety between the first appearance of spoilage odors ("maximum storage") and earliest appearance of toxin. When the storage temperature was 5.6°C a safety margin did not exist for samples treated with a dose of 2 kGy (toxicity after 28 days, spoilage odor after 30 days!).

Petrale sole fillets are a particularly good substrate for *C. botulinum* growth and toxin production. In haddock fillets toxin production occurred at a later time. The authors concluded that irradiated petrale sole fillets would not provide a health hazard provided that the product could be stored and distributed at a temperature of 3.3°C or below, while for haddock fillets a temperature of 5.6°C was considered sufficient.

Obviously, certain radicidized foodstuffs must be kept refrigerated, and in the case of fishery products temperature control during irradiation is of particular importance.

The same considerations apply to radurized foods. Irradiation as a means of extending refrigerated shelf life of meats, fish, chicken, and certain other products will be described in Chapter 9.

Because of the destruction of the more radiation-sensitive organisms and the partial survival of the more resistant species, radicidation and radurization will always cause changes in the composition of the spoilage flora. As an example, irradiation of vacuum-packaged ground beef with a dose of 3 kGy and with subsequent storage at 4°C trebled the shelf life in studies described by Holzapfel and Niemand [21]. Although lactobacilli constituted less than 1% of the initial microbial population, this changed to 5.5% directly after radurization and to over 90% within 5 days of refrigerated storage. The dominant lactobacilli were *L. sake* and *L. curvatus*. All potentially harmful bacteria were effectively eliminated.

This development of lactobacilli in radurized meat means additional protection for the consumer. The organisms are not harmful. They would help to prevent toxin formation from *C. botulinum*, which cannot occur below pH 4.5, and the acidic flavor would warn the consumer. For these reasons it was suggested that 1% glucose and an inactivated preparation of lactic acid bacteria, such as *Pediococcus cerevisiae*, be added to meat products before radurization, so as to increase lactic acid production during the spoilage process [47]. These authors found no toxin production in semi-preserved meat products inoculated with *C. botulinum* if the pH was below 5.4.

The older literature on radiation resistance of lactobacilli is contradictory. French authors reported D_{10} values of about 150 Gy, which would place *Lactobacillus* among the more radiation-sensitive organisms [48]. Whether they happened to use more sensitive strains or whether their use of nutrient broth rather than meat as a medium for the irradiation treatments was responsible for this high sensitivity is unclear. In contrast, Niven [49] had found D_{10} values of 820–1100 Gy for homofermentative and 270–400 Gy for heterofermentative lactobacilli (irradiation medium and temperature not indicated). The extensive recent studies of Holzapfel and co-workers [50] have clearly established the high radiation resistance of many strains of lactobacilli (see also 51). Interestingly, some of these strains are more radiation resistant during the log phase of growth than during the stationary phase [50].

The composition of the spoilage flora of foods treated with a nonsterilizing dose of radiation depends on several factors, particularly on the composition of the initial flora, on the kind of food (nutrients, pH, etc.), on the radiation dose, and on the atmosphere in the package (aerobic or anaerobic). However, certain trends are always observed, like the disappearance of the pseudomonads and of *Aeromonas* and the increasing proportion of *Moraxella-Acinetobacter*, of *Micrococcus*, and of *Lactobacillus* species. An example of work done in the author's laboratories [52] on freshwater trout is shown in Figure 7 and Table 6. The gutted fish were vacuum-packed individually in polyethylene pouches, irradiated with 10 MeV electrons and stored in a cold room under melting ice. Nonirradiated control samples had reached the spoilage level of 10^7 organisms/g after two weeks, while samples irradiated with 1 or 2 kGy had not quite reached that level after 4 weeks (Fig. 7). The changing composition of the microbial flora is indicated in Table 6. After 4 weeks *Bacillus* and *Proteus* species dominated in the control samples, *Micrococcus* in the

Table 6 Changes in the Microbial Flora of Irradiated and Nonirradiated Trout During Storage Under Melting Ice

	Percentage of microorganisms						
	Days of storage						
	0	14			28		
Microorganisms	Control	Control	1 kGy	2 kGy	Control	1 kGy	2 kGy
Pseudomonas I	0	5	0	0	0	0	0
Pseudomonas II	7	25	0	0	0	0	0
Pseudomonas III	15	0	0	0	0	0	0
Pseudomonas IV	8	20	0	0	3	0	0
Achromobacter-Moraxella	15	0	18	26	0	17	24
Flavobacterium	16	5	0	0	0	0	0
Proteus	6	11	0	0	16	0	0
Aeromonas	5	7	0	0	9	0	0
Micrococcus	5	5	35	45	10	31	41
Lactobacillus	2	5	24	18	7	30	22
Corynebacterium	2	5	10	11	9	10	22
Bacillus	8	7	8	0	20	7	0
Sarcina	0	0	0	0	12	3	2
Brevibacterium	0	0	0	0	12	0	0
Yeasts	11	5	5	0	2	2	0

Source: Ref. 52.

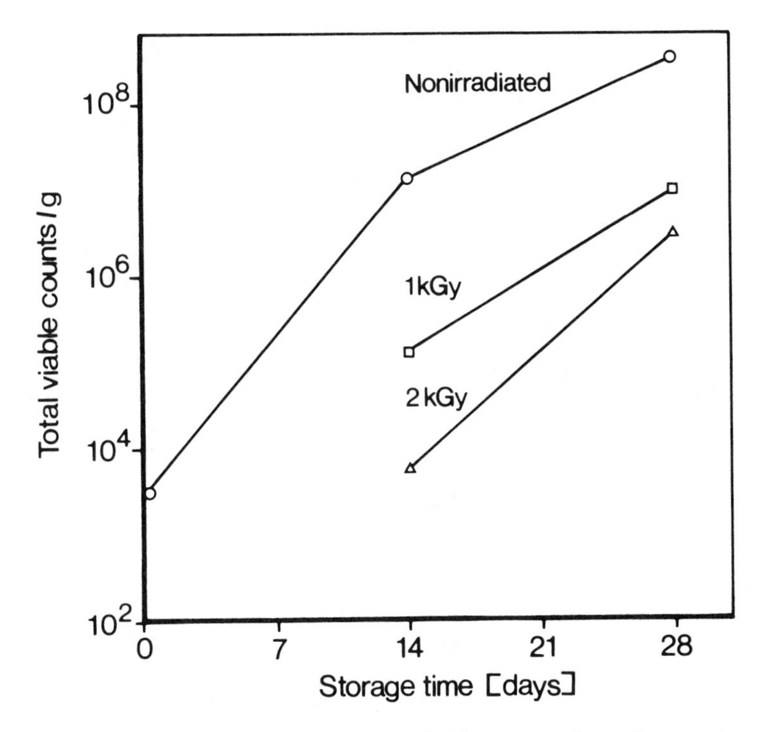

Figure 7 Total viable counts of microorganisms in vacuum-packed trout during storage at 0°C; unirradiated and irradiated with a dose of 1 or 2 kGy from 10 MeV electrons. (From Ref. 52.)

2 kGy samples, and *Micrococcus* and *Lactobacillus* in the 1 kGy samples. The spoilage odor in the control samples was putrid, while that of the irradiated samples, when they finally did spoil, was more "acidic." (The question of safety of nonsterilized irradiated products will be discussed in more detail in Chapter 6.)

In some areas knowledge of radiation microbiology is not as complete as it should be. No information appears to be available, for instance, on the radiation resistance of *Listeria monocytogenes*. The disease caused by this bacterium, listeriosis, is characterized by meningitis and septicemia. The disease can be very serious, especially in infants, pregnant women, and the aged. In recent years several epidemics of listeriosis with a high death rate have been identified in the Unites States and Canada. Cheeses, fermented sausages, and uncooked milk, meat, and vegetables were found to be carriers of the organism. Together with *C. botulinum* type E and *Yersinia enterocolitica*, *L. monocytogenes* belongs to the few pathogens which can survive and even grow at refrigeration temperature.

It would be very desirable to know the radiation dose required to eliminate this danger. [Note added in proof: Ongoing studies in several laboratories have indicated that the radiation resistance of *L. monocytogenes* is low.]

III. YEASTS AND MOLDS

What has been said about the factors that influence the radiation sensitivity of bacteria is generally also true for yeasts and molds. However, because of the difficulty of determining cell numbers in molds and yeasts, their radiation sensitivity is usually not expressed in the form of a D_{10} value. Instead, one can determine the dose required to prevent growth using the *end-point method*. In this method a large number of spores (their numbers can be determined) are inoculated into nutrient broth or into food samples, which are then irradiated at different dose levels. The samples are tested for presence or absence of survivors after incubation. The lowest dose giving no survival is regarded as the inactivation dose for the number of spores inoculated. The growth of molds can be visually observed in petri dishes. Besides determining the inactivation dose, one can measure the diameter or area of colonies, and by comparison with nonirradiated controls one can determine the dose required to cause, for instance, 50% growth reduction after 5 days of incubation.

Yeasts reproduce mainly by forming buds, or outgrowths of single cells, which form new cells—but they can also reproduce by forming spores. Molds grow by spreading out threadlike structures called hyphae. They can also produce spores. From the mass of budding yeast cells or of hyphae-producing molds, it is difficult to separate individual cells, hence the problem of plotting survival curves and determing D_{10} values.

When the effect of irradiation is expressed as reduction or suppression of growth, it is important to clearly indicate what was determined. The capacity for indefinite growth is lost at a much lower dose than the ability to germinate. For example, in the mold *Rhizopus stolonifer*, a dose of 5 kGy reduced the ability to form a colony to less than 1% but hardly affected the ability to germinate. A dose of 15 kGy was required to reduce germination to about 1%. A dose insufficient to inactivate a colony permanently may halt growth temporarily. After a few days normal growth may occur [53, 54]. The age of the cultures can influence radiation sensitivity considerably. While 3-week-old cultures of *Penicillium viridicatum* or *Aspergillus flavus* required a dose of 2 kGy for prevention of growth, a dose of 1 kGy was sufficient for 6-week-old cultures [55]. Under these circumstances it is not surprising that the experiments carried out by different authors to determine the radiation sensitivity of yeasts and molds have not led to identical results. There is agreement, however, that yeasts are generally more resistant than molds. The data presented in Table 7 indicate the dose required to prevent growth of an irradiated inoculum of spores within a specified time of postirradiation incubation. If we assume that an inactivation dose of 20 kGy for *Pullularia pullulans* means a reduction from initially 10^7 to zero cells per ml, a D_{10} value of roughly 3000 Gy can be estimated. This means a radiation resistance close to that of *Micrococcus radiodurans* (Table 2).

Instead of determining the dose required to prevent growth or germination, one can also determine the dose required to inhibit some biochemical function: in the case of fermenting yeasts, for example, the dose required to stop production of CO_2. Radiation-damaged cells surviving the treatment can gradually recover during postirradiation incubation. As indicated in Figure 8, a dose of 1.2 kGy, causing complete suppression of CO_2 production in *S. rosei* within the first 3 days after irradiation, had very little effect on total CO_2 production in 19 days [56].

As in the case of bacteria, results of determinations of radiation resistance are influenced by postirradiation environmental conditions. An example of the influence of the growth medium is shown in Figure 9. Irradiated *A. flavus* grew less well on Czapek agar than on three other agar media.

Some molds are of particular interest from a public health viewpoint because of their ability to produce mycotoxins. One of the most potent of these is aflatoxin, produced by *A. flavus* and *P. viridicatum*. Concerns that surviving cells of these species might have increased radiation resistance and/or an increased ability to produce aflatoxin after irradiation have led to extensive studies with various toxin-producing and non—toxin-producing strains of *A. flavus* and *P. viridicatum*. Overall, no increase in radiation resistance [60] or in ability to produce aflatoxin [61] was found in repeated cycles of irradiation and reculturing. (The effect of radiation on mycotoxin production will be further discussed in Chapter 6, Sec. III.)

Mycotoxins already formed are rather resistant to irradiation, as are toxins of bacterial origin. Inactivation of such toxins would require a dose much too high for practical food irradiation. Any suggestions that spoiled foods could be made marketable by radiation treatment must therefore be rejected.

IV. VIRUSES

Viruses occupy a position between living cells and nonliving biological molecules. They consist primarily of a nucleic acid, usually covered by a coat of protein. They can reproduce only by parasitizing a living cell, and each type of virus requires cells of specific

Table 7 Radiation Resistance of Some Species of Yeasts and Molds Irradiated at Ambient Temperature

Organism	Irradiation dose required to prevent growth (kGy)	Reference
Yeasts		
Saccharomyces rosei[a]	15	56
Saccharomyces cerevisiae[a]	18	56
Saccharomyces carlsbergensis[a]	15	56
Torulopsis stellata[a]	10	56
Rhodotorula glutinis[b]	10	57
Cryptococcus albidus[b]	10	57
Debaryomyces klöckeri[c]	7.5	57
Sporobolomyces pararoseus[c]	5.0	57
Candida krusei[d]	5.5	58
Candida tropicalis[d]	10	58
Pullularia pullulans[d]	20	58

Molds		
Penicillium viridicatum[e]	1.4	59
Aspergillus flavus[e]	1.7	59
Aspergillus niger[f]	2.5	58
Rhizopus nigricans[f]	2.5	58
Alternaria species[f]	6.0	58
Botrytis cinerea[f]	5.0	58
Cladosporium species[f]	6.0	58

[a]Initial cell number 1.5–3.0 × 10^6/ml. Gamma-irradiated and incubated (25°C, 19 days) in grape juice.
[b]Initial cell number 0.6–2.5 × 10^6/ml. Gamma-irradiated and incubated (17°C, 21 days) in grape juice.
[c]Initial cell number 0.6–2.5 × 10^6/ml. Gamma-irradiated and incubated (25°C, 21 days) in nutrient broth.
[d]Initial cell number 10^7/ml. Gamma-irradiated in phosphate buffer, subsequently incubated on malt extract agar at 27°C for up to 15 days.
[e]Initial number of conidia 4 × 10^6/ml. Electron-irradiated in 0.1% pepton + 0.1% Tween 80. Postirradiation incubation on Czapek agar, 5 days at 25°C.
[f]Initial number of conidia not given. Gamma-irradiated and subsequently incubated (27°C, 10 days) on malt extract agar.

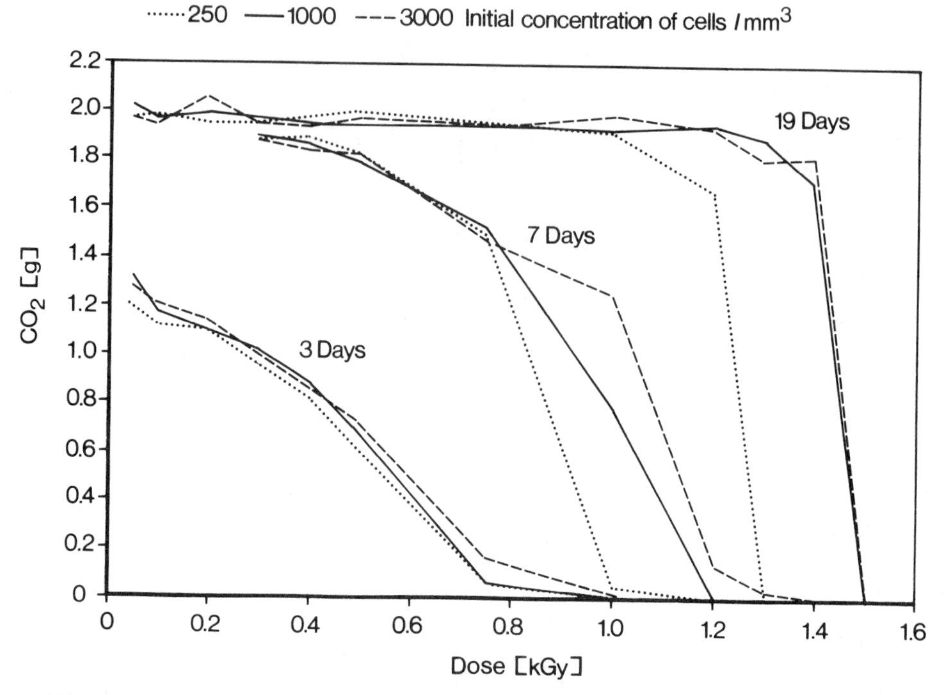

Figure 8 Effect of gamma radiation on cumulative CO_2 production by *Saccharomyces rosei* in grape juice at 3, 7, and 19 days after irradiation; incubation temperature 25°C. (From Ref. 56.)

organisms in which it can multiply. While plant viruses invade only plant cells, certain animal viruses can also propagate in human cells and are potentially dangerous for man.

When research on food irradiation began on a large scale in the 1950s, very little was known about foodborne viruses, and the radiation treatment was never intended to be used for the eradication of viruses. Due to the difficulties of propagating, harvesting, and detecting viruses, extensive studies on the radiation resistance of foodborne viruses were not available until Italian [62,63] and American [64] authors published their results in the mid-1960s and early 1970s. At that time food virology had been clearly established as a new science [65].

Viruses causing infectious hepatitis and poliomyelitis can definitely be transmitted through foods such as raw milk, the former also through shellfish from polluted waters and many other foods. Foot-and-mouth disease virus is highly infectious for animals, especially cattle, but transmission to man via meat of diseased animals is rare.

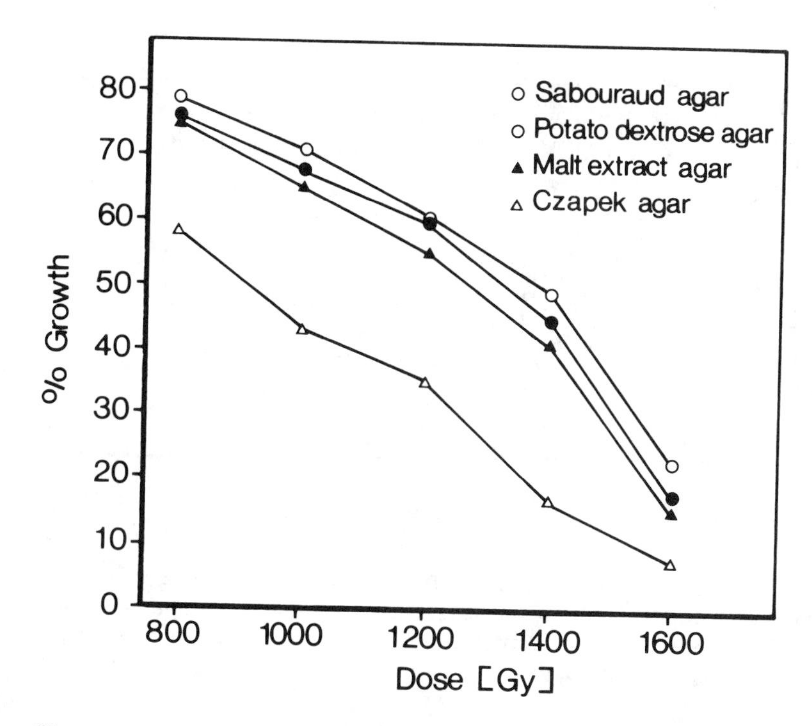

Figure 9 Influence of incubation medium on postirradiation growth of *Aspergillus flavus*, strain 373; initially 4×10^6 conidia/ml; 10 MeV electrons at ambient temperature; incubation at 25°C; growth indicated as % of control. (From Ref. 59.)

The role of foods in the transmission of some other viruses is less clear; echoviruses (diarrhea, especially in children; aseptic meningitis), reovirus (Colorado-tick fever), coxasackie viruses (aseptic meningitis, herpetic angina, myocarditis in young children), arboviruses (central nervous system diseases), and others [1].

Because of their dependence on living cells, viruses present in animal tissue cease to multiply either immediately or soon after slaughter, but they may survive for extended periods of time. Three of 12 market-purchased raw ground beef samples were reported to contain poliovirus and echovirus [66].

The rate of inactivation of viruses by irradiation is an exponential function of the radiation dose. Inactivation curves for foot-and-mouth disease virus are shown in Figure 10. D_{10} values are 4800 Gy for irradiation of an aqueous suspension and 6300 Gy for irradiation of a dry virus preparation [62]. Other viruses demonstrate a similarly high radiation resistance. The inactivation dose depends on

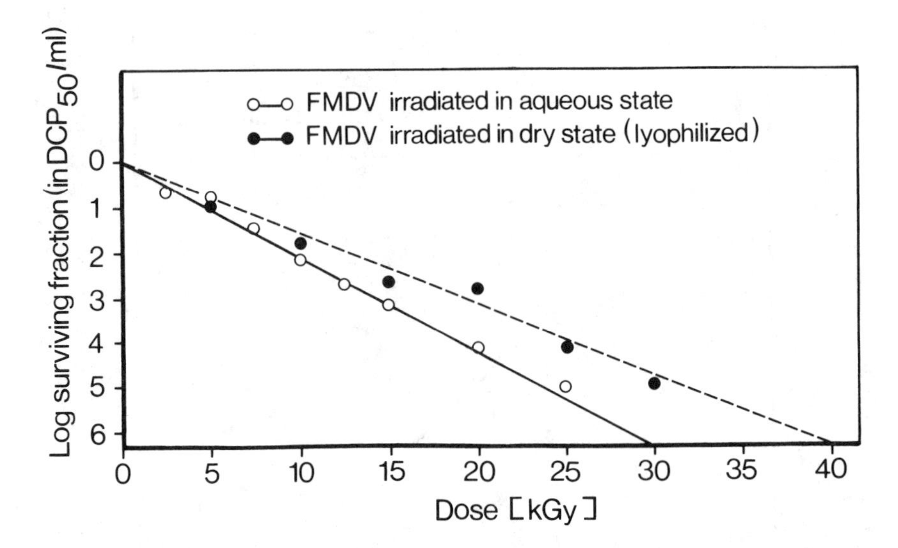

Figure 10 Radiation resistance of foot-and-mouth disease virus, gamma-irradiated at ambient temperature in aqueous suspension or in the dry state. The infective titre of the irradiated virus samples was determined by an in vitro method and is expressed as the cytopathogenic dose (DCP_{50}/ml). (Adapted from Ref. 63.)

the initial level of virus contamination. Clearly, the 10 kGy dose now considered as the upper limit in food irradiation cannot achieve more than about two log cycles, or 99% reduction of the number of viruses present. Radurized or radicidized foods must therefore be expected to still contain infectious viruses, the same as unheated dried, salted, or frozen foods. Because conventional heat processing will easily inactivate viruses, the combination of irradiation with a mild heat treatment (such as required for enzyme inactivation) can ascertain the absence of viable viruses.

V. PROTOZOA

Protozoa are one-celled organisms occurring in soil and water or as parasites in the bodies of animals. Many of the 30,000 species are of public health significance, for example plasmodium, which causes malaria.

In the context of food irradiation, only those species are of interest which are transmitted through food and are pathogenic to man.

Entamoeba histolytica is the causative organism of amoebic dysentery. It is acquired by man through consumption of fecally contaminated water or raw fruits and vegetables harboring the infective cysts. The amoeba invades the walls of the small intestine, causing ulcers and diarrhea. A dose of 250 Gy killed all viable cysts [67].

Toxoplasma gondii can parasitize all nucleated mammalian cells. It is found in many species, including man, cats, dogs, chickens, and many other animals. Infection in adults may go unnoticed in many cases or may cause rash, enlarged lymph nodes, fever, cough, headache, and other symptoms. The parasite can be passed from mother to child during pregnancy. The resulting congenital toxoplasmosis is serious, and infants who survive may have impaired vision and may be mentally deficient. An estimated 3300 babies born every year in the United States are infected. *T. gondii* occurs worldwide and is most common in warm, humid regions. The cyst form is highly resistant and can survive in shaded soil for about one year. The cysts can be transmitted orally through contact with infected animals or their feces and by consumption of improperly cooked meat. A radiation dose of 90 Gy was found to be sufficient to eliminate infectiousness of *T. gondii*, while a dose of 300 Gy was required to kill the parasites [68]. More recent work has indicated that a dose of 250 Gy is sufficient to make *T. gondii* in pig muscle noninfectious [69].

VI. PARASITIC HELMINTHS

Effects of radiation on parasitic helminths have been reviewed by King and Josephson [70]. Only some of the more important examples can be mentioned here.

Cysticercus bovis is the larval form found in cattle of a parasite causing taeniasis in man. When it occurs in man the parasite is called *Taeniarhynchus saginatus* (formerly *Taenia saginata*). It is found throughout the world and is acquired by eating uncooked or undercooked contaminated beef. The adult beef tapeworm lives in the small intestine of man, attached to the intestinal wall, where it may survive for many years, producing as many as 750,000 eggs a day. These are excreted with the feces, and where feces are deposited on grazing land, tapeworm eggs can reach the intestine of cattle. Here the eggs develop into oncospheres, which perforate the intestinal wall and migrate via blood and lymph systems into muscular tissues. Within several weeks of metamorphosis from oncosphere to cyst takes place. The life cycle of the parasite is completed when man consumes the contaminated beef and the cyst can develop into an adult tapeworm. Taeniasis frequently causes no symptoms. Abdominal pains, digestive disturbances, nausea, and weight loss may occur.

The disease is very widespread in some regions of the world. At slaughterhouses in certain parts of Central Africa cattle arrive with infection rates of 30–50%. According to WHO surveys, 20–30% of the human population are infected in some areas.

Cysts present in beef can be inactivated by deep freezing for 10 days at −10°C or 2 days at −35°C. Freezing is expensive in tropical regions and van Kooy and Robijns have tested irradiation as an alternative [71]. Using development in vitro as a criterion of viability, these authors recommended a dose of 3 kGy and a minimum storage of 7 days at 2°C postirradiation for inactivation. A study later carried out in Turkey showed that a relatively low dose of 0.4 kGy rendered *C. bovis* incapable of development in man [70].

Cysticercus cellulosa is the infective form, found in pork, of the tapeworm called *Taenia solium* when it occurs in man. As in the case of beef tapeworm, harboring pork tapeworm in the intestine is often not associated with any symptoms in man. However, if individuals having the adult pork tapeworm pass the eggs in their stool and contaminate their food with improperly washed hands, autoinfection may occur. In that case cysts may appear in the muscles, eyes, brain, and heart. This disease, called cysticercosis, can be very serious. Thorough cooking of pork inactivates the cysts. Studies carried out in South Africa have suggested that pork infested with *C. cellulosa* is fit for human consumption after irradiation with a dose of 200–600 Gy [72].

Trichinella spiralis is a parasitic nematode that localizes in the muscles of pigs and many wild mammals, especially carnivores. When meat containing the encysted larvae is eaten, the larvae are released in the intestinal tract. They burrow into the wall of the small intestine, where they mature and copulate, producing another generation of larvae. These pass through the lymphatic and circulatory systems to the muscle tissue, where they are encysted. During this period the symptoms of trichinosis occur: fever, edema, muscular pain. Heavy infections may result in permanent disability or death. There is presently no cure available for the disease.

In contrast to some other countries, no inspection program for *T. spiralis* in fresh pork exists in the United States. Although the prevalence rate for swine infections is low (an estimated 1 in 1000 in 1971) and although thorough cooking inactivates the cysts in contaminated meat, trichinosis still claims its victims. The Centers for Disease Control reported an average of about 120 cases per year during the 5-year period 1976–1980. In about 80% of the reported cases pork products such as raw smoked sausage or ham were incriminated. In recent years uneven heating of pork in microwave ovens was assumed to be responsible for several outbreaks of trichinosis.

Inactivation of *T. spiralis* by X-rays was observed by Schwartz in 1921, and this led to one of the first suggestions for a practical application of food irradition [73]. More recently, a cooperative project supported by the U.S. Department of Energy, the USDA Animal Parasitology Institute, and the National Pork Producers Council has evaluated the technical and commercial feasibility of producing trichina-safe pork by irradiation. The outcome was very favorable, and the only doubts that remained were related to "the hysteria that this food preservative process occasions among those who claim to represent the consumer in the USA" [74]. The dose required to render trichina-infected pork safe for human consumption was found to be 160 Gy. In order to provide safe "overkill" for all circumstances a dose of 300 Gy was suggested [75]. The cost of irradiation would add less than half a cent to the price of one pound of pork [76].

The dwarf tapeworm *Hymenolepis nana* does not need an intermediate host. Man acquires this parasite by eating cereals, dried fruits, and other foods which insect hosts have infested with the larval stage of the tapeworm. A dose of 370 Gy effectively prevented development to the egg-producing stage [77]. No information seems to be available on the radiation resistance of *Echinococcus granulosa*, another parasite which is of great importance in Africa and South America.

Man can be infected by the liver fluke, *Opisthorchis viverrini*, a parasitic flatworm, by eating contaminated raw, pickled, or smoked fish. The worms produce a chronic disease with symptoms varying from indigestion and stomach pain to diarrhea, enlarged liver, and swollen abdomen. Investigations in Thailand [78] have demonstrated that irradiation of contaminated fish with a dose of 500 Gy completely prevented the infectivity of the metacercariae of *O. viverrini* present in the fish flesh.

Although there is a need for further studies on the radiation behavior of a number of parasitic helminth, the information available to data indicates that probably all of these organisms are suppressed in their maturation by doses below 500 Gy. Much higher doses are required to achieve killing, but this is not required if the parasites lose their infectivity.

VII. EFFECTS OF IONIZING RADIATION ON INSECTS

Much of the work on grain pests has been carried out at the U.S. Department of Agriculture's Stored-Product Insects Research and Development Laboratory, Savannah, Georgia [79], where a pilot-scale bulk grain and packaged product irradiator was built in 1967, while much of the work on quarantine treatment of fruits was done at USDA's Hawaiian Fruit Flies Laboratory [80], also equipped with a pilot-scale irradiator in 1967. Important studies in radiation entomology were also carried out in the United Kingdom [81] and in many other countries [82].

The dose required to kill an insect depends primarily on age at time of treatment. Radiation sensitivity differs greatly in the different metamorphic stages of egg, larva, pupa, or adult. For example, the sensitivity of eggs of the yellow mealworm *Tenebrio molitor* L. is 250 times higher at the age of 0.5 days than at the age of 7.5 days. In general, radiation sensitivity is highest in the egg stage, lowest in the adult stage of development. Doses in the order of 1–3 kGy are required to achieve death at all stages within a few days. However, it is usually not necessary to aim at 100% mortality. Lower radiation doses are more economical and less damaging to the quality of the irradiated foods, and they suffice to sterilize the resident pupal or adult populations and to prevent emergence of eggs and larvae as adults.

Sterilizing doses for stored-product pests are listed in Table 8. They show that females are generally more sensitive than males and that considerable differences exist between the species. The moths (Lepidoptera) are more resistant to radiation than the mites (Acarina), which are more resistant than the beetles (Coleoptera). Sterilized insects will not produce offspring, and this is decisive for quarantine purposes. They will continue feeding and thus damaging the stored product, but it has been found that irradiated insects feed at a much reduced rate.

If a particular insect is to be eliminated, this insect's radiation sensitivity determines the required radiation dose. As indicated in Table 8, a dose of 50 Gy would suffice, for instance, to sterilize the females of *Tenebroides mauritanicus*, the black beetle with the common name cadelle. The males would not find fertile partners and their higher resistance therefore does not have to be taken into account. If a commodity is infested with many different insects, a dose of 500 Gy will control even the most resistant beetle species and immature moths. Some adult moths might remain fertile, but their few offspring would be sterile because of inherited genetic damage.

A dose of 250 Gy has been suggested as an effective quarantine treatment against fruit flies. It would prevent emergence of viable adults from eggs or larvae. Adults that might emerge from irradiated pupae would be sterile [83]. This assurance of sterility rather than complete kill would require a change in current U.S. quarantine regulations which are based on the concept of "probit 9 security." This means, in effect, that a treatment must result in at least 99.9968% mortality, or less than 1 survivor out of 31,250 [80].

The use of chemical insecticides is increasingly complicated by the development of resistance. Could insects also develop radiation resistance? Extensive tests have been made to answer this question,

Table 8 Comparative Sensitivity of Stored-Product Pests to the Sterilizing Effects of Gamma Radiation

Species	Sterilizing dose (Gy)		
	Male[a]	Female[a]	Both[b]
		Acarina	
Acarus siro L.	—	—	450
		Coleoptera	
Attagenus unicolor (Brahm)	175	175	—
Callosobruchus maculatus (F.)	70	70	—
Cathartus quadricollis (Guerin-Meneville)	300	200	—
Dermestes maculatus DeGeer	300	300	—
Gibbium psylloides (Czenpinski)	300	300	—
Gnathocerus maxillosus (F.)	200	200	—
Lasioderma serricorne (F.)	250	175	250
Latheticus oryzae Waterhouse	200	100	—
Oryzaephilus mercator (Fauvel)	—	—	200
O. Surinamensis (L.)	200	200	—
Palorus subdepressus (Wollaston)	400	400	—
Rhyzoperiha dominica (F.)	—	—	175
Sitophagus hololeptoides (La Porte)	100	100	—
Sitophilus granarius (L.)	100	100	—
S. oryzae (L.)	132	132	—
S. zeamais Motschulsky	100	100	—
Tenebrio molitor (L.)	150	50	—
T. obscurus (F.)	100	100	—
Tenebroides mauritanicus (L.)	100	50	—
Tribolium castaneum (Herbst)	200	200	—
T. confusum Jacquelin du Val	—	—	175
T. destructor Uyttenboogaart	100	100	—

Table 8 (Continued).

Species	Sterilizing dose (Gy)		
	Male[a]	Female[a]	Both[b]
	Coleoptera (continued)		
T. madens (Charpentier)	300	100	—
Trogoderma glabrum (Herbst)	250	132	—
T. inclusum (LeConte)	300	200	—
T. variabile (Ballion)	300	100	—
	Lepidoptera		
Cadra cautella (Walker)	1000	300	—
Plodia interpunctella (Hubner)	1000	1000	—
Sitotroga cerealella (Olivier)	1000	1000	—
Ephestia elutella (Hubner)	450	300	—

[a]The sex irradiated is specified.
[b]Unsexed irradiated populations exposed together.
Source: Ref. 79.

and no evidence of any pronounced increase in resistance to radiation has been found in any of the species studied. On the contrary, radiation-selected strains exhibited reduced fitness (higher percentages of sterility, lower fecundities, and decreased lifespans) [83].

VIII. LIVING PLANT TISSUES

The purpose of irradiating plant tissues is often to achieve a strong interference with the normal metabolic functions of that plant. Potatoes, onions, and garlic are irradiated to keep the tubers or bulbs from sprouting. Fruits may be irradiated with the intention of delaying the maturation process. It may be expected that a radiation dose which has such a pronounced effect on one physiological property will also have other, unintended effects.

Radiobiological changes in the main components of foods, especially the starch and sugar content, are too small to be of nutritional significance; changes in the level of vitamins may be of importance in nutritional evaluation, and such changes will be discussed in Chapter 7. A large number of other ingredients of irradiated living plant tissues have also been studied, primarily because of their contribution to the characteristic flavors and aromas of foods.

Kawakishi and co-workers studied the effect of irradiation on the development of characteristic aromatic substances in onions [84]. Others have described the effect of radiation on the ripening process in lemons [85], pears [86], mangoes [87–89], and papayas [90]. As indicative parameters of the ripening process, these studies measured ethylene and CO_2 production, oxygen consumption, the level of various carbohydrates, or identified characteristic aroma substances by gas chromatography.

Some of these changes develop slowly during storage. In an experiment designed to compare sprout inhibition by cold storage (4°C) with sprout inhibition by irradiation (storage at 15°C), organic acid levels in potatoes were not affected during the first days after irradiation. After 7 months of storage citric acid concentration had increased in the control samples and decreased in the irradiated samples [91], while malic acid had increased more in the irradiated samples (Table 9).

In some cases the effects of radiation on processes of cell metabolism, such as the pentose-phosphate cycle and the Embden-Meyerhof scheme, were traced individually. On the basis of this type of research on irradiated carrots, Massey and Bourke investigated how the rates of various stages in intermediary metabolism are speeded up or slowed down by irradiation [92].

The immediate cause of the biological phenomena observed after irradiation, such as inhibition of growth and sprouting, is probably the result of radiation damage to DNA, which causes changes in the activity of certain plant hormones and enzymes. The concentrations of the hormones gibberellic acid and indolyl acetic acid, which stimulate plant growth, are lower in irradiated seedlings than in nonirradiated ones—not because they are destroyed by irradiation, but because irradiation inhibits biosynthesis of these hormones [93].

Many of the physiological reactions of plants to irradiation are identical to their reactions to other stress factors. The increase in ethylene production following irradiation also occurs after mechanical damage and microbial infection. The same applies to the increase in phenylalanine-ammonialyase activity (PAL) (Fig. 11). Increased PAL activity causes increased conversion of phenylalanine into cinnamic acid and the phenol compounds produced from cinnamic acid, down to the hydroxycoumarin scopoletin. Increased phenol biosynthesis is a typical reaction of plant tissue to irradiation and other stress factors [94]. The following scheme indicates the pathway for the synthesis of phenols and coumarins.

D-glucose →→ HO—(shikimic acid ring)—COOH →→ (phenyl)—CH_2—CH(NH_2)—COOH

shikimic acid

phenylalanine

PAL ↓

(phenyl)—CH=CH—COOH

cinnamic acid

↓

HO—(phenyl)—CH=CH—COOH

p-coumaric acid

↓

HO, HO—(phenyl)—CH=CH—COOH

caffeic acid

↓

HO, H_3CO—(phenyl)—CH=CH—COOH

ferulic acid

↓

glucose—O, H_3CO—(phenyl)—CH=CH—COOH

glucosidoferulic acid

↓

glucose—O, H_3CO (coumarin)=O

scopolin

←

HO, H_3CO (coumarin)=O

scopoletin

Table 9 Effect of Gamma Irradiation (100 Gy) and 7 Months Storage on Organic Acid Levels in Potatoes, Variety Hansa

	mg/100 g		
Organic acid	Initial	After 7 months, control[a]	After 7 months, irradiated[b]
Phosphoric	2.6	7.0	12.2
Succinic	2.7	2.3	2.5
Malic	41.2	117.3	187
Pyroglutamic	8.3	9.2	25.1
Citric	208.3	266.7	158.5
Total	265.1	407.3	390.8

[a]Storage temperature 4°C.
[b]Storage temperature 15°C.
Source: Ref. 91.

The increase in scopoletin and scopolin levels in the flavedo of irradiated grapefruit is shown in Figure 12. This study is particularly interesting in the light of the hypothesis advanced some years ago by Kuzin on the formation of "radiotoxins" [95]. This term refers to substances, especially phenols [96], which have been identified in extracts from irradiated plants and are found to inhibit mitosis and growth in various test systems. As Table 10 shows, the cortex of irradiated potatoes was found to contain 50% more chlorogenic acid and 70% more caffeic acid than that of nonirradiated potatoes. It is worth noting that mitosis and growth are also inhibited in the plant test systems by extracts from nonirradiated plants, but to a lesser degree than by extracts from irradiated ones. As yet, there is no proof that irradiated plants contain any cytotoxic compounds not normally found in nonirradiated plants. In this respect, the use of the term "radiotoxins" is misleading.

It is important to be aware of the continuous changes in the composition of living plant tissue in irradiated and in unirradiated plant material. Determination of a plant constituent at one point in time may give a completely different result from that obtained at an earlier or later time. Indian authors who determined chlorogenic acid in potatoes found a steep increase within 6–24 hr after irradiation, followed by a decline. Three days after irradiation chlorogenic acid levels were actually lower in irradiated tubers (Fig. 13).

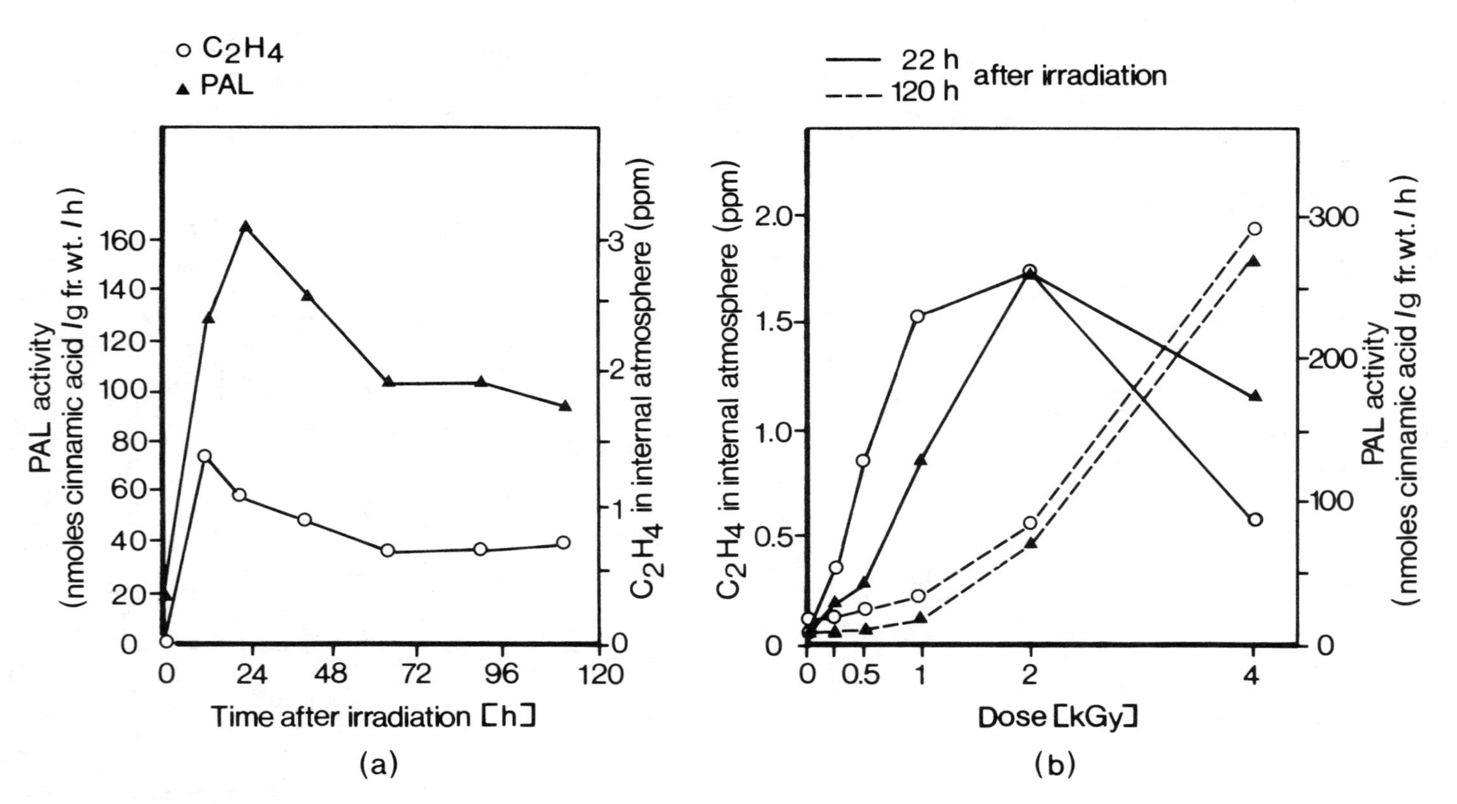

Figure 11 Ethylene (C_2H_4) production and phenylalanine ammonia-lyase (PAL) activity of irradiated grapefruit. (a) Plotted against time after irradiation with a dose of 2 kGy; (b) plotted against radiation dose, at 22 and 120 h after irradiation. (Adapted from Ref. 94.)

Table 10 Chlorogenic and Caffeic Acid (mg %) in the Cortex of Nonirradiated and Irradiated (150 Gy) Potato Tubers, 24 h After Irradiation

Experiment	Chlorogenic acid			Caffeic acid		
	Control	Experiment	% of level in control	Control	Experiment	% of level in control
1	15.7	30.0	191	1.61	2.75	171
2	16.2	23.3	143	1.88	2.90	155
3	16.0	22.0	138	1.74	2.80	160
4	19.7	24.8	126	171	3.32	195
Mean value	16.9 ± 1.1	25.0 ± 2.5	150	1.73 ± 0.7	2.90 ± 0.18	170

Source: Ref. 96.

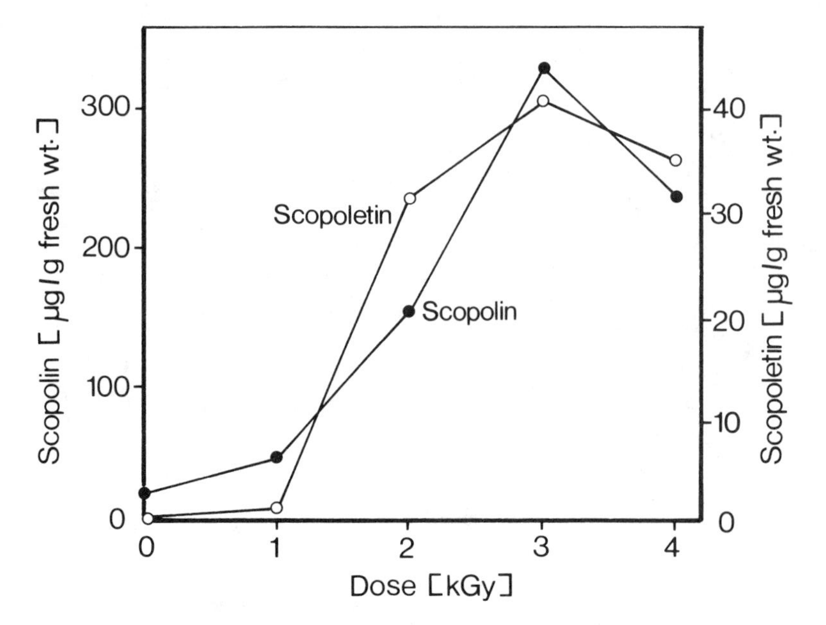

Figure 12 Scopoletin and scopolin levels in the flavedo of grapefruit 7 days after irradiation. (Adapted from Ref. 94.)

The increase in hydroxy-coumarins observed by Riov [94] in grapefruit flavedo is even more pronounced than the increase in phenols found by Kuzin in potatoes. Nine days after irradiation with 3 kGy, 160 µg scopoletin/g were found in the flavedo, 8.2 µg in nonirradiated controls. The natural occurrence of variable concentrations of scopoletin and other hydroxy-coumarins in carrots, celery, plums, apricots, and potatoes is well known and is not a peculiarity of irradiated products.

These radiation-stimulated physiological responses in living plant tissue differ greatly from the radiation chemical changes observed in nonliving foodstuffs. Radiation chemical effects increase with increasing dose, and they occur predominantly during irradiation. In contrast, the radiation-induced physiological effects do not show a linear dose dependence (Figs. 11b and 12), and they develop within hours or days or even months of the radiation treatment.

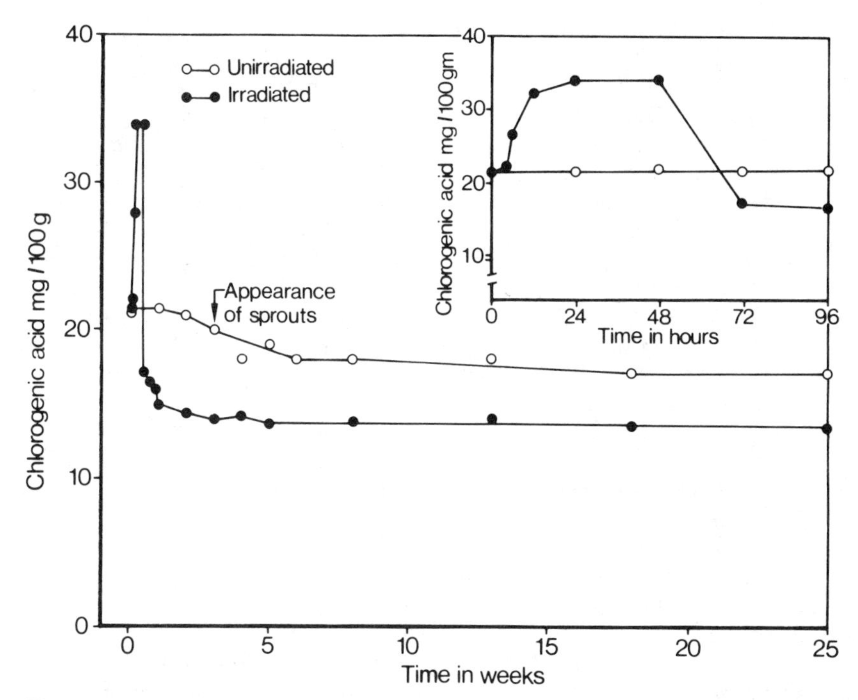

Figure 13 Variation in chlorogenic acid levels of gamma-irradiated (100 Gy) and unirradiated potatoes during storage. (Adapted from Ref. 97.)

REFERENCES

1. Grecz, N., D. B. Rowley, and A. Matsuyama, The action of radiation on bacteria and viruses, in *Preservation of Food by Ionizing Radiation*, vol. 2, E. S. Josephson and M. S. Peterson, eds. CRC Press, Boca Raton, FL, 1983. p. 167.
2. von Sonntag, C., Pulse radiolysis of nucleic acids and their base constituents: an updating review, *Radiat. Phys. Chem.*, 30:313 (1987).
3. von Sonntag, C., *The Chemical Basis of Radiation Biology*. Taylor and Francis, Philadelphia, PA, 1987.
4. Pollard, E. C., Phenomenology of radiation effects on microorganisms, in *Handbook of Medical Radiology*, vol. 2/2, L. Diethelm et al., eds., Springer, New York, 1966, p. 1.
5. Brynjolfsson, A., Chemiclearance of food irradiation: Its scientific basis, in *Combination Processes in Food Irradiation*, Proceedings of a Symposium held in Colombo, Nov. 1980, Internat. Atomic Energy Agency, Vienna, 1981, p. 367.

6. Stapleton, G. E., Factors modifying sensitivity of bacteria to ionizing radiations, *Bacteriol. Rev.*, 19:26 (1955).
7. Kampelmacher, E. H., Prospects of eliminating pathogens by the process of food irradiation, in *Combination Processes in Food Irradiation*, Proceedings of a Symposium held in Colombo, Nov. 1980. Internat. Atomic Energy Agency, Vienna, 1981, p. 265.
8. Ley, F. J., Application of radiation for the control of salmonellae in various foods, in *Food Irradiation*, Proceedings of a Symposium held in Karlsruhe, June 1966. Internat. Atomic Energy Agency, Vienna, 1966, p. 349.
9. Maxcy, R. B., and N. P. Tiwari, Irradiation of meats for public health protection, in *Radiation Preservation of Food*, Proceedings of a Symposium held in Bombay, Nov. 1972. Internat. Atomic Energy Agency, Vienna, 1973, p. 491.
10. Moussa, A. E., and J. F. Diehl, Combined effect of radiation dose, temperature during irradiation and water activity on the survival rate of three serotypes of Salmonella (in German) *Arch. Lebensmittelhyg.*, 30:157 (1979).
11. Härnulv, B. G., and B. G. Snygg, Radiation resistance of spores of *Bacillus subtilis* and *B. stearothermophilus* at various water activities, *J. Appl. Bact.*, 36:677 (1973).
12. Alper, T., Mechanisms of cell death due to ionizing radiation, in *Sterilization of Medical Products by Ionizing Radiation*, E. R. L. Gaughran and A. J. Goudie, eds., Multiscience Publication Ltd., Montreal, 1978, p. 9.
13. Goresline, H. E., M. Ingram, P. Macúch, G. Mocquot, D. A. A. Mossel, C. F. Niven, Jr., and F. S. Thatcher, Tentative classification of food irradiation processes with microbiological objectives, *Nature*, 204:237 (1964).
14. Matches, J. R., and J. Liston, Radiation destruction of *Vibrio parahaemolyticus*, *J. Food Sci.*, 36:339 (1971).
15. Koh, W. Y., C. T. Morehouse, and V. L. Chandler, Relative resistance of microorganisms to cathode rays. I. Nonsporeforming bacteria, *Appl. Microbiol.*, 4:143 (1956).
16. Tarkowski, J. A., S. C. C. Stoffer, R. R. Beumer, and E. H. Kampelmacher, Low dose gamma irradiation of raw meat. I. Bacteriological and sensory quality effects in artificially contaminated samples, *Int. J. Food Microbiol.*, 1:13 (1984).
17. Palumbo, S. A., R. K. Jenkins, R. L. Buchanan, and D. W. Thayer, Determination of irradiation D-values for *Aeromonas hydrophila*, *J. Food Protection*, 49:189 (1986).
18. Quinn, D. J., A. W. Anderson, and J. F. Dyer, The inactivation of infection and intoxication microorganisms by irradiation in seafood, in *Microbiological Problems in Food Preservation by Irradiation*, Panel Proceedings Series, Internat. Atomic Energy Agency, Vienna, 1967, p. 1.

19. Mossel, D. A. A., and H. Stegeman, Irradiation: An effective mode of processing food for safety, in *Food Irradiation Processing*, Proceedings of a Symposium held in Washington, D.C., March 1985. Internat. Atomic Energy Agency, Vienna, 1985, p. 251.
20. Licciardello, J. J., J. T. R. Nickerson, and S. A. Goldblith, Elimination of salmonella in poultry with ionizing radiation, in *Elimination of Harmful Organisms from Food and Feed by Irradiation*, Panel Proceedings Series, Internat. Atomic Energy Agency, Vienna, 1968, p. 1.
21. Holzapfel, W. H., and J. G. Niemand, The role of Lactobacilli and other bacteria in radurized meat, in *Food Irradiation Processing*, Proceedings of a Symposium held in Washington, D.C., March 1985. Internat. Atomic Energy Agency, Vienna, 1985, p. 239.
22. Anellis, A., D. Berkowitz, and D. Kemper, Comparative resistance of nonsporogenic bacteria to low-temperature gamma irradiation, *Appl. Microbiol.*, 25:517 (1973).
23. Ma, K., and R. B. Maxcy, Factors influencing radiation resistance of vegetative bacteria and spores associated with radappertization of meat, *J. Food Sci.*, 46:612 (1981).
24. Münzner, R., Studies on radiation resistance of strains of *Moraxella-Acinetobacter* isolated from marine fish (in German), *Arch. Lebensmittelhyg.*, 28:195 (1977).
25. Anellis, A., and R. B. Koch, Comparative resistance of strains of *Clostridium botulinum* to gamma rays, *Appl. Microbiol.*, 10: 326 (1962).
26. Schmidt, C. F., W. K. Nank, and R. V. Lechowich, Radiation sterilization of food. 2. Some aspects of the growth, sporulation and radiation resistance of spores of *Clostridium botulinum*, type E, *J. Food Sci.*, 27:77 (1962).
27. Roberts, T. A., Heat and radiation resistance and activation of spores of *Clostridium welchii*, *J. Appl. Bact.*, 31:133 (1968).
28. Christensen, E. A., The role of microbiology in commissioning a new facility and in routine control, in *Sterilization of Medical Products by Ionizing Radiation*, E. R. L. Gaughran and A. J. Goudie, eds., Multiscience, Montreal, 1978, p. 50.
29. Proctor, B. E., S. A. Goldblith, E. M. Oberle, and W. C. Miller, Jr., Radiosensitivity of *Bacillus subtilis* under different environmental conditions, *Rad. Res.*, 3:295 (1955).
30. Friedman, R. S., and N. Grecz, The role of heat resistance in thermorestoration of hydrated bacterial spores, *Acta Aliment.*, 2:209 (1973).
31. Maxcy, R. B., Irradiation of food for public health protection, *J. Food Protection*, 45:363 (1982).

32. Schmidt, C. F., Dose requirements for the radiation sterilization of food, in *Report of the European Meeting on the Microbiology of Irradiated Foods*. FAO and Internat. Association of Microbiological Societies, Paris, 1960, p. 2.
33. WHO, *The Technical Basis for Legislation on Irradiated Food*, Technical Report Series 316, Geneva, 1965.
34. Anellis, A., D. Berkowitz, C. Jarboe, and H. M. El-Bisi, Radiation sterilization of prototype military foods. 2. Cured ham, *Appl. Microbiol.*, 15:15 (1967).
35. Kreiger, R. A., O. P. Snyder, and I. J. Pflug, *Clostridium botulinum* ionizing radiation D-value determination using a micro food sample system, *J. Food Sci.*, 48:141 (1983).
36. Josephson, E. S., Radappertization of meat, poultry, finfish, shellfish and special diets, in *Preservation of Food by Ionizing Radiation*, vol. 3, E. S. Josephson and M. S. Peterson, eds. CRC Press, Boca Raton, FL, 1983, p. 231.
37. Anellis, A., D. B. Rowley, and E. W. Ross, Jr., Microbiological safety of radappertized beef, *J. Food Protection*, 42:927 (1979).
38. Thayer, D. W., R. V. Lachica, C. N. Huhtanen, and E. Wierbicki, Use of irradiation to ensure the microbiological safety of processed meats, *Food Technology* (Chicago), 40(4):159 (1986).
39. Farkas, J., and E. Andrassy, Increased sensitivity of surviving bacterial spores in irradiated spices, in *Fundamental and Applied Aspects of Spores*, G. J. Dring, D. J. Ellar, G. W. Gould, eds. Acad. Press, London, 1985, p. 397.
40. Ouwerkerk, T., Salmonella control in poultry through the use of gamma irradiation, in *Combination Processes in Food Irradiation*, Proceedings of a Symposium held in Colombo, Nov. 1980. Internat. Atomic Energy Agency, Vienna, 1981, p. 335.
41. Sang, F. C., M. E. Hugh-Jones, and H. V. Hagstad, Viability of *Vibrio cholerae* 01 on frog legs under frozen and refrigerated conditions and low dose radiation treatment, *J. Food Protection*, 50:662 (1987).
42. Farkas, J., Decontamination, including parasite control, of dried, chilled and frozen foods by irradiation, *Acta Aliment.*, 16:351 (1987).
43. Rowley, D. B., R. Firstenberg-Eden, E. M. Powers, G. E. Shattuck, A. E. Wasserman, and E. Wierbicki, Effect of irradiation on the inhibition of *Clostridium botulinum* toxin production and the microbial flora in bacon, *J. Food Sci.*, 48:1016 (1983).
44. Barbut, S., A. J. Maurer, and D. W. Thayer, Gamma-irradiation of *Clostridium botulinum* inoculated turkey frankfurters formualted with different chloride salts and polyphosphates, *J. Food Sci.*, 52:1137 (1987).
45. Dezfulian, M., and J. G. Bartlett, Effects of irradiation on growth and toxigenicity of *Clostridium botulinum* types A and B inoculated onto chicken skins, *Appl. Environm. Microbiol.*, 53: 201 (1987).

46. Eklund, M. W., Significance of *Clostridium botulinum* in fishery products preserved short of sterilization, *Food Technology* (Chicago), 36(12):107 (1982).
47. Riemann, H., W. H. Lee, and C. Genigeorgis, Control of *Clostridium botulinum* and *Staphylococcus aureus* in semi-preserved meat products, *J. Milk Food Technol.*, 35:514 (1972).
48. Dupuy, P., and O. Tremeau, Resistance of some strains of Lactobacillus to ionizing radiation (in French), *Int. J. Appl. Rad. Isotopes*, 11:145 (1961).
49. Niven, C. F., Jr., Microbiological aspects of radiation preservation of food, *Ann. Rev. Microbiol.*, 12:507 (1958).
50. Hastings, J. W., W. H. Holzapfel, and J. G. Niemand, Radiation resistance of lactobacilli isolated from radurized meat relative to growth and environment, *Appl. Environment. Microbiol.*, 52:898 (1986).
51. Ehioba, R. M., A. A. Kraft, R. A. Molins, H. W. Walker, D. G. Olson, and G. Subbaraman, Identification of microbial isolates from vacuum-packaged ground pork irradiated at 1 kGy, *J. Food Sci.*, 53:278 (1988).
52. Hussain, A. M., D. Ehlermann, and J. F. Diehl, Effect of radurization on microbial flora of vacuum-packaged trout (*Salmo gairdneri*), *Arch. Lebensmittelhyg.*, 27:197 (1976).
53. Behara, L., G. B. Ramsey, M. A. Smith, and W. R. Wright, Factors influencing the use of gamma radiation to control decay of lemons and oranges, *Phytopathology*, 49:91 (1959).
54. Behara, L., G. B. Ramsey, M. A. Smith, and W. R. Wright, Effects of gamma radiation on brown rot and Rhizopus rot of peaches and the causal organisms, *Phytopathology*, 49:354 (1959).
55. Malla, S. S., J. F. Diehl, and D. K. Salunkhe, In vitro susceptibility of strains of *Penicillium viridicatum* and *Aspergillus flavus* to beta-irradiation, *Experientia*, 23:492 (1967).
56. Fernandez, A., G. Stehlik, and K. Kaindl, The fungistatic effect of cobalt-60 gamma radiation on different concentrations of grape-juice yeasts. 1. Juice-fermenting yeasts. Seibersdorf Project Report SPR-4, Institute for Biology and Agriculture, Seibersdorf Reactor Centre, Austria, 1966.
57. Fernandez, A., G. Stehlik, and K. Kaindl, The fungistatic effect of cobalt-60 gamma radiation on different concentrations of grape-juice yeasts. 2. Nonfermenting yeasts. Seibersdorf Project Report SPR-14, Institute for Biology and Agriculture, Seibersdorf Reactor Centre, Austria, 1967.
58. Saravacos, G. D., L. P. Hatzipetrou, and E. Georgiadou, Lethal doses of gamma radiation of some fruit spoilage microorganisms, *Food Irradiation* (Saclay), 3(1/2):A6 (1962).
59. Münzner, R., Some factors affecting the radiosensitivity of molds (in German), *Arch. Mikrobiol.*, 64:349 (1969).

60. Münzner, R., and J. F. Diehl, Studies on the radiation sensitivity of cultures of *Aspergillus flavus* and *Penicillium viridicatum* repeatedly irradiated with electrons (in German), *Lebensm. Wiss. Technol.*, 2:44 (1969).
61. Frank, H. K., R. Münzner, and J. F. Diehl, Response of toxigenic and non-toxigenic strains of *Aspergillus flavus* to irradiation, *Sabouraudia*, 9:21 (1971).
62. Massa, D., Radiation inactivation of foot-and-mouth disease virus in the blood, lymphatic glands and bone marrow of the carcasses of infected animals, in *Food Irradiation*, Proceedings of a Symposium held in Karlsruhe, June 1966. Internat. Atomic Energy Agency, Vienna, 1966, p. 329.
63. Baldelli, B., Gamma radiation for sterilizing the carcasses of foot-and-mouth disease virus infected animals, in *Microbiological Problems in Food Preservation*, Panel Proceedings Series. Internat. Atomic Energy Agency, Vienna, 1967, p. 77.
64. Sullivan, R., A. C. Fassolitis, E. P. Larkin, R. B. Read, Jr., and J. T. Peeler, Inactivation of 30 viruses by gamma radiation, *Appl. Microbiol.*, 22:61 (1971).
65. Cliver, D. O., Transmission of viruses through foods, *Crit. Revs. Environ. Control*, 1:551 (1971).
66. Sullivan, R., A. C. Fassolitis, and R. B. Read, Jr., Method for isolating viruses in ground beef, *J. Food Sci.*, 35:624 (1970).
67. Schneider, C. R., Radiosensitivity of *Entamoeba histolytica* cysts, *Exp. Parasitol.*, 9:87 (1960).
68. Baldelli, B., K. A. Saravanos, M. Ambrosi, T. Frescura, and G. A. Polidori, Effects of gamma radiation on a strain of Toxoplasma isolated from dog (in Italian), *Parassitologia*, 13:105 (1971).
69. Dubey, J. P., R. J. Brake, K. D. Murrell, and R. Fayer, Effect of irradiation on the viability of *Toxoplasma gondii* cysts in tissue of mice and pigs, *J. Am. Vet. Med. Assoc.*, 187:304 (1985).
70. King, B. L., and E. S. Josephson, Action of radiation on protozoa and helminths, in *Preservation of Food by Ionizing Radiation*, vol. 2. E. S Josephson and M. S. Peterson, eds. CRC-Press, Boca Raton, FL, 1983, p. 245.
71. van Kooy, J. G., and K. G. Robijns, Gamma irradiation elimination of *Cysticercus bovis* in meat, in *Elimination of Harmful Organisms from Food and Feed by Irradiation*, Panel Proceedings Series. Int. Atomic Energy Agency, Vienna, 1968, p. 81.
72. Verster, A., T. A. du Plessis, and W. van den Neever, The eradication of tapeworms in pork and beef carcasses by irradiation, *Radiat. Phys. Chem.*, 9:769 (1977).
73. Schwartz, B., Effects of x-rays on trichinae, *J. Agric. Res.*, 20:845 (1921).

74. van Houweling, C. D., and D. Meisinger, The interest of the pork industry in the United States of America in irradiation, in *Food Irradiation Processing*, Proceedings of a Symposium held in Washington, D.C., March 1985. Internat. Atomic Energy Agency, Vienna, 1985, p. 281.
75. Brake, R. J., K. D. Murrell, E. E. Ray, J. D. Thomas, B. A. Muggenburg, and J. S. Sivinski, Destruction of *Trichinella spiralis* by low-dose irradiation of infected pork, *J. Food Safety*, 7:127 (1985).
76. Sivinski, J. S., and R. K. Switzer, Low-dose irradiation: a promising option for trichina-safe pork certification, in *Radiation Disinfestation of Food and Agricultural Products*, J. H. Moy, ed., Hawaii Institute of Tropical Agriculture & Human Resources, Univ. of Hawaii at Manoa, Honolulu, Hawaii, 1985, p. 181.
77. Onyango-Abuje, J. A., and C. J. Weinmann, Effects of gamma irradiation on the infectivity and development of *Hymenolepis nana* oncospheres (Cestoda: Cyclophyllidea), *Z. Parasitenk.*, 44:111 (1974).
78. Bhaibulaya, M., Effect of gamma rays on the metacercariae of liver fluke (*Opisthorchis viverrini*) infective stages of parasite caused by consumption of raw or semi-processed fish, *Food Irrad. Newsletter*, 9(2):8 (1985).
79. Tilton, E. W., and J. H. Brower, Ionizing radiation for insect control in grain and grain products, *Cereal Foods World*, 32: 330 (1987).
80. Burditt, A. K., Jr., Food irradiation as a quarantine treatment of fruits, *Food Technol.* (Chicago), 36(11):51 (1982).
81. Cornwell, P. B., *The Entomology of Radiation Disinfestation of Grain*. Pergamon Press, Elmsford, NY, 1966.
82. Moy, J. H., ed., *Radiation Disinfestation of Food and Agricultural Products*, Proceedings of an Internat. Conference held in Honolulu, Hawaii, Nov. 1983. Hawaii Inst. Trop. Agric. and Human Resources, Univ. of Hawaii at Monoa, Honolulu, 1985.
83. Tilton, E. W., and A. K. Burditt, Jr., Insect disinfestation of grain and fruit, in *Preservation of Food by Ionizing Radiation*, vol. 3, E. S. Josephson and M. S. Peterson, eds. CRC Press, Boca Raton, FL, 1983, p. 215.
84. Kawakishi, S., K. Namiki, H. Nishimura, and M. Namiki, Effects of gamma irradiation on the enzyme relating to the development of characteristic odor of onions, *J. Agr. Food Chem.*, 19:166 (1971).
85. Maxie, E. C., H. L. Rae, I. L. Eaks, and N. F. Sommer, Studies on radiation-induced ethylene production by lemon fruits, *Rad. Botany*, 6:445 (1966).
86. Maxie, E. C., N. F. Sommer, C. J. Muller, and H. L. Rae, Effect of gamma radiation on the ripening of Bartlett pears, *Plant Physiol.*, 41:437 (1966).

87. Dennison, K. A., and E. M. Ahmed, Irradiation effects on the ripening of Kent mangoes, *J. Food Sci.*, 32:702 (1967).
88. Cuevas-Ruiz, J., H. D. Graham, and R. A. Luse, Gamma radiation effects on biochemical components of Puerto Rican mangoes, *J. Agr. Univ. Puerto Rico*, 56(1):26 (1972).
89. Khan, I., A. Sattar, M. Ali, and A. Muhammed, Some physiological and biochemical changes in irradiated mangoes, *Lebensm. Wiss. Technol.*, 7:25 (1974).
90. Hilker, D. M., and R. L. Young, Effect of ionizing radiation on some nutritional and biochemical properties of papaya, *Hawaii Farm Sci.*, 15:9 (1966).
91. Thomas, P., S. Adam, and J. F. Diehl, Role of citric acid in the after-cooking darkening of gamma-irradiated potato tubers, *J. Agric. Food Chem.*, 27:519 (1979).
92. Massey, L. M., Jr., and J. B. Bourke, Some radiation-induced changes in fresh fruits and vegetables, in *Radiation Preservation of Foods*, Advances in Chemistry Series 65. American Chemical Society, Washington, D.C., 1967, p. 1.
93. Ananthaswamy, H. N., K. K. Ussuf, P. M. Nair, U. K. Vakil, and A. Sreenivasan, Role of auxins on the reversal fo radiation-induced growth inhibition in plants, *Rad. Res. Revs.*, 3:429 (1972).
94. Riov, J., S. P. Monselise, R. Goren, and R. S. Kahan, Stimulation of phenolic biosynthesis in citrus fruit peel by gamma irradiation, *Rad. Res. Revs.*, 3:417 (1972).
95. Kuzin, A. M., *Radiation Biochemistry*, Moscow, 1962. (English edition: Israel Program for Scientific Translations, Jerusalem, 1964.)
96. Kuzin, A. M., E. G. Plyshevskaya, and V. A. Kopylov, On the role of the systems orthophenols—orthoquinones in the initial radiation-induced processes in an organism (quoted: Nucl. Sci. Abstr. 20, No. 1647 (1966), *Izvestiya Akademii Nauk SSR, Seriya Biologicheskaya*, (4):507 (1965).
97. Thomas, P., A. N. Srirangarajan, and S. R. Pawdal-Desai, A. S. Ghanekar, S. G. Shirsat, M. B. Pendharkar, P. M. Nair, and G. B. Nadkarni, Feasibility of radiation processing for post-harvest storage of potatoes under tropical conditions, in *Food Preservation by Irradiation*, vol. 1, Proceedings of a Symposium held in Wageningen, November 1977. Int. Atomic Energy Agency, Vienna, 1978, p. 71.

5

Radiological and Toxicological Safety of Irradiated Foods

With regard to irradiated foods, considerations of safety for consumption involve four aspects: radiological safety, toxicological safety, microbiological safety, and nutritional adequacy. The first two will be discussed in this chapter, the other two in the next chapter.

I. ARE IRRADIATED FOODS RADIOACTIVE?

The terms "radiation" and "radioactivity" are easily confused by the layman, and consumers who are first confronted with the idea of eating irradiated foods often express fears that such foods might be radioactive and therefore harmful. This confusion became particularly apparent in May 1986, when the media reported about the radioactive contamination of crops in many European countries caused by the radioactive cloud emanating from the ruins of the nuclear reactor at Chernobyl in the Soviet Union. These reports alternately spoke of radioactive foods, radiated foods, radiating foods or irradiated foods as if these were synonymous. Only two of these terms are correct, the first and the last, and they mean different things. Radioactive foods are foods containing radioactive isotopes, or in more modern terminology radionuclides. Irradiated foods are foods that have been treated with ionizing radiation.

Are irradiated foods radioactive? There is no simple yes or no answer to this question because all foods are radioactive. Naturally radioactive elements are found in all foods, primarily isotopes of potassium (^{40}K), carbon (^{14}C), and hydrogen (^{3}H or tritium). Natural trace constituents such as radium (^{226}Ra), thorium (^{228}Th), radioactive lead (^{210}Pb), and polonium (^{210}Po) are present in most foods. The foods consumed by an adult in one day contain about 150–200 Bq (becquerel) of natural radioactivity, which means that 150–200 numclear disintegrations per second take place in that quantity of food.

The question to be asked is, therefore, do foods become more radioactive when they are irradiated? The answer depends on the radiation energy and the radiation dose used. It is possible to convert stable, nonradioactive elements to radioactive isotopes by treating them with radiation of sufficiently high energy. Each stable element requires a certain minimal radiation energy (threshold energy) to make it radioactive. How much radioactivity will be produced at a particular energy level above this threshold depends on the radiation dose.

The maximum energy of gamma radiation emitted by the commonly used radioactive sources ^{60}Co (1.33 MeV) and ^{137}Cs (0.66 MeV) is too low to induce radioactivity in the constituent elements of foods. Electrons and photons of higher energy can be produced by electron accelerators and by X-ray machines. Measurable amounts of short-lived radioactive isotopes can be present in foods irradiated with 14 MeV electrons, and radioactive isotopes having half-lives of hours or days can be produced when electron energies of higher than 20 MeV are used. X-rays are more efficient in inducing radioactivity than electrons of the same energy. To be on the safe side, the FAO/IAEA/WHO Joint Expert Committee on Irradiated Foods (JECFI) has recommended 10 MeV as the maximum permissible energy for electron generators and 5 MeV for X-rays [1]. Under this condition induced radioactivity has never been observed in foods treated with a radiation dose in the practically useful range of up to 50 kGy. The U.S. Food and Drug Administration [2], the U.K. Advisory Committee on Irradiated and Novel Foods [3], and the Codex Alimentarius Commission [4] have all adopted the 10 MeV limit for electron radiation and the 5 MeV limit for X-radiation, as have the governments of most countries that have permitted irradiation of certain foodstuffs.

There is a considerable safety margin between these approved energy levels and those capable of inducing measurable radioactivity [6]. The question of induced radioactivity is really the least controversial issue in the debate about the safety of irradiated foods. In its new regulations on food irradiation (final rule of April 18, 1986), FDA firmly stated [2]:

> Because no evidence has been submitted to contradict FDA's finding that the irradiation of food does not cause the food to become radioactive, no further discussion of this issue is necessary.

II. CAN IRRADIATED FOODS BECOME TOXIC?

As described in Chapter 3, the radiation energy absorbed by irradiated foods causes various chemical reactions in the food. Many reaction products are formed, albeit in very small amounts. How much of each radiolytic product is present in an irradiated food depends primarily on the radiation dose applied, but also on modifying factors such as water content and presence or absence of oxygen. The nature of these chemical changes was largely unknown when the first experiments on food irradiation were carried out. In this respect food irradiation was in the same position as any new food processing method. When microwave heating was first introduced nobody knew for sure what chemical changes would take place in the food. Were they exactly the same as those produced by conventional heating, or did the microwave field produce particular breaks and recombinations in molecules different from those produced by traditional cooking? Although nobody knew the answer, microwaved food never caught the attention of regulatory agencies in the way irradiated food did. It was simply assumed that no significant amounts of harmful products were present in microwaved food. The same attitude prevailed when, more recently, extrusion cooking was introduced in the food industry. This process consists of extruding a dough under very high pressure (3000 lb per sq inch and more, or 21 MPa expressed in SI units) through a small orifice. Many snack foods are produced in this way. Heat results from the pressure and friction, and it was not inconceivable that this combination of pressure and heat would cause previously unknown chemical changes in the starch, protein, or other molecules present in the extruded material. But again nobody thought there was a need for special studies to examine the wholesomeness of extruded foods. Not so with irradiated foods.

Probably because of associations such as radiation sickness, radiation cancer, and atomic bombs, the word radiation carries a very negative image. Although heat can be just a deadly as radiation, the latter causes much more concern, probably because we can feel heat but cannot feel radiation. From the beginning of systematic studies in this area in the late 1940s, irradiated foods were considered to require careful toxicological investigation before this process could be used in food manufacturing.

III. ANIMAL FEEDING STUDIES

Some animal feeding studies with foods treated with X-rays had already been carried out in the 1920s. Although these studies are now of historical interest only, it is fascinating to read how firmly Groedel and Schneider concluded in 1926 that irradiation did not produce any toxic factors in animal diets [7]. They had observed clearly deleterious effects of feeding irradiated plant foods to mice and guinea pigs and interpreted these as being due to destruction of vitamins by the X-ray treatment. They summarized their study with the words, "The formation of a particular 'poison' in food under the influence of irradiation can be excluded."

The first animal feeding study with an electron-irradiated diet was reported by Da Costa and Levenson of the Massachusetts Institute of Technology in 1951 [8], and a multitude of such studies, both with electron-irradiated and gamma-irradiated diets, were carried out in the following years, mainly in the United States and in the United Kingdom [9, 10].

In the United States a legal requirement for safety testing of irradiated foods was introduced on September 6, 1958, when the Food Additives Amendment to the Food, Drug and Cosmetic Act was passed by Congress. The law required that all chemicals added to food, as well as foods treated by radiation, had to be tested for possible toxicity. Risk estimations of food additives are primarily based on animal feeding studies. Groups of the test animals receive, in their diet or drinking water, the substance to be tested at different dose levels.* Adverse effects, such as growth depression, liver damage, and reduced reproductive capacity, will usually begin to show up at a certain dose level. The next lower level, which has produced no adverse effect, is called the *no-effect level*. In setting up legal limits for the use of a food additive it is normal practice to regulate intake of that substance in man to no more than 1% of the no-effect level observed in the most sensitive animal species. This hundredfold safety factor takes into account the uncertainties of extrapolating results from animals to man and the presence in the human population of persons who may be particularly sensitive to the substance under consideration, i.e., pregnant women, babies, the sick, and the old.

While this approach has worked well for the safety evaluation of food additives, it has caused many difficulties when applied to the testing of foods. It is simply not possible to incorporate 100 times as much meat or potato or onion into a rat diet as is contained in the average human diet. Moreover, with foods irradiated in the practically useful dose range of up to 50 kGy, it has never been possible to produce significant toxic effects attributable to irradiation. When unfavorable effects were occasionally found, they were shown to be due to nutritional inadequacies, especially due to feeding too high a level of the test food [11]. It is not surprising to see rats

---

*In this context "dose" means the amount of test substance fed daily to the test animals, usually expressed in g/kg of body weight.

get sick when their diet contains 30% onions—regardless of whether the onions have been irradiated or not.

Even though there is a great deal of data from long-term feeding studies on irradiated food that demonstrates a no-effect situation, this evidence is of rather limited value to the toxicologist unless it can be related to the dosage of potential toxicants, either in direct or in indirect terms. If a clear-cut, radiation-related effect cannot be found even at the highest level of feeding, a no-effect level cannot be established. The experimenter does not know what the target organs might be or which laboratory values require particularly careful monitoring. It is not satisfactory for the evaluating toxicologist to conclude after a 2- or 3-year feeding study that nothing unusual has been found in any of the groups. Perhaps the level of irradiated food was just below the effect level? If man consumed such food in quantity, would the safety factor perhaps be only two or five, rather than the usually demanded 100? This problem is not unique to irradiated foods. It applies to the testing of foods in general, as contrasted to the testing of single compounds such as food additives, pesticides, or drugs.

Using higher radiation doses to provide a higher safety margin is only possible to a very limited extent. Chemical changes may take place at the higher dose levels that do not occur or are insignificant at the lower dose level. Essential nutrients may be partly or wholly destroyed or the food may become unpalatable at the higher dose level. Such effects would certainly confound experimental results.

Other test methods have supplemented the traditional animal feeding studies. Assays using bacterial cultures or mammalian cell cultures to test for possible cytotoxic and mutagenic properties of irradiated foods (designated in vitro tests) have achieved increasing sophistication since they were introduced some 20 years ago. Workers in the field of food irradiation were actually among the first to make extensive use of such tests for toxicological evaluation [12, 13]. While mutagenic effects of some irradiated sugar solutions in various in vitro systems have been well established [14], no such effects were found in vivo (i.e., in studies on animals) [15]. No convincing evidence of mutagenicity of any irradiated substrate has been observed under in vivo conditions in mammalian systems.

IV. THE CHEMICLEARANCE APPROACH

The more knowledge we have of the kinds and amounts of radiolysis products present in an irradiated food, the less we need animal feeding studies. However, even when the most sophisticated modern methods of analysis are used, one can never be sure that all radiation-induced compounds are detected. As a matter of fact, it is very unlikely that this goal will ever be reached with foods of complex composition, such as meat. A complete or nearly complete balance sheet of radiolytic products might be achieved with foods of simple, homogeneous composition, such as corn starch or sucrose. But even then, chemical data alone will not suffice to make a risk evaluation. Among dozens or hundreds of reaction products identified in an irradiated food, there will certainly be some whose toxicological properties have never been tested. It helps little to know that pork irradiated at a dose of 10 kGy contains 0.1 mg/kg of a substance X, if the toxicologists do not know how harmful or harmless substance X is.

What is needed under these circumstances is a combination of animal feeding studies and of chemical tests. Lehmann and Laug [16] of the Food and Drug Administration suggested in 1954, in one of the earliest descriptions of the principles of wholesomeness testing of irradiated foodstuffs:

> Chemical and physical tests should always precede animal tests, for the advance clues may enable the investigator to plan the animal experiment with greater intelligence and insight. In some instances even, animal experiments may be omitted when a radiation product has been adequately characterized.

Not much attention was paid to this advice in the early phase of food irradiation research. During the 1950s and 1960s, toxicological evaluation of irradiated foods was predominantly based on the results of animal feeding studies, although many chemical studies were carried out even then. However, it became more and more obvious that this exclusive reliance on feeding studies was unsatisfactory. In 1975, Diehl and Scherz [17] in Germany proposed the estimation of radiolytic products as an additional basis for evaluating the wholesomeness of irradiated foods, a concept also favored by Taub and co-workers in the United States [18]. The FAO/IAEA/WHO Joint Expert Committee on Irradiated Foods at its meeting in 1976 accepted this approach and proposed [19]:

> The analyses of radiolytic products that have been carried out so far have removed much of the previous uncertainty . . . .
> The general principal of radiation chemical reactions, as revealed by analytical studies, will reduce considerably the extent to which toxicological testing is needed and will simplify the testing procedures.

The term "chemiclearance" was introduced by Basson of South Africa [20], and the concept was applied to evaluating the wholesomeness of irradiated fruits [21]. This general approach was also the basis of reports prepared by the Select Committee on Health Aspects of

Irradiated Beef, a committee created by the Life Science Research Office of the Federation of American Societies for Experimental Biology. The committee evaluated the toxicological significance of over 100 compounds identified in irradiated beef and concluded that there were no grounds to suspect that these radiolysis products, present at a level of μg per kg, would constitute any hazard to the consumer [22]. The usefulness of chemical studies was also emphasized by an advisory committee of the FDA appointed in 1979 to recommend those toxicological requirements appropriate for assessing the safety of irradiated foods. Basing its recommendations on radiation chemistry, the committee concluded that foods irradiated at dose levels up to 1 kGy and foods comprising no more than 0.01% of the daily diet (e.g., spices) irradiated at 50 kGy or less can be considered safe for human consumption without any toxicological testing. The anticipated low levels of human exposure to any possible unique radiolytic products generated in irradiated foods under these conditions did not warrant expensive tests with thousands of laboratory animals as it could be predicted that such tests would not provide useful information [23].

V. A LONG HISTORY OF SAFETY STUDIES

The results of safety studies—animal feeding studies, in vitro tests, chemical investigations—carried out in various laboratories, in the United States and in other countries, have been periodically evaluated, and the tenor of these evaluations has been very reassuring. Kraybill [24] concluded in 1959 that irradiated foods

> have not been found to be toxic to experimental animals or man. In animal feeding experiments, with a broad spectrum of foods tested, longevity, reproduction and lactation performance in general are the same as for animals maintained on diets of unirradiated food. Investigations conducted to the present time with mice and rats have not revealed any evidence as to radiation induced carcinogens.

The Food and Drug Administration approved irradiation of wheat and wheat products for insect control with a dose of 0.2–0.5 kGy in 1963, and in the same year cleared sterilization of vacuum-packed canned bacon with a dose of 45–56 kGy. Permission for irradiation of white potatoes for inhibition of sprouting with a dose of 0.05–0.1 kGy followed in 1964. Reviewing 10 years of wholesomeness research sponsored by the U.S. Army at a cost of $ 6.1 million, the Surgeon General concluded in 1965 that [25]:

> Food irradiated up to absorbed doses of 5.6 megarads with a cobalt-60 source of gamma radiation or with electrons with energies up to 10 million electron volts has been found to be wholesome; i.e., safe, and nutritionally adequate.

Instead of 5.6 megarads we would now say 56 kGy, but other than that, the tremendous research efforts made since then have only reconfirmed this conclusion again and again.

Similar progress has been achieved in other countries. A Working Party on Irradiation of Food established by the British Government concluded in 1964 [10]:

> No evidence has yet accrued from these long-term investigations that the treated food is in any way deleterious to the species employed in the test . . . . No evidence of any carcinogenic activity of the irradiated food has been recorded.

Irradiation of potatoes was approved in the Soviet Union in 1958, followed by Canada in 1960.

All seemed to go well until a severe setback occurred in 1968, when the FDA not only refused to act on a petition of the Army for approval of radiation-sterilization of ham but also rescinded its approval of irradiation of bacon granted in 1963. The reasons given by FDA were that the animal feeding studies carried out some 10 years earlier did not meet the more stringent requirements for food additive testing now considered as state of the art. Significant deficiencies in the design and execution of the older studies were noted. To explain the change in the agency's attitude, Daniel Banes, then Associate Commissioner for Science of the FDA, declared in the course of Congressional Hearings [26]:

> The questions we now ask about the effects . . . on the reproduction process and on metabolic systems and the biochemistry of the body are far more advanced than they were 8 or 10 years ago . . . . The field of science is not static. It is constantly changing and expanding. We apply our best judgment based on the facts available to us at any given time.

In response, the Army embarked on a massive program of testing the safety of radiation-sterilized beef. It involved 1500 dogs, 27,000 rats, and 20,000 mice [27]. When most of the funds allowed for this program had been spent, the Army's supervisors of this project were shocked to find out that its major contractor, Industrial Biotest Labs (IBT), was following sloppy laboratory practices and even falsified data [28]. Consequently, all of the data provided by IBT had to be considered as useless—another major disaster in the history of food irradiation.

In the meantime, representatives of international agencies and of various national governments had been debating ways of joining forces and of settling the question of toxicological safety of irradiated foods in a concerted worldwide effort. It had become obvious that it was unnecessary and unwise to duplicate the costly animal feeding studies in different countries. In order to coordinate and rationalize these various efforts, the International Project in the Field of Food Irradiation was created in 1970 (see Chapter 1). The Project established its headquarters at the Federal Research Center for Nutrition, Karlsruhe, Federal Republic of Germany. Project directors were toxicologists of worldwide reputation: Roy Hickman (1971–1973), David Clegg (1974–1975), both delegated from the Canadian Health and Welfare Department, and Peter Elias (1975–1982), formerly with the British Ministry of Health.

Feeding studies contracted by the Project involved a range of commodities irradiated at dose levels of up to 10 kGy. The selection of the foodstuffs was based on a consideration of the interest likely to be accorded to the product as a staple food entering international trade, its usefulness to developing countries, and its technological and economic suitability for radiation preservation by doses of up to 10 kGy. The work was limited to this dose range because most of the applications that are of practical interest require a radiation dose of not more than 10 kGy. The only country in which there was much interest in using higher doses, such as those needed for sterilization, was the United States. The very comprehensive study on radiation-sterilized beef sponsored by the Army was proceeding at that time, and thus it seemed unnecessary for the Project to expand its own program into the range above 10 kGy.

FDA's criticism of some of the older feeding studies was fully considered in the design of the International Project's investigations. The research programs also included screening tests for mutagenicity and the evaluation of chemical investigations. During its 12 years of existence the Project produced 67 Technical Reports and 4 Activity Reports. Two extensive monographs were published in book form [29, 30]. One of the reports published in scientific journals described a comprehensive program of in vivo mutagenicity testing [31]. None of the studies have given any indication of the presence of radiation-induced carcinogens or other toxic factors. The Project was terminated in 1982 because the member countries found that it had fulfilled its purpose of clearly answering the question of wholesomeness of foods irradiated in the dose range of up to 10 kGy.

Before the International Project was initiated and while it proceeded, toxicological studies were also carried out within national research programs. They are much too numerous to be listed here but a few will be mentioned because their experimental approach differed from the usual procedure of feeding irradiated foods to test animals.

Instead of feeding the irradiated food as whole, some authors fed that portion of the food which was suspected as possibly harmful so as to increase the chances of detecting slight effects on the health of the animals. For example, the possible formation of toxic or carcinogenic compounds in the fat portion of irradiated bacon was tested by Dixon and co-workers [32] by feeding the lipid fraction of bacon to mice in a long-term study. Since suspected lipid oxidation products might be formed by the combined action of irradiation and frying, lipids of fried bacon were fed. The bacon was sliced and fried in the regular manner until thoroughly cooked, and the fat was separated from the meat. A diet was prepared containing 20% of this fat. No significant differences were observed between mice consuming the lipids of unirradiated bacon or of bacon irradiated at a dose of 56 kGy.

A further concentration of suspected radiolytic products was achieved by French authors testing the safety of irradiated starch under the guidance of the well-known toxicologist R. Truhaut. They prepared a mixture of nine compounds that had been identified in irradiated starch (formic acid, hydrogen peroxide, methyl alcohol, acetaldehyde, formaldehyde, glycolaldehyde, glyceraldehyde, malonaldehyde, and glyoxal) and fed 0.3 g of this mixture per kg of body weight daily to rats. This intake corresponds to a quantity of radiolytic products 800 times greater than that taken up by a baby consuming 30 g of irradiated starch per day. No toxic effect was found even at this highly exaggerated dose level [33].

For physiological reasons the irradiated food to be tested should not constitute more than a certain percentage of the test animal's total diet. For commodities such as potatoes or rice the level chosen is usually 30%, but this may be far too high when materials such as onions or spices are to be tested. Another approach to the wholesomeness testing of irradiated foods is the feeding of irradiated complete diets. If irradiation produced toxic substances these should be easier to detect when 100% of the diet is irradiated, rather than only 30%. Many generations of animals have been raised on such completely irradiated diets, and again no carcinogenic or other toxic effects were found. In one of the early investigations of this kind a diet compounded of nine food items, all irradiated to a dose of 56 kGy was fed to rats for 2 years [34]. A multigeneration study with rats fed a whole diet irradiated at two dose levels (2 and 25 kGy) was described by a group of scientists in India [35], long- and short-term feeding trials with poultry fed a diet irradiated at 35 kGy were carried out in Canada [36], while a comparative hog feeding study with radiation-sterilized (50 kGy) or heat-sterilized feed was done in the Netherlands [37]. The full text of Barna's long-term study on over 5000 rats receiving a diet irradiated at 25 or 45 kGy is available only in Hungarian, but a reference to this work has appeared in an English-language report [38].

As a matter of fact, hundreds of thousands of laboratory animals used in the testing of drugs and other chemicals have been raised and maintained on irradiated diets because these have been found to be superior to heat-treated or fumigated diets [39–41].

Critics of food irradiation sometimes argue that animal feeding studies are not carried out under realistic conditions. Whereas, presumably, the test animals are completely protected from harmful environmental influences, man is exposed to many environmental contaminants. It is argued that harmful effects of irradiated foods may only become manifest when animals or man are simultaneously exposed to many environmental contaminants. This criticism is not justified because the test animals consume diets which inevitably contain the same traces of environmental contaminants present in man's food, the water they drink is usually tap water, and the air they breathe is much the same as in an air-conditioned office or living room. Nevertheless, the question of possible combination effects has been studied in some of the early rat feeding studies, when irradiated foods were administered together with carcinogenic croton oil. No co-carcinogenic effect of irradiated foods (doses up to 100 kGy!) was found [42]. More recent work failed to show any evidence for a co-mutagenic effect of an irradiated diet (dose 45 kGy) fed to mice together with the mutagen cyclophosphamide [43].

Could irradiation possibly increase the allergenic properties of irradiated foods? This question was investigated by testing proteins from irradiated milk (doses up to 93 kGy) in sensitized guinea pigs. Irradiation actually reduced the allergenic properties of the milk proteins [44]. The immune response of rats maintained on a diet irradiated at a dose of 10 kGy was found to be unaffected [45].

The fact that irradiation causes the formation of free radicals and that these are quite stable in dry foods [46, 47] has often been mentioned as a reason for special caution with irradiated dry foods. A long-term feeding study especially designed to look for possible effects of a diet containing a high free radical concentration (irradiated with electrons at a dose of 45 kGy) was carried out at the Federal Research Center for Nutrition in West Germany. No indication of harmful effects was found with regard to tumor formation [48] or mutagenic effects [49]. Nine generations of rats were continuously fed this diet without any indication of toxic effects [50].

Both because of the extent of the investigation and because of the high dose (58 kGy) applied, a feeding study with radiation-sterilized chicken meat recently carried out in the United States will also be mentioned. The work was performed by Raltech Scientific Services, first under the sponsorship of the U.S. Army, later under the sponsorship of USDA. The design included five groups of animals receiving a) electron-irradiated, b) gamma-irradiated, c) heat-sterilized, and d) enzyme-inactivated (blanched) chicken meat stored frozen. A control group e) was fed a 100% rodent or dog chow diet. Long-term feeding studies were carried out with dogs, rats, and mice. Teratology studies (tests for possible damage to embryo or fetus during gestation) were conducted with mice, hamsters, rats, and rabbits. Additional tests utilized the fruit fly *Drosophila melanogaster*. The complete set of 12 reports describing these studies, comprising 35,769 pages, is available on microfiche from the National Technical Information Service, 5285 Port Royal Rd., Springfield, VA, 22161. More than 230,000 eviscerated broilers were used to produce the 134 metric tons of chicken meat needed for these studies. The whole enterprise took 7 years and cost some $8 million [51–53].

To give an idea of the complexity of such investigations, the experimental plan for the mouse feeding study with irradiated chicken meat is shown in Figure 1. The first generation of progeny obtained from the breeding stock (or P generation) is called F_0. Their offspring are called F_1, and so on. F_{1A} and F_{1B} are produced by two matings of the F_0 generation. The numbers of animals shown in Figure 1 refer only to one diet. Since five diets were fed, these num-

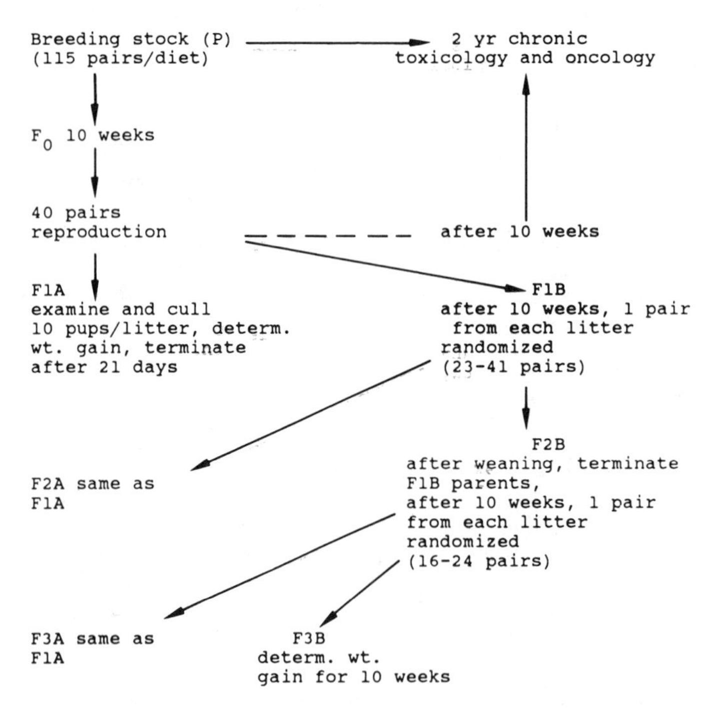

Figure 1 Protocol for testing irradiated chicken meat in a mouse chronic toxicity, oncogenicity, and multigeneration reproductive study. Albino CD-1 mice were used. (Adapted from Ref. 53.)

bers must be multiplied by five to estimate the total number of mice used in the chicken meat study.

Although the initial evaluation of results suggested the possibility that diets a) and b) had produced testicular tumors and some other adverse effects, the final evaluation by the FDA did not confirm any adverse effects from the ingestion of irradiated chicken meat [2]. We shall return to this issue shortly.

VI. DISSATISFACTION WITH EXPERT OPINIONS

"What's so great about science?" asked philosopher Paul Feyerabend in his deliberately provocative book *Science in a Free Society* in 1978. Earlier he had demanded (in *Against Method*, 1975) that laypersons should supervise science, declaring that scientific rationality was only one kind, and in many respects an inferior kind, of rationality. Theodore Roszak in *The Making of a Counter Culture* (1968) even maintained that, to the extent today that we are scientific, we are not truly human. Adherents of this counterculture will hardly be prepared to listen to scientific arguments about food irradiation—or about any other aspect of science.

Even among those who do not subscribe to the intellectual revolt against science there are probably quite a few sceptics who find summary conclusions like "none of the studies have indicated adverse effects" quite unsatisfactory. They may wish to see the original data so that they can come to their own conclusions. Unfortunately, this is not possible without making a book like this unreadable and unaffordable. During an animal feeding study hundreds of data are registered for each individual animal: body weight during the whole life time, results of hematologic studies (leucocyte count, erythrocyte count, hemoglobin concentration, hematocrit value, prothrombin time, mean corpuscular volume, and possibly other data), results of clinical blood chemistry studies (such as blood glucose, serum alkaline phosphatase, serum glutamic-pyruvic transaminase, blood urea nitrogen, total protein, albumin, globulins, sodium, potassium, calcium, chloride), urine analysis (glucose, albumin, pH, specific gravity, presence of occult blood, etc.). Usually these values must be determined several times during the lifetime of the animal, e.g., at 3, 6, 12, 18, and 24 months of age. When the animals die, organ weights are determined (adrenal glands, brain, caecum, gonads, heart, kidneys, liver, spleen, thyroid gland, uterus), and various tissues are examined histopathologically. The post mortem studies are designed to give an overall description of each animal's health status (are there symptoms of necrosis or fibrosis or bile duct proliferation in the liver? nephrosis, nephritis, pyelitis in the kidneys? myocarditis in the heart? parasites in the intestines? cystitis in the bladder? keratitis or retinal degeneration in the eyes? chronic inflammation in the forestomach? chronic murine pneumonia in the lungs? diffuse or focal degeneration in the testes? increased extramedullary hematopoiesis in the spleen?—to name just a few of the signs the pathologists will look for). The search for tumors (neoplasms) and their characterization (as fibroadenoma, fibrosarcoma, adenocarcinoma, hemangioma, lymphosarcoma, etc.) is an important part of post moretem examinations. In the reproduction phase of such studies data on fertility, number of offspring born alive or dead (with malformations?), number of off-glands, brain, caecum, gonada, heart, kidneys, liver, spleen, thyroid gland, uterus), and various tissues are examined histopathologically. The post mortem studies are designed to give an overall description of each animal's health status (are there symptoms of necrosis or fibrosis or bile duct proliferation in the liver? nephrosis, nephritis, pyelitis in the kidneys? myocarditis in the heart? parasites in the intestines? cystitis in the bladder? keratitis or retinal degeneration in the eyes? chronic inflammation in the forestomach? chronic murine pneumonia in the lungs? diffuse or focal degeneration in the testes? increased extramedullary hematopoiesis in the spleen?—to name just a few of the signs the pathologists will look for). The search for tumors (neoplasms) and their characterization (as fibroadenoma, fibrosarcoma, adenocarcinoma, hemangioma, lymphosarcoma, etc.) is an important part of post mortem examinations. In the reproduction phase of such studies data on fertility, number of offspring born alive or dead (with malformations?), number of offspring weaned, birth weight, weaning weight, and some other observations have to be registered. The protocol often requires other data, for instance, on food consumption, kidney function tests, liver function tests, or behavioral studies.

Data obtained on animals receiving the irradiated test diet must be compared with those obtained on animals receiving the same but unirradiated diet and those obtained on control animals receiving a standard laboratory ration. Often two groups of animals were fed the test diet irradiated at two different dose levels. About 100 animals are usually assigned to each group—50 males and 50 females. Tests including reproduction studies may result in a total population of several thousand animals. The final summary report may require hundreds of pages, the raw data (laboratory protocols) may comprise thousands of pages.

The evaluation of results requires medical knowledge, experience with toxicity tests, and considerable know-how in statistics. Much confusion and misinformation can result when persons not having these qualifications draw unjustified conclusions from the data. A typical (simplified) case might look like this:

Among 100 rats receiving the irradiated test diet, 10 tumors may have been found, among 100 rats receiving the unirradiated test diet, only 5 tumors. The inexperienced observer would be inclined to conclude that the irradiated food caused a 100% increase in cancer rate. But a look at the long-term experience with that particular strain of rat in that particular laboratory may show that 10 tumors in 100 animals has been the normal average overy many years. The 100 animals receiving the standard laboratory ration may even have developed 11 or 12 tumors. The experienced scientist would perhaps conclude in that case that the low rate of only 5 tumors in the group receiving the unirradiated test diet was accidental and that there was no indication of an increased tumor rate due to the feeding of the irradiated diet.

In reality a much more complex array of data has to be considered: what kind of tumors were observed? what was the tumor incidence among males, what among females? what was the tumor incidence in the following generation (or generations, if it was a multigeneration test)? was there a dose dependence (if test food irradiated at two or more dose levels was fed)? The results of all the other tests (hematology, etc.) must also be considered. Obviously, the evaluation of such animal feeding studies is a very difficult task that should be left to the specialists—although friends of consumer advocacy may not like this conclusion. The author wishes to add that he is not a toxicologist and that he also relies on the work of expert groups when results of toxicity tests have to be evaluated.

All that has been said refers to data gathered when *one* food is tested in *one* animal species. Since many irradiated foods have been investigated, usually on several animal species (mice, rats, dogs, monkeys), and many of these studies have been repeated in different laboratories, a staggering amount of information has to be evaluated. As an illustration, when the FAO/IAEA/WHO Joint Expert Committee (JECFI) met in 1980 it has to evaluate those toxicological studies which had been completed since the Committee's previous meeting in 1976: reports on 17 chronic studies (lifetime, multigeneration), 70 short-term studies (90-day feeding or in vivo mutagenicity tests), and 35 in vitro studies, plus a dozen tests with drosophila or other insect species—not to mention nutritional, microbiological, and chemical studies. Even a large group of experienced toxicologists cannot master this task in the usual 1-week meeting, unless a secretariat consisting of several scientists has examined and refereed all the reports beforehand.

VII. SOME RESULTS CAUSED CONCERN

While the vast majority of all animal feeding studies have not indicated harmful effects of irradiated foods, some results have necessitated careful reevaluations. Sometimes this was due to problems of statistics. When the animals on the irradiated test diet do better than those on the control diet, it is assumed that the apparent detrimental effect of the control diet is due to statistical fluctuations, but when a detrimental effect appears in some of the animals eating the test diet, it makes the test diet suspect. Subsequent analysis or repeat studies have as a rule shown that the apparent detrimental effects were due to biological variability or faulty experimental design or evaluation.

For example, Monsen thought he had seen a particular type of lesion in heart muscle of mice receiving an irradiated diet [54]. The study was repeated on almost 5000 mice of the same strain, and not a single case of such a lesion was found [55]. From 125,000 tissue and 800,000 serial heart sections prepared, in addition to detailed breeding, genetic, and necropsy reports, it was concluded that irradiation could not be the cause of the heart lesion originally observed.

Another reason for presumed toxic effects may be an insufficient vitamin supply in the irradiated diet—already recognized by Groedel and Schneider in 1926 [7]. Rats receiving a diet containing 35% radiation-sterilized beef developed internal bleeding (haemorrhagic diathesis) after long-term feeding, and this caused considerable concern at the time. It was later shown that the vitamin K level in this diet was very low even before irradiation, and that the high-dose irradiation destroyed enough of this to cause symptoms of vitamin K deficiency. Addition of vitamin K to the diet completely prevented the bleeding [56, 57].

A German group [58] reported in 1976 that pigs consuming fish meal irradiated at 7 or 14 kGy had a higher rate of mitosis (cell division) in the crypt region of the jejunum compared with control animals. Three years later a repeat experiment was reported which demonstrated that such increases in the rate of mitosis also occurred in control groups, purely as a matter of chance. Reexamination of the records of the earlier study confirmed a marked heterogeneity of the natural mitosis rate in pigs. The authors now concluded that neither the first nor the second study had shown significant differences between control animals and those consuming irradiated fish meal [59].

As mentioned earlier, the Raltech study with radiation-sterilized chicken meat, probably the most comprehensive safety evaluation ever carried out with an irradiated food—or with any food, also caused considerable concern initially. Another setback for food irradiation seemed to threaten when the preliminary evaluation of data indicated that mice fed radiation-sterilized chicken meat showed an increase in testicular tumors, increased kidney damage (glomerulonephropathy), and decreased survival. In addition, male dogs fed radiation-sterilized chicken seemed to have lower body weights than dogs fed the frozen control diet. Reexamination of histopathological slides by the National Toxicology Program's Board of Scientific Counselors did not confirm an increased incidence of testicular tumors in mice fed irradiated chicken, and FDA scientists concluded that neither glomerulonephropathy or survival in mice nor body weight in dogs were affected by irradiation of chicken meat in the diet [2, 60].

Another study frequently cited as evidence that irradiated foods may be harmful is a report of increased polyploidy in malnourished Indian children who consumed freshly irradiated wheat for 4 to 6 weeks. This effect was not seen when wheat stored for 12 weeks after irradiation was used [61]. The story of starving children having to eat irradiated wheat for experimental purposes is ideally suited for heartbreaking press reports, and it has, not surprisingly, become a focal point in the argumentation of activist groups against food irradiation. This story therefore deserves a closer look. The following evaluation is largely based on the report of a working party from the National Food Agency of Denmark [62].

Polyploidy means the occurrence, as detected under the microscope, of cells containing twice or more the normal number of chromosomes. Human cells normally have 46 chromosomes. If they are polyploid they have 92 chromosomes and occasionally 138 or even more. The incidence of polyploid cells varies to a certain extent between individuals, and even in one individual from day to day. It also varies from organ to organ within one individual. Normally there will be about 0.1–1% of polyploid cells in the bone marrow and among the lymphocytes in the circulating blood; levels of 3 or 4% are occasionally observed in healthy individuals. Other organs, such as the liver, may have a higher percentage of polyploid cells, and this is not necessarily an indication of abnormality. Although some authors have reported large increases in the percentage of polyploid cells in cancer, viral infections, and senility (levels of 20% or even more), it is not possible to assign a specific health significance to polyploidy. Certain substances not considered to be carcinogens are known to cause an increased incidence of polyploidy. In short, the biological significance of polyploidy is not known.

To get a clear perspective of the Indian report it is important to know about two technical problems in the execution of such studies. First, since polyploid cells are rare it is essential that enough cells are observed before any valid conclusions can be drawn. When the incidence is 1%, an observer counting only 100 cells may accidentally find no polyploid cell at all, or he may find 2 or even 3 among the 100. Obviously, many thousands of cells have to be counted in order to see the effect of a treatment, for instance a particular diet. Second, it is sometimes very difficult to recognize polyploid cells. If two normal (diploid) cells happen to be superimposed on the microscope slide, they look very much like one polyploid cell.

In one report it was noted that of two independent observers looking at the same slides, one recognized 34 and the other 9 polyploid cells. Under these circumstances the observers can easily be influenced subconsciously by a knowledge of the slide's origin. If they know the slide stems from a treatment group they will expect to find more polyploid cells and—quite unwittingly—they will look harder and will detect more polyploid cells. Such studies must therefore be carried out "blind," i.e., the observers must not be able to recognize the origin of the slides they look at.

In the study with malnourished children, 0% polyploid cells was found in the reference group of five children (i.e., the group getting non-irradiated wheat), 0.8% in the group receiving irradiated wheat for 4 weeks, 1.8% in the group ingesting irradiated wheat for 6 weeks. Only 100 cells were counted for each child. The percentages thus relate to the discovery of 0 polyploid cells in five children (500 cells), 4 polyploid cells in five children (500 cells), and nine polyploid cells in five children (500 cells). Unfortunately, the authors have reported only the group results, not those of the individual children. It could be, for example, that all the polyploid cells in one group were found in one child. If this was the case, any conclusion regarding a possible dietary effect is impossible. It is not possible on the basis of the data presented to validate the authors' claim that the differences between groups are statistically significant.

It is very strange that no polyploid cells at all were found among the 500 cells counted in the reference group. The puzzling thing about this study is not the normal (0.8%) or slightly elevated (1.8%) incidence of polyploidy in the groups of children receiving irradiated wheat, but rather the below normal (0%) incidence in the reference group. No indication is given that the evaluations were carried out blind. Regrettably, the authors did not fully describe the diet given to the children. They mention only that it contained 20 g wheat/kg body weight. Surely the children did not eat raw grains of wheat. Did they consume whole wheat flour? Cooked? Baked? Since this information is not given, nobody can exactly repeat this study. Any report that cannot be repeated because of insufficient description of methods is scientifically worthless.

There were no differences between the three groups of children with regard to chromosomal aberrations like breaks, gaps, and deletions. The incidence of these aberrations at the height of starvation and after treatment was similar. This is surprising in view of the report of Armendares et al. [63] who found that malnourished Mexican children exhibited a high incidence of chromosomal aberrations (12–21%) in lymphocytes relative to the background incidence of chromosomal aberrations in lymphocytes of well-fed children (2–4%).

Similar experimental and statistical objections must be directed against the studies on mice [64], rats [65] and monkeys [66] conducted in the same laboratory at the National Institute of Nutrition, Hyderabad. In contrast, a rat feeding study carried out at the Bhabha Atomic Research Centre, Bombay, with freshly irradiated wheat in which the incidence of polyploidy was determiend by counting 3,000 cells from each animal showed no effect of consuming the irradiated wheat [67]. Other studies that also cound not find adverse effect of feeling freshly irradiated wheat will be mentioned in Section 5.VIII. Since—more than 10 years later—the biological significance of an elevated incidence of polyploidy is still unknown and because of the serious deficiencies of the studies reporting increased polyploidy as a result of consuming irradiated wheat, the claim that the studies performed at the Hyderabad institute indicated harmful effects of food irradiation is unsubstantiated.

In order to investigate the reasons for the discrepancy between the results reported by the researchers in Hyderabad and in Bombay a committee of experts was appointed by the Indian Goverment in October 1975. The Committee's report, submitted in July 1976, was very critical of the work of the Hyderabad authors, who were described as "victims of preconceived notions," their data being "not only mutually contradictory but also at variance with well established facts of biology." P. C. Kesavan of Jawaharlal Nehru University, New Delhi, one of the members of the government committee, concluded that the experiments carried out at the Hyderabad institute were not well designed and the results imprecise, and that the data failed to demonstrate any mutagenic potential of irradiated wheat [68]. National and international committees of experts and regulatory agencies which had to make decisions about the safety of irradiated foods have all come to the same conclusion, although they have phrased their evaluations politely so as to avoid offending the Indian colleagues [2, 19, 62, 69]. Regrettably, this does not prevent some of the critics of food irradiation from quoting the Hyderabad studies as an indication or even as proof that irradiated foods are dangerous.

VIII. WARNINGS FROM OPPONENTS OF FOOD IRRADIATION

The conclusion that foods irradiated in accordance with established practice are toxicologically safe has recently been challenged by advocates of the anti-nuclear movement. Wheras most of these challenges have been pronounced in the form of oral statements or of leaflets handed out at political rallies, an article written by Richard Piccioni, senior staff scientist with Accord Research and Educational Associates, a New York-based anti-nuclear advocacy organization, has appeared in a respected journal [70].

Piccioni claims that the irradiation of foods "exposes the consumer to a whole new range of carcinogens" and implies that this increases the risk of getting cancer. He concludes that the actions of government agencies which have permitted irradiation of certain foodstuffs "have been unlawful and dangerous to public health." According to Piccioni, "Examination of the scientific literature reveals a large number of research reports which attest to the presence of carcinogenic or mutagenic activity in irradiated foods and food components." He supports this claim with 14 references.

Four of these references relate to in vitro studies on *Salmonella typhimurium* [14, 71] or on cell cultures [72, 73], three more refer to studies on the fruit fly *Drosophila melanogaster* [74, 75, 76]. While none of these publications describe carcinogenic effects of irradiated foods or food components, they do show mutagenic effects of irradiated solutions of glucose and of some other food constituents. Indeed, this caused concern about the safety of irradiated foods when these results were published years ago. Piccioni fails to mention, however, that other investigations have been carried out which have removed those concerns.

Firstly, the mutagenic effects observed with irradiated solutions of single compounds such as glucose were not observed when actual foods were irradiated, even when high doses of irradiation were used. Of the great number of studies which support this conclusion only a few can be mentioned as examples: in vitro studies with *Salmonella typhimurium* or other microbial assay systems did not show mutagenic effects of irradiation in fresh vegetables [77], onions [78], mango fruits [79], onion powder [80, 81], cod fillets [82], dates, cod fillets and chicken meat [83], spices [84], and chicken meat [53]. In vitro studies with mammalian cell cultures demonstrated no mutagenic activity of irradiated dates, cod fillets and chicken meat [85]. When the drosophila assay was used, no indication of increased mutagenicity was found in irradiated onion powder [86] dates [31], and beef and ham [87]. It is well known from many studies and can be well explained on the basis of established principles of radiation chemistry that radiation effects are more pronounced in dilute solutions than in complex mixtures of compounds, such as foodstuffs [88]. Den Drijver and coworkers [89] have shown that the compounds primarily responsible for in vitro mutagenicity of irradiated sugar solutions are not present in measurable amounts when the sugars are irradiated as constituents of fruits or fruit juices.

It should be pointed out that many compounds naturally occurring in unirradiated foods have been found to exert mutagenic effects in vitro or in drosophila. For instance, thymidine, a compound present in most foods, was found to be mutagenic when fed to drosophila; irradiation actually reduced the mutagenic potential of thymidine [90]. Whereas in vitro and drosophila tests are valuable for pre-screening large numbers of samples, extrapolation of results to man is doubtful and is actually misleading when opposing results have been obtained under in vivo conditions in mammals.

Secondly, numerous mutagenicity tests with irradiated food constituents and irradiated foodstuffs have been carried out under in vivo conditions in mammals, and in no case have effects been observed that could be clearly related to the radiation treatment. Instead of a complete listing, the following publications may again serve as examples: glucose irradiated in the dry state [91] or in solution [71, 92, 93] exhibited no mutagenic effects when tested in mice or rats. When irradiated dates, cod fillets or chicken meat were fed to mice, rats and Chinese hamsters no indication of increased mutagenicity was observed with regard to chromosome aberrations in bone marrow, micronucleus test, sister-chromatid exchange in bone marrow and in spermatogonia, and DNA metabolism in spleen cells [31]. Irradiation did not increase mutagenic effects of fresh onions in mice [94] or of onion powder in Chinese hamsters and in three strains of mice [80]. A variety of in vivo tests with rats showed no increased mutagenicity of irradiated spices as compared to unirradiated spices [84, 95, 96]. Consumption of irradiated fish by mice [97] or rats [98, 99] caused no observable mutagenic effects.

Whereas in the studies just mentioned the radiation doses were usually in the range of below 10 kGy and the irradiated foodstuff constituted only a minor part of an otherwise unirradiated diet, a number of in vivo investigations were carried out with animal diets that were 100% irradiated, using high radiation doses (25–50 kGy). Again, no evidence of radation-related mutagenicity was observed [49, 93, 100–104]. Since irradiated food would replace only a small fraction of an individual's total diet, feeding a wholly irradiated diet to animals enhances the safety factor inherent in the evaluation; using a higher radiation dose than the one contemplated in practice provides additional assurance.

Looking back at Piccioni's first 7 references it is obvious that he has selected studies which seemed to support his claims but do not stand up to closer examination. Moreover, Piccioni failed to mention the work of numerous authors who used much more relevant mammalian in vivo systems and who found no evidence of mutagenic—much less carcinogenic—effects of irradiated foodstuffs or whole diets. So far Piccioni's text is as interesting for what it doesn't say as for what it does.

What other "evidence" is presented by Piccioni? Four more of the publications on which he bases his claim of carcinogenic activity in irradiated foods are the discredited reports of increased incidence of polyploidy in groups of malnourished children, mice, rats and monkeys consuming freshly irradiated wheat [61, 64, 65, 66]. As mentioned earlier, government agencies and expert committees charged with evaluating the experimental evidence concerning the safety of irradiated foods have repeatedly concluded that these studies have not demonstrated harmful effects of the ingestion of irradiated foods in general or of freshly irradiated food in particular.

Piccioni has once more omitted to mention important evidence that runs counter to his preconceived notions. When freshly irradiated wheat was fed to rats and adequate numbers of cells were evaluated to permit proper statistical evaluation, no increase in the frequency of polyploid cells was observed [67]. Whereas the significance of polyploidy is unknown and an increased incidence of polyploidy cannot be considered as an indication of mutagenic or carcinogenic activity in the diet, the dominant lethal assay is recognized as a valid in vivo mutagenicity test. The feeding of freshly irradiated wheat did not induce any dominant lethal mutations in rats [105]. Another group of Indian authors concluded that the feeding of irradiated wheat (storage time not indicated) caused no genetic or cytogenetic effects in mice [106]. The sister-chromatid exchange (SCE) test is another indicator of mutagenicity. Feeding freshly irradiated wheat to monkeys for a period of 12 months did not affect the SCE frequency in blood lymphocytes [107].

The most extensive attempt to replicate the studies which had led to reports of increased polyploidy as a result of feeding freshly irradiated wheat were carried out at the request of the International Project in the Field of Food Irradiation [108]. In these studies, wheat freshly irradiated at a dose of 0.75 kGy caused no elevated polyploidy in bone marrow, no increase in dominant lethal mutations and no response in the micronucleus test in three assays with rats.

Piccioni lumps another publication together with the studies on freshly irradiated wheat [109]. This paper, however, is not concerned with irradiated wheat. The purpose of irradiating wheat is the destruction of insects and this can be achieved with a relatively low dose of radiation. The Indian authors who thought they had observed increased polyploidy had used a dose of 0.75 kGy in all of their studies. In contrast, Renner [109] used a radiation-sterilized stock ration treated with high doses of radiation. He did find a slight increase in polyploidy in the Chinese hamsters to which this ration was fed—but the incidence was still below 0.5%, even when the radiation dose was 100 kGy! Renner expressly concluded: "These results are not interpreted as a mutagenic effect of the irradiated diet." Other feeding studies in which he used assay methods more suitable for mutagenicity testing and found no evidence of radiation-induced mutagenicity were mentioned earlier [49, 93, 101].

Piccioni mentions two more papers which supposedly confirm his claim of carcinogenicity. One [22] refers to the presence of benzene, "a known carcinogen" in irradiated meat. What Piccioni fails to mention is that benzene was found in parts per billion quantities when beef was irradiated at high dose levels. He also fails to mention that benzene is found in many natural, unirradiated foods including fish, vegetables, nuts and dairy products. In non-irradiated eggs it is present in many times larger concentrations than in irradiated meat. In the heat sterilization process benzene is produced at about the same level as in the radiation sterilization process [60].

Piccioni then refers to work of Gower and Wills who observed oxidation of benzo(a)pyrene to its active carcinogenic form when mixed with unsaturated fatty acid preparations and irradiated [110].

Gower and Wills have irradiated mixtures of starch and polyunsaturated fats. In some experiments they added benzo(a)pyrene (BP) to this mixture at a concentration of 10 mg/kg (37.5 nmoles/g). Not surprisingly, they found evidence of peroxidation in the polyunsaturated fats during storage, and evidence of accelerated peroxidation in irradiated samples. When BP was present in the mixture, oxidation products of PB, presumably quinones, were formed—again accelerated by irradiation. Absence of oxygen or presence of antioxidants inhibited both lipid peroxidation and BP oxidation. The authors imply (without presenting evidence) that the unidentified oxidation products of BP are more toxic to man than BP itself.

It is well known that ionizing radiation, like heat, UV light or traces of certain metals, can induce the formation of lipid peroxides from polyunsaturated fats in the presence of an oxygen atmosphere. However, it is not justifiable to equate the experimental model of a starch/lipid mixture with a real food. The model lacks other essential elements present in food, especially proteins and natural antioxidants, which inhibit the oxidation process. Moreover, the effect of starch in the model mixture would be to create an unusually large surface, making the lipid even more susceptible to oxidation in an oxygen atmosphere.

The formation of oxidation products from BP in the presence of lipid peroxides is also not surprising, but again the conditions chosen by Gower and Wills are quite atypical of real foods. A concentration of 10 mg BP/kg is 10,000 times higher than the highest level of BP allowed in smoked foodstuffs according to German food law. The finding of elevated levels of oxidation products of BP in a starch/lipid mixture after irradiation, and in one experiment also in irradiated herring flesh with 10 mg BP/kg added, is not proof that such products are formed in any food containing a realistic level of BP. The results do not justify the conclusion that irradiation of foods produces carcinogenic activity.

The basis of decisions to classify a substance as carcinogenic or not is the long-term study in several species of animals—except in the few cases where epidemiological experience in man has provided this basis. As mentioned earlier many long-term animal feeding studies with a large variety of irradiated foods fed to many species of test animals are available, and they have in no case demonstrated a treatment-related increase in cancer rate.

Finally, Piccioni refers to a claim made by Russian authors in 1972 that alcoholic extracts of freshly irradiated potatoes induced dominant lethal mutations in mice [111], and he darkly hints that the FAO/IAEA/WHO Joint Expert Committee paid insufficient attention to this finding. In fact, several groups tried without success to replicate the work reported from Russia: a dominant lethal test in mice carried out by Canadian authors [112], a micronucleus test in rats reported from Germany [113] and a study employing several in vitro and in vivo tests completed in Japan [114] gave no indication of mutagenic effects of extracts of freshly irradiated potatoes. Since these were more extensive, more statistically valid studies, it must be concluded that the original finding [111] was artefactual. Once more, Piccioni has conveniently overlooked those publications which contradict his conclusions.

Nothing remains of his allegations. What does remain is the impression that his article is not an attempt to inform objectively about food irradiation. It is an attempt to add fuel to the fires of controversy regarding this method of processing. By carefully omitting to mention the massive evidence contradicting his point of view, Piccioni misleads the reader.

IX. RELEVANCE TO HUMAN HEALTH

All evidence based on animal studies can be criticized using the argument that there are physiological differences between man and test animals. Indeed, an experiment not demonstrating deleterious effects of a certain diet for an animal species cannot prove the absence of such effects in man if such a diet were consumed by man. However, the often heard demand for proof or even "absolute proof" of the absence of untoward health effects of a diet (or a food additive or drug) is unrealistic. Absolute safety does not exist. In this context the FDA [2] quotes a Senate report stating:

> Safety requires proof of a reasonable certainty that no harm will result from the proposed use of an additive. It does not—and cannot—require proof beyond any possible doubt that no harm will result under any conceivable circumstances.

Or in the words of a Canadian Government report on food irradiation [69]:

> The nature of toxicology as used in the regulatory process is such that it can never offer guarantees of absolute safety under all circumstances. What can be offered is assurance that under specified conditions there is no evidence to suggest hazard to humans.

However, many years of experience have shown that evaluations based on long-term studies on several animal species do provide a good basis for extrapolation to man. The fact that a few food additives were first permitted and later withdrawn is not a valid argument against animal experiments. Additives removed from the permitted list in recent years had been traditionally considered as safe, without much toxicological testing. When they were finally tested they turned out to be unsafe. They were removed from the list as a precautionary measure, not because they had caused disease in man, but because the animal studies showed that a certain risk for man could not be excluded. Examples are coumarin and safrole, natural substances once widely used as flavoring agents.

One may even say that modern toxicological methods are supercritical, that they exaggerate the risks. The fate of the sweeteners cyclamate and saccharin in the United States would suggest this. Cyclamate was banned in 1969. Most toxicologists now agree that this action was based on a false alarm and that the ban should be lifted. In many other countries this sweetener was never removed from the shelves. Animal studies indicating a carcinogenic risk of saccharin would have caused the banning of saccharin by the FDA in 1977 had not the Congress intervened with a special law allowing the continued use of this only remaining sweetener. Today there is general agreement that the amounts of saccharin consumed by man do not increase the risk of cancer.

Studies on man should be more convincing than studies on animals—but they also have their drawbacks. If they are carried out under well-controlled conditions in a metabolic ward they are very expensive and cannot comprise large numbers of persons. Statistical significance of results obtained with small numbers is doubtful. On the other hand, if human studies are not carried out under closely controlled conditions, they are also subject to criticism. How can one be sure that persons living at home really adhere to a prescribed experimental diet? Some study designs may provide a sufficient number of participants and satisfactory control, for example with groups living and eating in dormitories or in military service. But these studies are necessarily limited to a few weeks or months at best, while long-term animal studies can cover the complete lifetime of the test animals and of their progeny in several generations. For ethical reasons the particularly sensitive groups of a population, babies and pregnant women, cannot be included in any controlled human study.

In view of these limitations it is not surprising that not many human studies with irradiated foods have been carried out. Kraybill reported that clinical and laboratory examinations failed to reveal any significant alterations in groups of 10 volunteers consuming either nonirradiated foods or foods irradiated at a dose of 30 kGy [115]. In all seven such studies, each carried out with 9 to 10 volunteers, irradiated items supplied 35–100% of the energy intake [116].

More recently, eight experiments involving several hundred volunteers consuming irradiated foods for periods of 7 to 15 weeks were carried out in the People's Republic of China [117–121]. Clinical tests failed to discern significant differences between control groups and test groups. In view of the aforementioned Indian reports which had indicated an effect of freshly irradiated wheat on polyploidy in lymphocytes, this question received particular attention. The average incidence of polyploidy was 0.36 before beginning the consumption of irradiated diets and 0.26% at the end of the study period, a nonsignificant difference. The Chinese authors, who emphasized that the microscopic evaluations had been carried out blind, concluded that consumption of irradiated diets had no effect on polyploidy.

One experience with human consumption of irradiated foods should be mentioned here which, although it is not aimed at testing the safety of such foods, can contribute to the discussion. That is the use of radiation-sterilized foods in the diet of severely ill patients. A number of hospitals in the United States and the United Kingdom use irradiated foods for patients who have to be kept in a completely sterile environment because of their susceptibility to bacterial or viral infections. These may be leukemia patients who have received whole body irradiation or organ transplant recipients who receive immunosuppressive medication. For weeks or sometimes months they can be fed only sterilized foods. Supplementing heat-sterilized foods with radiation-sterilized items can provide more varied, more palatable, nutritious menus for these patients. At the Fred Hutchinson Cancer Research Center, Seattle, for instance, irradiated foods have been used for this purpose since 1974 with excellent results. Irradiated items constitute about 25% of the foods selected by this group of patients at Fred Hutchinson [122]—again, this cannot be considered as a proof of safety of irradiated foods, but it does contribute to the overall evaluation.

One can surely say that the safety for consumption of no other food processing method has been as thoroughly studied as that of irradiation. For good reasons the FAO/IAEA/WHO Joint Expert Committee on Irradiated Food, at its meeting in Geneva in 1980, came to the conclusion that further toxicological testing of foods irradiated in the dose range of up to 10 kGy, is not required. Since then the results of the Raltech study, the very extensive toxicological evaluation of chicken meat irradiated at a dose of 58 kGy, have become available [53]. Since foods other than meat and meat products are hardly of interest in the dose range above 10 kGy, it appears difficult to justify *any* further animal feeding studies on irradiated foods.

REFERENCES

1. WHO, *Wholesomeness of Irradiated Food*, World Health Organization, Technical Report Series 659, Geneva, 1981.
2. FDA, Irradiation in the production, processing, and handling of food, *Federal Register*, 51:13376 (1986).

3. HMSO, *Report on the Safety and Wholesomeness of Irradiated Foods*, Advisory Committee on Irradiated and Novel Foods, Her Majesty's Stationery Office, London, 1986.
4. FAO/WHO, Codex General Standard for Irradiated Foods and Recommended International Code of Practice for the Operation of Radiation Facilities for the Treatment of Foods, Codex Alimentarius Commission, vol. XV, Ed. 1, Rome, 1984.
5. Becker, R. L., Absence of induced radioactivity in irradiated foods, in P. S. Elias and A. J. Cohen, eds., *Recent Advances in Food Irradiation*, Elsevier Biomedical, Amsterdam, 1983, p. 285.
6. Miller, A., and P. E. Jensen, Measurements of induced radioactivity in electron- and photon-irradiated beef, *Int. J. Appl. Rad. Isotopes*, 38:507 (1987).
7. Groedel, F. M., and E. Schneider, Experimental studies on the question of the biological effect of X-rays (in German), *Strahlentherapie*, 23:411 (1926).
8. da Costa, E., and S. M. Levenson, Effect of diet exposed to capacitron irradiation on the growth and fertility of the albino rat, U.S. Army Medical Res. and Nutrit. Lab. Rept. No. 89, 1951.
9. Kraybill, H. F., and L. A. Whitehair, Toxicological safety of irradiated foods, *Ann. Rev. Pharmacol.*, 7:357 (1967).
10. HMSO, Report of the Working Party on Irradiation of Food, Her Majesty's Stationery Office, London, 1964.
11. Elias, P. S., The wholesomeness of irradiated food, *Ectoxicol. Environment. Safety*, 4:172 (1980).
12. Schubert, J., Mutagenicity and cytotoxicity of irradiated foods and food components, *Bull. World Health Organ.*, 41:873 (1969).
13. Kesavan, P. C., and M. S. Swaminathan, Cytotoxic and mutagenic effects of irradiated substrates and food material, *Rad. Botany*, 11:253 (1971).
14. Wilmer, J., H. Leveling, and J. Schubert. Mutagenicity of gamma-irradiated oxygenated and deoxygenated solutions of 2-deoxy-D-ribose and D-ribose in *Salmonella typhimurium*, *Mutat. Res.*, 90:385 (1981).
15. Varma, M. B., K. P. Rao, S. D. Nandan, and M. B. Rao, Lack of clastogenic effects of irradiated glucose in somatic and germ cells of mice, *Mutation Res.*, 169:55 (1986).
16. Lehmann, A. J., and E. P. Laug, Evaluating the safety of radiation-sterilized foods, *Nucleonics*, 12:52 (1954).
17. Diehl, J. F., and H. Scherz, Estimation of radiolytic products as a basis for evaluating the wholesomeness of irradiated foods, *Int. J. Appl. Rad. Isotopes*, 26:499 (1975).
18. Taub, I. A., P. Angelini, and C. Merritt, Jr., Irradiated food: validity of extrapolating wholesomeness data, *J. Food Sci.*, 41: 942 (1976).

19. WHO, *Wholesomeness of Irradiated Food,* World Health Organization, Technical Report Series 604, Geneva, 1977.
20. Basson, R. A., Chemiclearance, *Nuclear Active,* 17:3 (1977).
21. Basson, R. A., M. Beyers, and A. C. Thomas, A radiation chemical approach to the evaluation of the possible toxicity of irradiated fruits. 1. The effect of protection by carbohydrates, *Food Chem.,* 4:131 (1979).
22. FASEB, *Evaluation fo the Health Aspects of Certain Compounds Found in Irradiated Beef,* Life Sciences Research Office, Federation of American Societies for Experimental Biology, Bethesda, MD, 1977, with Supplements I and II, 1979.
23. Pauli, G. H., and C. A. Takeguchi, Irradiation of foods—an FDA perspective, *Food Revs. Internat.,* 2:79 (1986).
24. Kraybill, H. F., Safety in the operation of radiation sources and use of irradiated foods, *Int. J. Appl. Rad. Isotopes,* 6: 233 (1959).
25. Congress of the United States, *Radiation Processing of Foods,* Hearings before the Subcommittee on Research, Development and Radiation of the Joint Committee on Atomic Energy, June 9 and 10, 1965, p. 105.
26. Congress of the United States, *Status of the Food Irradiation Program,* Hearings before the Subcommittee on Research, Development and Radiation of the Joint Committee on Atomic Energy, July 18 and 30, 1968, pp. 100, 103.
27. Josephson, E., A. Brynjolfsson, and E. Wierbicki. The use of ionizing radiation for preservation of food and feed products, in *Radiation Research—Biomedical, Chemical and Physical Perspectives,* Nygaard et al., eds. Acad. Press, New York, 1975, p. 96.
28. Comptroller General of the United States, *The Department of the Army's Food Irradiation Program—Is it worth continuing?* U.S. General Accounting Office PSAD, 1978.
29. Elias, P. S., and A. J. Cohen, eds., *Radiation Chemistry of Major Food Components.* Elsevier, Amsterdam, 1977.
30. Elias, P. S., and A. J. Cohen, eds., *Recent Advances in Food Irradiation.* Elsevier Biomedical, Amsterdam, 1983.
31. Renner, H. W., U. Graf, F. E. Wurgler, H. Altmann, J. C. Asquith, and P. S. Elias, An investigation of the genetic toxicology of irradiated foodstuffs using short-term test systems. Part 3. In vivo tests in small rodents and in Drosophila melanogaster, *Food Chem. Tox.,* 20:867 (1982).
32. Dixon, M. S., D. L. Moyer, L. J. Zeldis, and R. W. McKee, Influence of irradiated bacon lipids on body growth, incidence of cancer and other pathologic changes in mice, *J. Food Sci.,* 26:611 (1961).
33. Truhaut, R., and L. Saint-Lèbe, Different approaches to the toxicologic evaluation of irradiated starch (in French), in *Food*

Preservation by Irradiation (Proceedings Sympos. Wageningen, Nov. 1977), Internat. Atomic Energy Agency, Vienna, 1978, vol. 2, p. 31.

34. Kraybill, H. F., M. S. Read, and T. E. Friedemann, Wholesomeness of gamma-irradiated foods fed to rats. *Fed. Proc.*, 15:933 (1956).
35. Aravindakshan, M., R. C. Chaubey, P. S. Chauhan, and K. Sundaram, Multigeneration feeding studies with an irradiated whole diet, in *Food Preservation by Irradiation* (Proc. Sympos., Wageningen, Nov. 1977) Internat. Atomic Energy Agency, Vienna, 1978, vol. 2, p. 41.
36. Cox, C., N. Nikolaiczuk, and E. S. Idziak, Poultry feed radicidation. 2. Long and short term poultry feeding trials with irradiated poultry feeds, *Poultry Sci.*, 53:619 (1974).
37. van Kooij, J. G., Chemical and biological evaluation of the nutritive value of heat-sterilized and radappertized feed mixtures, in *Decontamination of Animal Feeds by Irradiation*. Int. Atomic Energy Agency, Panel Proceedings Series, Vienna, 1979, p. 89.
38. Nádudvari, I., Experience of radiation treatment of laboratory and farm animal feeds in Hungary, in *Decontamination of Animal Feeds by Irradiation*. Int. Atomic Energy Agency, Panel Proceedings Series, Vienna, 1979, p. 33.
39. Adamiker, D., Irradiation of laboratory animal diets. A review, *Versuchstierkunde*, 18:191 (1976).
40. Ley, F. J., Radiation processing of laboratory animal diet, *Radiat. Phys. Chem.*, 14:677 (1979).
41. Tsuji, K., Low-dose cobalt 60 irradiation for reduction of microbial contamination in raw materials for animal health products, *Food Technol.* (Chicago), 37(2):48 (1983).
42. Teply, L. J., and B. E. Kline, Wholesomeness and possible carcinogenicity of irradiated foods, *Federation Proc.*, 15:927 (1956).
43. Renner, H. W., Search for a combined effect of a chemical mutagen and radiation-sterilized feed by mutagenicity test and reproduction study in mice (in German), *Fd. Cosmet. Toxicol.*, 13:427 (1975).
44. Kraybill, H. F., R. O. Linder and M. S. Read, T. M. Shaw, and G. J. Isaak, Effect of ionizing radiation on the allergenicity of milk protein, *J. Dairy Sci.*, 42:581 (1959).
45. Huismans, J. W., Effects of irradiated semi-synthetic diets on the immune response and some other parameters in male rats, International Project in the Field of Food Irradiation Technical Report IFIP-R35, Karlsruhe, 1975.
46. Diehl, J. F., and S. Hofmann, Electron spin resonance studies on radiation-preserved foods. I. Influence of radiation dose on spin concentration (in German), *Lebensm. Wiss. Technol.*, 1:19 (1968).

47. Diehl, J. F., Electron spin resonance studies on radiation-preserved foods. II. The influence of water content on spin concentration (in German), *Lebensm. Wiss. Technol.*, 5:51 (1972).
48. Renner, H. W., and D. Reichelt, On the wholesomeness of high concentrations of free radicals in irradiated foods (in German), *Zentralbl. Vet. Med.* B, 20:648 (1973).
49. Renner, H. W., T. Grünewald, and W. Ehrenberg-Kieckebusch, Mutagenicity test of irradiated foods by dominant-lethal assay (in German), *Humangenetik*, 18:155 (1973).
50. Renner, H. W., Long-term feeding study for testing the wholesomeness of an irradaited diet with a high content of free radicals, 2nd report (in German), Reports of the Federal Research Center for Food Preservation, Karlsruhe, 1974/1.
51. Wierbicki, E., Technological and irradiation conditions for radappertization of chicken products used in the U.S. Army Raltech toxicology study, in *Food Irradiation Processing* (Proceedings Sympos. Washington, DC, 4–8 March 1985), Internat. Atomic Energy Agency, Vienna, 1985, p. 79.
52. CAST, Ionizing Energy in Food Processing and Pest Control: 1. Wholesomeness of Food Treated with Ionizing Radiation. Council for Agricultural Science and Technology, Report no. 109, Ames, Iowa, 1986.
53. Thayer, D. W., J. P. Christopher, L. A. Campbell, D. C. Ronning, R. R. Dahlgren, G. M. Thomson, and E. Wierbicki, Toxicology studies of irradiation-sterilized chicken, *J. Fd. Protection*, 50:278 (1987).
54. Monsen, H., Heart lesions in mice induced by feeding irradiated foods, *Federat. Proc.*, 19:1029 (1960).
55. Thompson, S. W., R. D. Hunt, J. Ferrell, E. D. Jenkins, and H. Monsen, Histopathology of mice fed irradiated foods, *J. Nutrit.*, 87:274 (1965).
56. Johnson, B. C., M. S. Mameesh, V. C. Metta, and P. B. Rama Rao, Vitamin K nutrition and irradiation sterilization, *Fed. Proc.*, 19:1038 (1960).
57. Matschiner, J. T., and E. A. Doisy, Jr., Vitamin K content of ground beef, *J. Nutrit.*, 90:331 (1966).
58. Giese, W., U. Reusse, C. Messow, and J. Priebe, Pasteurization of fish meal by irradiation. 2. Studies on the harmlessness of feeding fattening pigs fish meal pasteurized by irradiation (in German), *Zbl. Vet. Med.*, B23:769 (1976).
59. Reusse, U., C. Messow, and R. Geister, Pasteurization of fish meal by irradiation. 3. The question of increased rates of mitosis after feeding radiation-pasteurized fish meal to pigs (in German), *Zbl. Vet. Med.*, B26:500 (1979).
60. Brynjolfsson, A., Wholesomeness of irradiated foods: a review , *J. Food Safety*, 7:107 (1985).

61. Bhaskaram, C., and G. Sadasivan, Effects of feeding irradiated wheat to malnourished children, *Am. J. Clin. Nutrit.*, 28:130 (1975).
62. LST, *Irradiation of Food,* Report by a Danish Working Group, National Food Agency, Publication no. 120, Søborg, 1986.
63. Armendares, S., F. Salamanca, and S. Frenk, Chromosome abnormalities in severe protein calorie malnutrition, *Nature,* 232: 271 (1971).
64. Vijayalaxmi, Genetic effects of feeding irradiated wheat to mice, *Can. J. Genet. Cytol.*, 18:231 (1976).
65. Vijayalaxmi and G. Sadasivan, Chromosomal aberration in rats fed irradiated wheat, *Int. J. Radiat. Biol.*, 27:135 (1975).
66. Vijayalaxmi, Cytogenetic studies in monkeys fed irradiated wheat, *Toxicology,* 9:181 (1978).
67. George, K. P., R. C. Chaubey, K. Sundaram, and A. R. Gopal-Ayengar, Frequency of polyploid cells in the bone marrow of rats fed irradiated wheat, *Fd. Cosmet, Toxicol.*, 14: 289 (1976).
68. Kesavan, P. C., Indirect effects of radiation in relation to food preservation: facts and fallacies, *J. Nucl. Agric. Biol.*, 7:93 (1978).
69. Health and Welfare Canada, Comprehensive Federal Government Response to Report of the Standing Committee on Consumer and Corporate Affairs on the Question of Food Irradiation and the Labeling of Irradiated Foods, Ottawa, Sept. 1987.
70. Piccioni, R., Food irradiation: contaminating our food, *Ecologist,* 18:48 (1988).
71. Aiyar, A. S., and V. Subba Rao, Studies on mutagenicity of irradiated sugar solutions in *Salmonella typhimurium, Mutation Research,* 48:17 (1977).
72. Wilmer, J. W. G. M., and A. T. Natarajan, Induction of sister-chromatid exchanges and chromosome aberrations by gamma-irradiated nucleic acid constituents in CHO cells, *Mutation Research,* 88:99 (1981).
73. Shaw, M. W., and E. Hayes, Effects of irradiated sucrose on the chromosomes of human lymphocytes in vitro, *Nature,* 211: 1254 (1966).
74. Swaminathan, M. S., S. Nirula, A. T. Natarajan, and R. P. Sharma, Mutations: incidence in *Drosophila melanogaster* reared on irradiated medium, *Science,* 141:637 (1963).
75. Parkash, O., Mutagenic effect of irradiated DNA in *Drosophila melanogaster, Nature*, 214:611 (1967).
76. Rinehart, R. R., and F. J. Ratty, Mutation in *Drosophila melanogaster* cultured on irradiated food, *Genetics,* 52:1119 (1965).
77. van Kooij, J. G., H. B. Leveling, and J. Schubert, Application of the Ames mutagenicity test to food processed by physi-

cal preservation methods, in *Food Preservation by Irradiation*, Proceedings of a Symposium held in Wageningen, 21–25 Nov. 1977, IAEA, Vienna 1978, vol. 2, p. 63.
78. Hattori, Y., M. Mori, F. Kaneko, and A. Matsuyama, Mutagenicity tests of irradiated onions by *Escherichia coli* mutants in vitro, *Mutat. Res.*, 60:115 (1979).
79. Niemand, J. G., L. den Drijver, C. J. Pretorius, and H. H. van der Linde, A study of the mutagenicity of irradiated sugar solutions: Implications for the radiation preservation of subtropical fruits, *J. Agric. Food Chem.*, 31:1016 (1983).
80. Münzner, R., and H. W. Renner, Mutagenicity testing of irradiated onion powder, *J. Food Sci.*, 46:1269 (1981).
81. Farkas, J., and E. Andrassy, A study of possible mutagenicity of irradiated onion powder by Salmonella mammalian-microsome tests, *Acta Aliment.*, 10:209 (1981).
82. Joner, P. E., B. Underdal, and G. Lunde, Mutagenicity testing of irradiated cod fillets, *Lebensm. Wiss. Technol.*, 11:224 (1978).
83. Phillips, B. J., E. Kranz, P. S. Elias, and R. Münzner, An investigation of the genetic toxicology of irradiated foodstuffs using short-term test systems. I. Digestion in vitro and the testing of digests in the *Salmonella typhimurium* reverse mutation test, *Food Cosmet. Toxicol.*, 18:371 (1980).
84. Farkas, J., E. Andrássy, and K. Incze, Evaluation of possible mutagenicity of irradiated spices, *Acta Aliment.*, 10:129 (1981).
85. Phillips, B. J., E. Kranz, and P. S. ELias, An investigation of the genetic toxicology of irradiated foodstuffs using short-term test systems. II. Sister-chromatid exchange and mutation assays in cultured Chinese hamster ovary cells, *Food Cosmet. Toxicol.*, 18:471 (1980).
86. Mittler, S., and M. I. Eiss, Failure of irradiated onion powder to induce sex-linked recessive lethal mutations in *Drosophila melanogaster*, *Mutat. Res.*, 104:113 (1982).
87. Mittler, S., Failure of irradiated beef and ham to induce genetic aberrations in drosophila, *Int. J. Radiat. Biol.*, 35:583 (1979).
88. Diehl, J. F., Radiolytic effects in foods, in *Preservation of Food by Ionizing Radiation*, E. S. Josephson and M. S. Peterson, eds. CRC Press, Boca Raton, FL, 1982, vol. 1, p. 279.
89. den Drijver, L., H. J. van der Linde, and C. W. Holzapfel, High-performance liquid chromatographic determination of D-arabino-hexos-2-ulose (D-glucosone) in irradiated sugar solutions. Application of the method to irradiated mango, *J. Agric. Food Chem.*, 34:758 (1986).
90. Parkash, O., Thymidine teratogenesis and mutagenesis in *Drosophila melanogaster*, *Experientia*, 23:859 (1967).

91. Varma, M. B., K. P. Rao, S. D. Naudan, and M. S. Rao, Lack of clastogenic effects of irradiated glucose in somatic and germ cells of mice, *Mut. Res.*, 169:55 (1986).
92. Schubert, Pan and Wald, unpublished results quoted in J. Schubert [12].
93. Münzner, R., and H. W. Renner, Mutagenicity testing of irradiated laboratory animal diet by the host mediated assay with *S. typhimurium* G 46 (in German), *Int. J. Radiat. Biol.*, 27:371 (1975).
94. Goud, S. N., O. S. Reddi, and P. P. Reddy, Mutagenicity testing of irradiated onions in mice, *Int. J. Radiat. Biol.*, 41: 347 (1982).
95. Barna, J., Genotoxicity test of irradiated spice mixture by dominant lethal test, *Acta Aliment.*, 15:47 (1986).
96. Farkas, J., and E. Andrassy, Prophage lambda induction (inductest) of blood of rats fed irradiated spices, *Acta Alim.*, 10: 137 (1981).
97. Chaubey, R. C., M. Aravindakshan, P. S. Chauhan, and K. Sundaram, Mutagenicity evaluation of irradiated Indian mackerel in Swiss mice, in *Food Preservation by Irradiation*, Proceedings of a Symposium held in Wageningen, 21–25 November 1977, IAEA, Vienna, 1978, vol. 2, p. 73.
98. Zaitsev, A. N., and I. N. Osipova, Study on the mutagenic properties of irradiated fish in a toxicity trial (in Russian), *Voprosy Pitanyia*, 53(4):53 (1981).
99. Zaitsev, A. N., and N. B. Maganova, Effects of feeding gamma-irradiated fish on the embryogenesis and chromosomes of white rats (in Russian), *Voprosy Pitanyia*, 61(6):61 (1981).
100. Eriksen, W. H., and C. Emborg, The effect of pre-implantation death of feeding rats on radiation-sterilized food, *Int. J. Radiat. Biol.*, 22:131 (1972).
101. Münzner, R., and H. W. Renner, Mutagenicity testing of irradiated laboratory animal diet by the host mediated assay with *S. typhimurium* TA 1530 (in German), *Zbl. Vet. Med. B*, 23:117 (1976).
102. Leonard, A., M. Wilcox, and W. Schietecatte, Mutagenicity tests with irradiated food in the mouse, *Strahlentherapie*, 153: 349 (1977).
103. Chauhan, P. S., M. Aravindakshan, A. S. Ayar, and K. Sundaram, Dominant lethal mutations in male mice fed gamma-irradiated diet, *Fd. Cosmet. Toxicol.*, 13:433 (1975).
104. Chauhan, P. S., M. Aravindakshan, A. S. Ayar, and K. Sundaram, Studies on dominant lethal mutations in third generation rats reared on an irradiated diet, *Int. J. Radiat. Biol.*, 28:215 (1975).

105. Chauhan, P. S., M. Aravindakshan, N. S. Kumar, V. S. Rao, A. S. Ayar, and K. Sundaram, Evaluation of freshly irradiated wheat for dominant lethal mutations in Wistar rats, *Toxicology*, 7:85 (1977).
106. Reddi, O. S., P. P. Reddy, D. N. Ebenezer, and N. V. Naidu, Lack of genetic and cytogenetic effects in mice fed on irradiated wheat, *Int. J. Radiat. Biol.*, 31:589 (1977).
107. Murthy, P. B. K., SCE in monkeys fed irradiated wheat, *Fd. Cosmet. Toxicol.*, 19:523 (1981).
108. Tesh, J. M., E. S. Davidson, S. Walker, A. K. Palmer, D. D. Cozens and J. C. Richardson, Studies in Rats Fed a Diet Incorporating Irradiated Wheat, International Project in the Field of Food Irradiation, Technical Report Series IFIP-R 45, Karlsruhe 1977.
109. Renner, H. W., Chromosome studies on bone marrow cells of Chinese hamsters fed a radiosterilized diet. *Toxicology*, 8:213 (1977).
110. Gower, J. D., and E. D. Wills, The oxidation of benzo(a)pyrene mediated by lipid peroxidation in irradiated synthetic diets, *Int. J. Radiat. Biol.*, 49:471 (1986).
111. Kopylov, V. A., I. N. Osipova, and A. M. Kuzin, Mutagenic effects of extracts from gamma-irradiated potato tubers on the sex cells of male mice (in Russian), *Radiobiologiya*, 12:58 (1958).
112. Levinsky, H. V., and M. A. Wilson, Mutagenic evaluation of an alcoholic extract from gamma-irradiated potatoes, *Fd. Cosmet. Toxicol.*, 13:243 (1975).
113. Hossain, M. M., J. W. Huismans, and J. F. Diehl, Mutagenicity studies on irradiated potatoes and chlorogenic acid; micronucleus test in rats, *Toxicology*, 6:243 (1976).
114. Ishidate, M., Jr., K. Yoshikawa, T. Sofuni, S. Iwahara, and T. Sibuya, Mutagenicity studies on alcohol extracts from gamma-irradiated potatoes. Tests for biological activities in bacterial and mammalian cell systems, *Radioisotopes* (Tokyo), 30:662 (1981).
115. Kraybill, H. F., Nutritional and biochemical aspects of food preserved by ionizing radiation, *J. Home Econ.*, 50:695 (1958).
116. Plough, I. C., E. L. Bierman, L. M. Levy, and N. F. Witt, Human feeding studies with irradiated foods, *Federation Proc.*, 19:1052 (1960).
117. Anonymous, Study on safety evaluation of diet consisting of 35 kinds of irradiated foods for human consumption (in Chinese), *J. Radiat. Res. Radiat. Processing*, 5:38 (1987).
118. Dai, Y., Safety evaluation of irradiated foods in China, in Proceedings of an FAO/IAEA Seminar on Practical Application of Food Irradiation in Asia and the Pacific, Shanghai, 7–11 April 1986, IAEA-TECDOC-452, Vienna, 1988, p. 162.

119. Jin, W., and J. Yuan, Safety evaluation of 35 kinds of irradiated food for human consumption, IAEA-TECDOC-452, Vienna, 1988, p. 163.
120. Han, C., S. Guo, M. Wang, Z. Liu, W. Cui, and Y. Dai, Feeding trial of diet mainly composed of irradiated foods in human volunteers, IAEA-TECDOC-452, Vienna, 1988, p. 202.
121. Brynjolfsson, A., Results of feeding studies of irradiated diets in human volunteers: summary of the Chinese studies, *Food Irrad. Newsletter*, 11:33 (1987).
122. Aker, S. N., On the cutting edge of dietetic science, *Nutrition Today*, 19(4):24 (1984).

6

Microbiological Safety of Irradiated Foods

When foods are treated with a nonsterilizing dose of radiation, some microorganisms will survive. As described in Chapter 4, different species or strains of microorganisms differ in their radiation resistance. The surviving microflora will therefore differ in composition from the flora present before the radiation treatment. This raises a number of important questions related to public health:

Could the selective effect of radiation on the microbial flora result in a higher health risk, e.g., when harmless organisms are less radiation resistant than certain pathogenic species, so that a more pathogenic population of microorganisms survives?

Could mutations in the surviving population convert nonpathogenic organisms into pathogenic organisms or less virulent strains into more virulent strains? Could irradiation stimulate toxin formation in toxin-producing bacteria or moulds?

Could repeated sublethal treatments with radiation lead to increased radiation resistance?

Could the diagnostic characteristics of microorganisms be changed as a result of irradiation so that the species or strain cannot be correctly identified?

Could the outer appearance of a spoiled food, possibly one containing a radiation-resistant toxin, be improved by irradiation so that the consumer would not be warned by the usual danger signs, such as typical spoilage odor? Is irradiation only a "cosmetic" treatment?

While some of these concerns could be quickly dispelled, years of research in many laboratories have been devoted to finding answers to some of the other questions.

I. SELECTIVE EFFECTS

Fortunately, most food poisoning pathogens are quite sensitive to radiation. A dose of 5 kGy will reduce a population of *Salmonella* serotypes, *Staphylococcus aureus*, *Shigella*, *Escherichia coli*, *Brucella*, and *Vibrio* species by at least 6 log cycles, which means that not more than one in one million organisms originally present will survive the radiation treatment. These pathogens will not have a selection advantage in irradiated foods [1]. The importance of *Campylobacter jejuni* [2], *Aeromonas hydrophila* [3], and *Yersinia enterocolitica* [4] as human pathogens has been firmly established in recent years, and they have also been found to have a low tolerance for radiation.

In contrast, some spore-forming pathogens are more resistant to radiation than most other microorganisms. *Clostridium botulinum*, *Clostridium perfringens*, and *Bacillus cereus* belong to this group. Conceivably, irradiation with a nonsterilizing dose will largely eliminate the normally competing flora, thus creating more favorable conditions for the germination, growth and toxin production of the surviving spores. This possible hazard has been a prime consideration of microbiologists studying the process of food irradiation. As discussed in Chapter 4, Sec. II, perishable foods irradiated with a nonsterilizing dose require certain precautions, such as the use of curing salts and/or low storage temperature.

The problems of the surviving microflora are clearly different according to the nature of the food and its associated microbial species. Foods too dry to permit growth of microorganisms, or frozen foods, present no problems—they simply carry fewer microorganisms, of all kinds, after irradiation than before. Similarly, acid foods do not support hazardous species of bacteria and will not give rise to problems through selective destruction by irradiation. The same applies to refrigerated foods, with very few exceptions, of which *C. botulinum* type E in seafoods is the most important.

Studies in which high numbers of this organism were inoculated into fish meat showed production of toxin during storage at temperatures down to about 3°C [5]. Since temperature in household refrigerators may occasionally be as high as 8°C, possible microbiological hazards of fish irradiation required close scrutiny.

An outbreak of botulism in the Federal Republic of Germany in 1970 causing a number of deaths was traced to the ingestion of hot-smoked trout [6]. Before this outbreak little was known of *C. botulinum* in freshwater fish, but subsequent examination of farmed trout and the fish farms showed the presence of *C. botulinum* E [6, 7]. A group in the author's laboratories at the Federal Research Center for Nutrition therefore studied the effect of radiation and storage temperature on toxin formation in trout inoculated with *C. botulinum* E. Some of the results are presented in Figure 1. Storage life was judged by the odor of the raw fish. A group of test

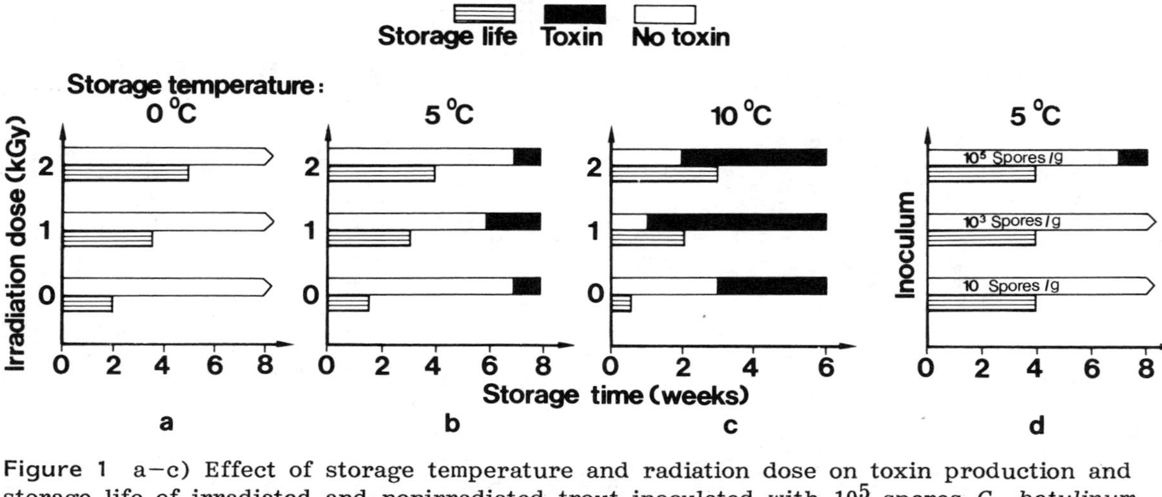

Figure 1 a–c) Effect of storage temperature and radiation dose on toxin production and storage life of irradiated and nonirradiated trout inoculated with 10^5 spores *C. botulinum* E per g fish. d) Effect of size of the incoulum of type E spores in trout irradiated with a dose of 2 kGy and stored at 5°C. (From Ref. 8.)

persons smelled the fish daily and when they agreed that a sample smelled spoiled this was considered as the end of storage life. Toxin production was determined at weekly intervals by the mouse test. As the figure indicates, no toxin was produced at 0°C. Irradiation with 2 kGy increased storage life to 5 weeks as compared to 2 weeks in unirradiated control samples. At 5°C, toxin was found after 6 weeks of storage in the 1 kGy sample and after 7 weeks in the 2 kGy and the control sample. Spoilage was observed much earlier. At 10°C, irradiated samples showed toxin production before odor indicated the end of storage life. At that storage temperature the hazard of botulism is indeed increased by radurization.

The high inoculum of 10^5 spores per gram used in these experiments exaggerated the possible hazard. When 10^3 spores per gram were inoculated, no toxin was found in irradiated fish even after 8 weeks of storage at 5°C, while the fish had reached the end of its storage life after 4 weeks (Fig. 1d). It was concluded that the radurization of trout at doses of about 1 or 2 kGy and a storage temperature below 5°C was safe with regard to a possible risk of botulism [8]. Similar conclusions were reached by other authors who studied toxin formation in various species of marine fish [9]. A storage temperature of 3°C or less for irradiated marine fish was usually recommended.

It should be pointed out that irradiation is by no means unique with regard to selective effects on the microflora. Nonsterilizing heat treatments select heat-resistant organisms and this is the reason why in countries where both cooked and raw ham are marketed, more cases of staphylococcal food poisoning are caused by ham sold cooked than sold raw. Similarly, vacuum-packaging favors growth of anaerobic species, salting favors survival of salt-resistant (halophilic) species, and refrigeration fabors cold-tolerant (psychrophilic) species. Thus, *Yersinia enterocolitica*, a species of psychrophilic bacteria, can grow at 4°C in meat, poultry, and milk and has been recognized as the cause of several outbreaks of food poisoning associated with consumption of refrigerated food.

It is also noteworthy that *C. botulinum* E and its toxin are quite sensitive to heat. Domestic cooking procedures will generally kill the vegatative cells of *C. botulinum* and will inactivate any toxin already present in the food [10]. As irradiated fish would normally be cooked before consumption, a hazard from irradiation would be even less likely. In contrast, smoked fish is consumed raw and the botulism hazard of smoked fish is much more real. (The radiation resistance of other types of *C. botulinum* was discussed in Chapter 4).

In summary, the selective effect of irradiation on microorganisms requires certain precautions in the handling of perishable foods treated with a nonsterilizing dose. The possible hazard is not greater than that of perishable foods treated by other nonsterilizing processes.

II. MUTATIONS

It is well known that exposure to ionizing radiation can increase the rate of mutation in bacteria and other organisms. Conceivably, mutations could spontaneously convert nonpathogenic organisms into pathogenic ones or less virulent strains into more virulent ones, and irradiation could increase the chance of this occurring.

When Ingram and Farkas surveyed the scientific literature, they found no indication that such an event had ever been observed. On the contrary, there were some reports of loss of virulence and infectivity as a result of radiation treatment [1], and this has been amply confirmed by more recent studies [11]. Irradiated populations usually exhibit reduced fitness. They become more exacting in their environmental requirements and more sensitive to heat, drying, and antimicrobial agents. Because of their reduced fitness, survivors of sublethal radiation treatment do not compete well with the nonirradiated organisms present in practical situations. Mutants produced by irradiation therefore usually disappear unnoticed.

The use of irradiation to produce microbial strains for certain industrial purposes is, of course, well known. In this way strains of *Penicillium notatum* have been developed which produce enormously increased yields of penicillin in comparison with the parent strains, and there are many other examples of this kind: Industrial production of many vitamins and amino acids utilizes mutated strains of microorganisms and in many cases the desired mutant has been obtained from irradiated samples. However, in all these cases, screening of the mutants has been designed to select only the strains desired. Recovery conditions after irradiation were carefully chosen to enable the desired mutant to survive.

It should be noted that treatments other than with ionizing radiation also can increase the rate of mutation. Among these are ultraviolet radiation (as present in sunlight), food preservatives, and heat. Genetic changes can be brought about even by various conditions of drying [1]. There is nothing unique about the mutagenic properties of ionizing radiation.

Mutations and their effects on the properties and characteristics of organisms are unpredictable. No amount of research can exclude the possibility that a new highly pathogenic strain will arise from a harmless parent strain, irradiated or not. Again, demands for absolute safety cannot be fulfilled. However, there is agreement among experts who work in this field that such an event is most unlikely. The British microbiologist Gould [12] has summarized his evaluation thus:

> Mutants are most often "crippled," i.e. far less able to survive than the organisms they were derived from, and numerous practical studies have confirmed that the induction of mutations in micro-organisms has no significance in food irradiation.

III. TOXIN PRODUCTION

Some food-spoiling molds are known to produce toxins. A well-known representative of these mycotoxins is aflatoxin, which is produced by *Aspergillus flavus* and *Aspergillus parasiticus*. Concern was caused by reports indicating increased production of aflatoxin when spores of *A. flavus* [13, 14] or *A. parasiticus* [15] or cultures derived from such spores were irradiated. Similar observations were made concerning production of ochratoxin by *Aspergillus ochraceus* [16]. However, results were not clear-cut, increased or decreased production of toxin being observed in different experiments. When Ingram and Farkas [1], eminent British and Hungarian microbiologists, reviewed some of the early studies in this field, they concluded that the results provided convincing illustrations of the power of irradiation to diminish aflatoxin production, without any convincing illustrations of an increase. With one exception [17], later studies have found less aflatoxin in irradiated than in unirradiated samples [18–20].

There are probably several reasons for these discrepancies. Firstly, as pointed out by Frank and co-workers [21], *Aspergillus* belongs to the heterocaryotic fungi, which means that the genetic information contained in different nuclei of a mycelium from one strain is not identical. Through a process called anastomosis the mycelium of one strain can exchange some nuclei with the mycelium of another strain and thus aquire some of the latter's properties. This explains the high variability and selective adaptability of this class of fungi. It also explains certain changes of properties when a strain is subcultured. The occasional occurrence, during repeated reculturing, of colonies having higher or lower aflatoxin-producing ability—as compared to the parent strain—is not surprising under these conditions and is quite unrelated to irradiation.

Secondly, the size of the inoculum has a decisive influence on toxin formation, and this was not taken into account in many of these studies. As demonstrated with *A. parasiticus* [22] and with *A. flavus* [23], a medium inoculated with a high number of spores will develop less aflatoxin than one inoculated with a lower number of spores. A reduction in the number of spores by about 4 log cycles, either by simple dilution or by irradiation, caused a twofold increase in toxin production by *A. parasiticus* [22] and an up to 12-fold increase in toxin production by *A. flavus* [23]. Toxin production is apparently suppressed when the number of inoculated spores per unit volume of substrate exceeds a certain optimum.

Experiments in which irradiation stimulated toxin formation were usually done with a massive inoculum of spores. In a real situation such conditions can hardly occur. When an Indian group of authors [24] studied spontaneous formation of aflatoxin in wheat (i.e., without infecting the wheat with an inoculum of spores) storing the

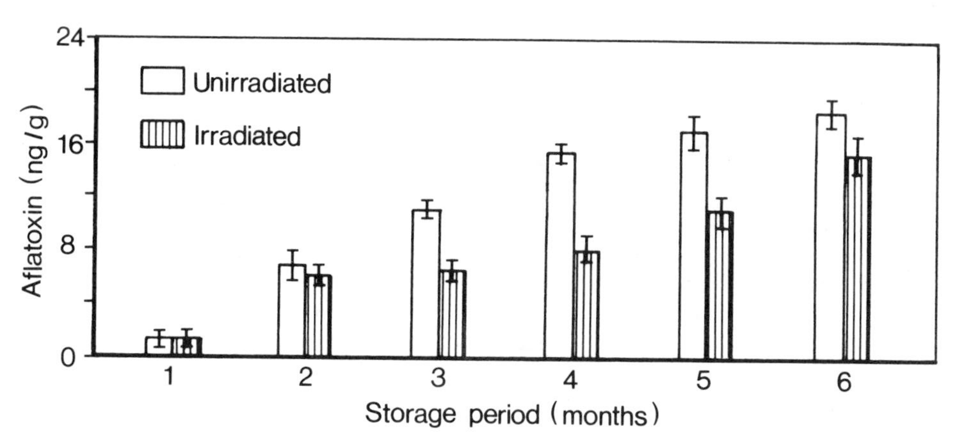

Figure 2 Aflatoxin levels in control and gamma-irradiated (0.2 kGy) wheat during storage at 90% relative humidity, 28°C. (From Ref. 24, with permission.)

wheat at 90% relative humidity at 28°C—conditions not unusual in tropical countries—they found either the same or lower levels of aflatoxin in irradiated samples when compared with unirradiated controls, as demonstrated in Figure 2.

A different experimental design was used by Priyadarshini and Tulpule [25], who irradiated various foods, which were subsequently heat-sterilized, inoculated with spores of *A. parasiticus*, and incubated. Aflatoxin formation was higher than in unirradiated control samples. The same authors suggested in a later publication that irradiation caused a higher level of free fatty acids in these foods and that this may have stimulated the formation of aflatoxin [26]. Whatever the explanation, no practical situation is conceivable in which a food would be first irradiated, then heat-sterilized and then stored under conditions permitting infection and molding.

Effects of many repeated cycles of sublethal irradiation and growth on various strains of *A. flavus* were studied in the author's institute [21]. The results showed that such treatment more frequently led to a complete loss or to a decrease of aflatoxin production than to a gain or increase.

In summary, there are no indications that any food irradiated and stored under conditions prevailing in practice would be at risk of increased formation of mycotoxins.

Effects of irradiation on toxin production in certain bacteria have also been studied and no alarming results were found. Several attempts at the former Food and Drug Directorate, Canada, now the Canadian Health Protection Branch, to produce toxic strains from nontoxic strains of *C. botulinum* by means of irradiation have not met with success (Erdman, as quoted by Idziak [27]).

IV. INCREASED RADIATION RESISTANCE?

Microorganisms have an amazing capacity to adapt to changing environmental conditions. Certain species or strains have developed a remarkable resistance to heat, cold, high salt concentration or antibiotics. Not surprisingly, some can also develop high resistance to radiation. This has been demonstrated in many experiments, of which only a few typical examples will be mentioned.

The effect of treating several species important in food hygiene to repeated doses of 2 kGy of gamma radiation alternating with periods of growth (22 hours at 35°C) was examined by a Canadian group [28]. They found radiation resistance to be increased by factors up to twofold when *E. coli, Streptococcus faecalis, S. aureus*, and *C. botulinum* were exposed to up to 12 cycles of irradiation and culturing of surviving cells. However, they found no increased radiation resistance in *Salmonella gallinarum*. Observations with *C. botulinum* indicated that the increased radiation resistance acquired by recycling vegetative cells may be passed on to the spores and transmitted through the spores to subsequent vegetative cells. In other experiments [29] radiation resistance was not transferred from vegetative cells to spores of *Bacillus pumilus*.

Twelve cycles of 2.5 kGy interspersed with 24–48 hours of growth in a poultry meat base at 37°C doubled the radiation resistance of *Salmonella anatum, enteritidis*, and *give*, and increased that of *S. typhimurium* by 50% [30]. Severalfold increases of radiation resistance were observed after 20-cycle irradiation of several other species of Salmonella [31].

The basis of the increased resistance to radiation may either be the selection effect of the recycling treatment alone or mutations and selection of the resistant mutants. Proponents of the selection hypothesis assume that individual cells in a population differ greatly in their radiation resistance. A cycle of repeated sublethal irradiations and growth phases would destroy the less resistant cells and would permit only the most resistant cells to survive and multiply. Other authors believe that a substantial increase in radiation resistance can only be explained on the basis of mutations. The most important mechanism by which cells can defend themselves against damage caused by ionizing radiation is the ability to repair, and therefore tolerate, radiation-induced lesions in the genetic material, DNA.

Different species of microorganisms respond differently to recycling treatments, and even different strains of the same species can respond differently [32]. The failure of attempts to increase radiation resistance in *S. gallinarum* [28] has already been mentioned.

Ley and co-workers actually noted reduced resistance after recycling of *S. typhimurium* [33]. Münzner and Diehl irradiated cultures of seven strains of *Penicillium viridicatum* and six of *A. flavus* with electrons at different dose levels of up to 2.4 kGy. All the strains of *P. viridicatum* and three of *A. flavus* became less resistant at all dose levels. In no case was the lethal dose increased [34].

It should be noted that all the recycling experiments were carried out under unrealistic conditions. Under industrial conditions of food irradiation no situation is conceivable where a population of microorganisms would be repeatedly resuscitated after sublethal irradiation by providing an optimal growth medium and optimal temperature. The appearance of genotypes with higher radiation resistance in some of these recycling experiments is, therefore, more academic interest than of practical significance. Some food irradiation plants and many other radiation-emitting sources have been in operation for many years now, and no evidence for increased occurrence of resistant strains in the environment of these plants has been found.

Radiation resistance can be compared with heat resistance. Under very special environmental conditions, particularly in waters of hot springs and geysers, bacteria have been found which tolerate temperatures of several hundred °C. But these cannot survive in other, colder environments. In food canning it is, therefore, not necessary to take the existence of such highly heat-tolerant species into account. In the same way, the fact that very radiation-tolerant strains have been obtained under special laboratory conditions has no bearing on practical food irradiation.

Maxcy [35] has summarized his extensive experience in this field by stating:

> The production of a resistant culture requires a highly specific set of laboratory conditions. The pure culture is not in competition with bacteria in the wild state where there would be a reversion to the common wild form through survival of the fittest. This set of highly theoretical conditions would not be expected in an industrial environment.

V. CHANGES IN DIAGNOSTIC CHARACTERS

If treatment with ionizing radiation changed the typical properties of pathogenic microorganisms to such an extent that they could no longer be correctly identified, this could have serious consequences for public health. This question has therefore received considerable attention. Changes in morphological, biochemical, or serological properties may appear after irradiation, especially after cyclic reirradiation. As mentioned earlier, survivors of sublethal irradiation are usually more demanding in their growth requirements. For example, Robern and Thatcher recycled a strain of *E. coli* which originally did not require amino acids, purines, or pyrimidines for growth. After 6 cycles of irradiation leucine, methionine and proline were required, after 12 cycles 8 amino acids plus uracil or cytosine [36]. Similarly, recycling of *B. pumilus* was accompanied by an increased requirement for specific amino acids [29]. The irradiation treatment apparently caused sufficient genetic damage so that the enzymes critical in the synthesis of certain amino acids were altered to such an extent that they were either nonfunctional or lost.

However, in all these instances the majority of other characters remained unchanged so that identification was still possible. Ingram and Farkas, who have closely examined this issue, consider radiation-induced changes in diagnostic characters "less serious than might be feared" [1]. They point out that the changes are usually temporary and can be reversed by a period of resuscitation. They also call attention to the fact that radiation has been used in medicine for many decades without causing any of the microbiological problems discussed with regard to food irradiation—diagnostic difficulties, increased virulence, increased radiation resistance, and so on. The fact that widespread medical use of radiation (unlike antibiotics) has caused no trouble from altered or more virulent forms of microorganisms is the best possible indication that such difficulties are not probable with food. Compared with doses used in food irradiation, the radiation doses normally used for diagnosis or therapy may be small, but the total amount of radiation used since the days of Röntgen and the Curies must be very large.

VI. IS IRRADIATION A "COSMETIC" TREATMENT?

Some critics of the food irradiation process have claimed that the object of irradiation is to improve the appearance of spoiled food, that it is only a "cosmetic" treatment designed to make a spoiled food marketable. They fear that irradiation will make the deception of consumers easier.

Such fears are unfounded. Spoilage of foods is accompanied by more or less pronounced changes of odor, taste, and visual appearance. While irradiation can reduce or eliminate the spoilage flora and the pathogenic organisms that may be present in a spoiled food, it cannot improve its sensory properties.

Because treatment with a nonsterilizing dose of radiation changes the microflora present in an irradiated food, this food may exhibit changed spoilage characteristics when it finally spoils. Other nonsterilizing treatments have similar effects. For instance, pasteurized milk will often not turn sour when it spoils, unlike unpasteurized milk. The pasteurizing of milk is not a cosmetic treatment. It protects consumers from infection with pathogens of which milk is a typical carrier, such as those causing tuberculosis or brucellosis. Freshly slaughtered poultry can be infected with Salmonella species. Irradiation of such poultry would protect the consumer from salmonellosis—it would not make a spoiled chicken marketable.

In the words of the authoritative review of Ingram and Farkas [1]:

> Irradiation pasteurisation produces definite and large improvements in the microbiological status of foods and feeds, by eliminating most pathogens and extending storage life. The associated risks seem relatively small and are still not demonstrated, except in the case of fish and sea-foods where storage below 3°C is recommended. The balance of advantage to public health seems heavily in favour of such a process, as in heat pasteurisation or cooking of food.

Additional experimental evidence supporting this conclusion has been summarized by Farkas [37].

REFERENCES

1. Ingram, M., and J. Farkas, Microbiology of foods pasteurised by ionising radiation, *Acta Aliment.*, 6:123 (1977).
2. Lambert, J. D., and R. B. Maxcy, Effect of gamma radiation on *Campylobacter jejuni*, *J. Food Sci.*, 49:665 (1984).
3. Palumbo, S. A., R. K. Jenkins, R. L. Buchanan, and D. W. Thayer, Determination of irradiation D-values for *Aeromonas hydrophila*, *J. Food Protect.*, 49:189 (1986).
4. El-Zawahry, Y. A., and D. B. Rowley, Radiation resistance and injuring of *Yersinia enterocolitica*, *Appl. Environ. Microbiol.*, 37:50 (1979).
5. Eklund, M. W., and F. T. Poysky, The significance of non-proteolytic *Clostridium botulinum* types B, E and F in the development of radiation-pasteurized fishery products, in *Preservation of Fish by Irradiation*. Int. Atomic Energy Agency, Panel Proceedings Series, Vienna, 1970, p. 125.
6. Bach, R., S. Wenzel, G. Müller-Prasuhn, and H. Gläsker, Trout from ponds as a carrier of *Clostridium botulinum* and cause of botulism. Part 3 (in German), *Arch. Lebensmittelhyg.*, 22:107 (1971).
7. Wenzel, S., R. Bach, and G. Müller-Prasuhn, Trout from ponds as a carrier of *Clostridium botulinum* and cause of botulism. Part 4 (in German), *Arch. Lebensmittelhyg.*, 22:131 (1971).

8. Hussain, A. M., D. Ehlermann, and J. F. Diehl, Comparison of toxin production by *Clostridium botulinum* type E in irradiated and unirradiated vacuum-packed trout (*Salmo gairdneri*), *Arch. Lebensmittelhyg.*, 28:1 (1977).
9. Eklund, M. W., Significance of *Clostridium botulinum* in fishery products preserved short of sterilization, *Food Technol.*, (Chicago), 36(12):107 (1982).
10. Hobbs, G., *Clostridium botulinum* and its importance in fishery products, *Adv. Food Res.*, 22:135 (1976).
11. Farkas, J., *Irradiation of Dry Food Ingredients*. CRC Press, Boca Raton, FL, 1988, p. 57.
12. Gould, G. W., Food irradiation—Microbiological aspects, *Institute of Food Science and Technology*, 19:175 (1986).
13. Jemmali, M., and A. Guilbot, Influence of irradiation of spores of *A. flavus* on the production of aflatoxin (in French), *Compt. Rend. Acad. Sci. Paris*, 269 Ser. D:2271 (1969).
14. Applegate, K. L., and J. R. Chipley, Increased aflatoxin G_1 production by *Aspergillus flavus* via gamma irradiation, *Mycologia* 65:1266 (1973).
15. Bullerman, L. B., and T. E. Hartung, Effect of low dose gamma irradiation on growth and aflatoxin production by *Aspergillus parasiticus*, *J. Milk Food Technol.*, 37:430 (1974).
16. Paster, N., R. Barkai-Golan, and R. Padova, Effect of gamma radiation on ochratoxin production by the fungus *Aspergillus ochraceus*, *J. Sci. Food Agric.*, 36:445 (1985).
17. Schindler, A. F., A. N. Abadie, and R. E. Simpson, Enhanced aflatoxin production by *Aspergillus flavus* and *Aspergillus parasiticus* after gamma irradiation of the spore inoculum, *J. Food Protection*, 43:7 (1980).
18. Ogbadu, G., Influence of gamma irradiation on aflatoxin B_1 production by *Aspergillus flavus* growing on some Nigerian foodstuffs, *Microbios*, 27:19 (1980).
19. Chang, H.-G., and P. Markakis, Effect of gamma irradiation on aflatoxin production in barley, *J. Sci. Fd. Agric.*, 33:559 (1982).
20. Rodriguez, M., and Y. A. Rodriguez, Reduction of aflatoxin formation in peanuts by gamma-irradiation (in Spanish), *Ciencia y Technica en la Agricultura, Veterinaria*, 5:103 (1983).
21. Frank, H. K., R. Münzner, and J. F. Diehl, Response of toxigenic and non-toxigenic strains of *Aspergillus flavus* to irradiation, *Sabouraudia*, 9:21 (1971).
22. Sharma, A., A. G. Behere, S. R. Padwal-Desai, and G. B. Nadkarni, Influence of inoculum size of *Aspergillus parasiticus* spores on aflatoxin production, *Appl. Environm. Microbiol.*, 40:989 (1980).
23. Odamtten, G. T., V. Appiah, and D. I. Langerak, Influence of inoculum size of *Aspergillus flavus* link on the production of

B_1 in maize medium before and after exposure to combination treatment of heat and gamma radiation, *Int. J. Food Microbiol.*, 4:119 (1987).
24. Behere, A. G., A. Sharma, S. R. Padwaldesai, and G. B. Nadkarni, Production of aflatoxins during storage of gamma-irradiated wheat, *J. Food Sci.*, 43:1102 (1978).
25. Priyadarshini, E., and P. G. Tulpule, Aflatoxin production on irradiated foods, *Fd. Cosmet. Toxicol.*, 14:293 (1976).
26. Priyadarshini, E., and P. G. Tulpule, Effect of graded doses of gamma-irradiation on aflatoxin production by *Aspergillus parasiticus* in wheat, *Fd. Cosmet. Toxicol.*, 17:505 (1979).
27. Idziak, E. D., Effect of radiation on micro-organisms, *Internat. J. Radiat. Steriliz.*, 1:45 (1973).
28. Erdman, I. E., F. S. Thatcher, and K. F. McQueen, Studies on the irradiation of microorganisms in relation to food preservation. Parts 1 and 2, *Canad. J. Microbiol.*, 7:199, 207 (1961).
29. Parisi, A., and A. D. Antoine, Increased radiation resistance of vegetative *Bacillus pumilus*, *Appl. Microbiol.*, 28:41 (1974).
30. Idziak, E. S., and K. Incze, Radiation treatment of foods. I. Radiation of fresh eviscerated poultry, *Appl. Microbiol.*, 16:1061 (1968).
31. Licciardello, J. J., J. T. R. Nickerson, S. A. Goldblith, C. A. Shannon, and W. W. Bishop, Development of radiation resistance in Salmonella cultures, *Appl. Microbiol.*, 18:24 (1969).
32. Gaden, E. L., Jr., and E. J. Henley, Induced resistance to gamma irradiation in *Escherichia coli*, *J. Bact.*, 65:727 (1953).
33. Ley, F. J., T. S. Kennedy, K. Kawashima, D. Roberts, and B. C. Hobbs, The use of gamma radiation for the elimination of *Salmonella* from frozen meat, *J. Hyg.*, 6:293 (1970).
34. Münzner, R., and J. F. Diehl, Studies on the radiation sensitivity of cultures of *Aspergillus flavus* and *Penicillium viridicatum* repeatedly irradiated with electrons (in German), *Lebensm. Wiss. Technol.*, 2:44 (1969).
35. Maxcy, R. B., Significance of residual organisms in foods after substerilizing doses of gamma radiation: A review, *J. Food Safety*, 5:203 (1983).
36. Robern, H., and F. S. Thatcher, Nutritional requirements of mutants of *Escherichia coli* resistant to gamma-irradiation, *Canad. J. Microbiol.*, 14:711 (1968).
37. Farkas, J., Decontamination, including parasite control of dried, chilled and frozen foods by irradiation, *Acta Aliment.*, 16:351 (1987).

7

Nutritional Adequacy of Irradiated Foods

As has been pointed out (see Chapters 3 and 5), treatment of foods with ionizing radiation causes changes in the chemical composition of foods, and this can also affect the nutrient value. The nature and extent of these changes depend on the composition of the food, on the radiation dose, and on modifying factors such as temperature and presence or absence of oxygen. Numerous investigations have been carried out to study the nutritional adequacy of irradiated foods under various conditions, and several reviews of this work are available [1–5].

I. MAIN COMPONENTS OF FOODS (MACRONUTRIENTS)

Carbohydrates, proteins, and fats are the main components of foods. (In most foods water is also a main component, but it does not require discussion in this context.) These three components provide energy and serve as building blocks for the growth and maintenance of the body. Animal feeding studies have shown that irradiation of foods at any dose level that is of practical interest, i.e., up to about 50 kGy, will not impair these functions of the main components.

Chemical analysis does show effects of radiation on carbohydrates, fats, and proteins which increase with increasing dose of radiation (see Chapter 3), but even in the dose range of 10–50 kGy, these are so small and so unspecific that attempts to develop analytical methods capable of detecting whether a food has been irradiated or not have met with limited success.

While fats and carbohydrates in food serve primarily as sources of energy, proteins additionally provide essential amino acids, which the human organism needs to build up its own proteins. Particular attention has therefore been paid to possible effects of radiation on the amino acid composition of proteins. No significant destruction of essential amino acids has been observed in irradiated beef [76], fish [7], or many other foodstuffs, even when sterilizing doses of radiation were used. The good growth observed in various animal species fed different kinds of irradiated feeds supports the conclusion that digestibility and biological value of proteins are essentially unchanged by treatment with radiation doses even in the range of 50 kGy [8]. In several chick feeding assays irradiation was found to improve the nutritional value of feed proteins in wheat bran [9], rye [10], field beans [11], and lentils [12]. Even extremely high radiation doses did not adversely affect protein quality: 210 kGy in the study with beans [11], 180 kGy in the study with lentils [12].

II. VITAMINS

The literature on vitamin losses caused by irradiation is somewhat contradictory. Some reports overestimate the losses because they are based on results obtained with pure solutions of vitamins. The radiation-induced loss of a vitamin is much greater in pure solution than in a food irradiated with the same dose. For example, vitamin B_1 in aqueous solution (0.25 mg/100 ml) showed 50% loss after irradiation with a dose of 0.5 kGy, while irradiation of dried whole egg (B_1 content 0.39 mg/100 g) with the same dose caused less than 5% destruction [13]. This is due to the mutually protective action of various food constituents on each other.

Some reports possibly underestimate the losses because analysis was carried out only immediately after irradiation, although radiation-induced losses of some vitamins in some foods may continue during storage, especially if oxygen is not excluded. An example of this phenomenon is shown in Figure 1. Rolled oats irradiated with a dose of 1 kGy and stored in the presence of air had lost 20% vitamin E (alpha-tocopherol) immediately after irradiation and around 45%, as compared to the unirradiated sample, after 6–8 months of storage. When the same material was irradiated and stored in airtight packaging in an atmosphere of nitrogen, losses of vitamin E were negligible, even after 8 months of storage [14].

Another factor that is often not sufficiently considered is the possibility of additional vitamin losses during cooking. As shown in Figure 2, the loss of vitamin B_1 (thiamin) in rolled oats caused by heating for 10 min in an oven set at 100°C was always greater in irradiated samples. After a storage period of 6 months, for example, the unheated, irradiated sample had 50% less vitamin B_1 than the

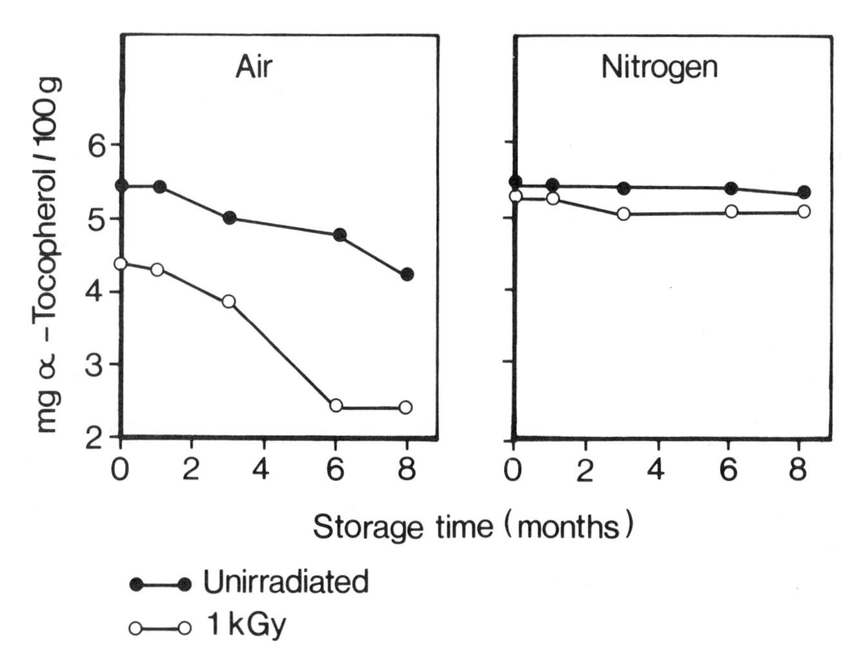

Figure 1 Changes in α-tocopherol (vitamin E) level of rolled oats, unirradiated or irradiated at 1 kGy with 10 MeV electrons, during storage at ambient temperature in air or nitrogen atmosphere (samples sealed in airtight Hostaphan/polyethylene bags). (From Ref. 14.)

unirradiated control, while after the heat treatment the irradiated sample had almost 70% less than the unirradiated control [13a]. As in the case of vitamin E these losses were much reduced by exclusion of air (not shown in Fig. 2). Some of the data presented in Figure 2 again demonstrate an important effect of storage on radiation-induced thiamin losses. Egg powder, for instance, showed practically no loss of thiamin when analysis was carried out immediately after irradiation, but a substantial loss after 8 months.

It should not be assumed that vitamin losses are always potentiated by the combined effects of irradiation and heating. Products other than rolled oats in Figure 2 do not clearly show such potentiation. Another example of a study of the combined effects of irradiation and cooking is shown in Table 1. Nicotinic acid and riboflavin are obviously much more stable than thiamin.

The combination of irradiation and cooking may even result in improved vitamin retention. The required cooking time for dry legumes can be considerably shortened by irradiation. As shown in

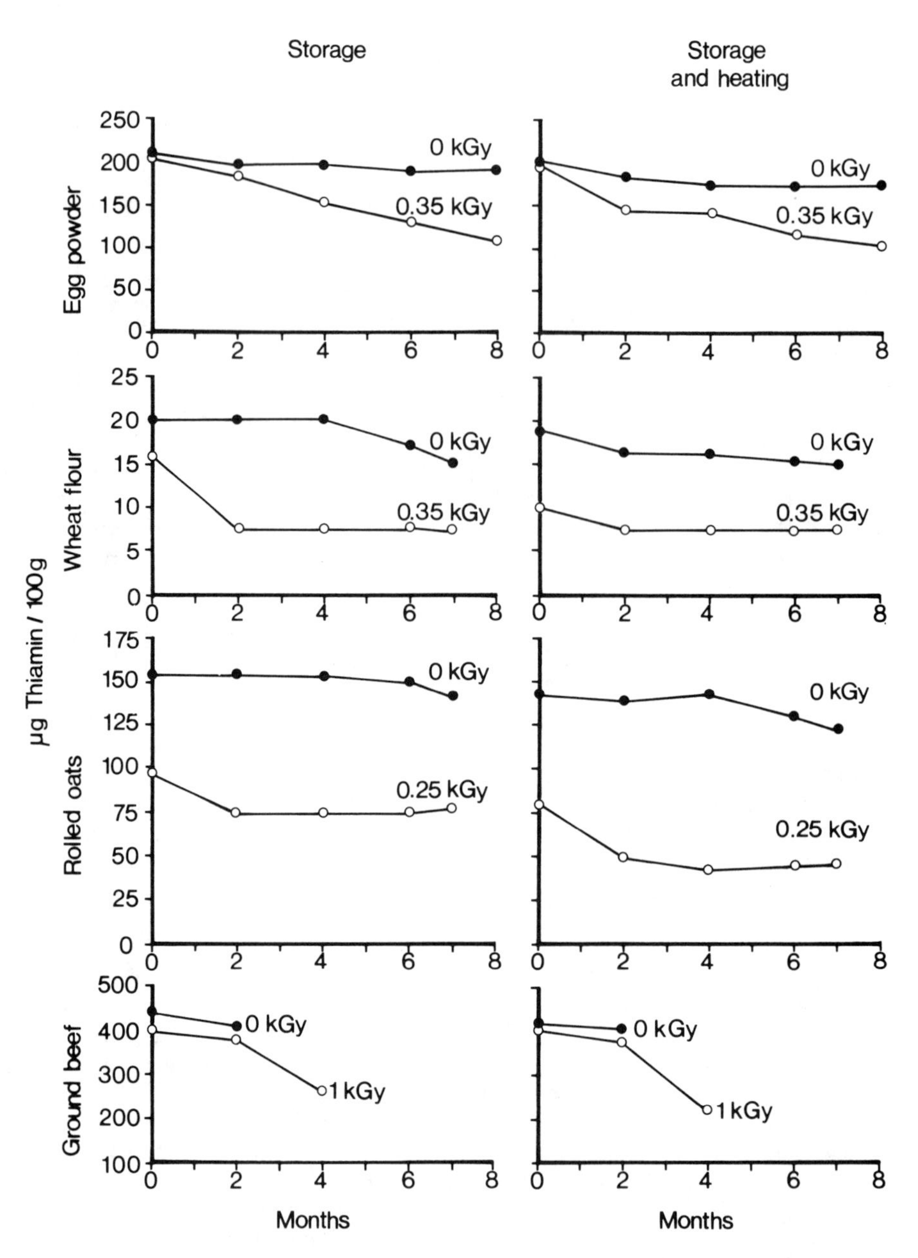

Figure 2 Changes in thiamin levels of four foodstuffs, unirradiated or irradiated, as a result of storage or storage plus heating (10 MeV electrons, dose indicated in graph, storage in polyethylene bags without exclusion of air, ambient storage temperature, except for minced pork 0°C; heat treatment 10 min in 100°C oven). (From Ref. 13a.)

Table 1 Effect of Radiation (6 kGy) and Cooking (4 min at 15 lb/in^2) on Some B-Complex Vitamins in Cod

Treatment	Vitamin		
	Nicotinic acid	Riboflavin	Thiamin
Nonirradiated raw (control) vitamin level (μg/g[a])	166	1.98	2.67
Irradiated, raw			
vitamin level (μg/g[a])	165	1.87	1.41
% of control	99	94	53
Nonirradiated, cooked			
vitamin level (μg/g[a])	160	1.8	2.4
% of control	96	91	90
Irradiated, cooked			
vitamin level (μg/g[a])	161	1.67	1.22
% of control	97	84	46

[a]Fat-free dry matter.
Source: Ref. 15.

Table 2, Sreenivasan observed that red gram (*Cajanus cajan*), a staple food in India and a chief source of protein, retained more thiamin, riboflavin, and niacin after irradiation (10 kGy) and cooking than the unirradiated cooked control [16]. Presumably the shorter cooking time of the irradiated samples caused less vitamin destruction during cooking.

Vitamin B_1 and E, the most radiation-sensitive vitamins, cannot only be protected by exclusion of oxygen but also by irradiation at low temperature. This is demonstrated in Figure 3. Treatment of minced pork with a dose of 50 kGy in the presence of air at ambient temperature destroyed alpha-tocopherol completely. When irradiation was carried out at 0°C the loss was 75%, and at −30°C it was 55%.

Table 2 Effect of Radiation and Cooking on Retention of B Vitamins in Red Gram

Treatment	% Retention		
	Riboflavin	Thiamin	Niacin
Irradiated (10 kGy, uncooked)	98.7	92.7	93.3
Control (cooked)	88	76.1	83.3
Irradiated (10 kGy, cooked)	95	82.7	89.3

Retention of vitamins was calculated on 100% basis in control unirradiated samples which contained (μg/g air-dry basis) 6.2 riboflavin; 4.6 thiamin; and 46 niacin.
Source: Ref. 16.

Similarly, loss of thiamin in egg powder irradiated with a dose of 10 kGy was about 50% when irradiated at ambient temperature and about 20% when irradiated at −30°C [14].

Although in general vitamin losses increase with increasing radiation dose, irradiation of foods with high radiation doses often requires conditions of treatment which minimize undesirable sensory effects. These conditions also reduce vitamin losses. For example, the process for radiation-sterilization of meat developed at the Natick Laboratories of the U.S. Army involved exclusion of oxygen (vacuum-packing) and irradiation at about −30°C. Under these conditions thiamin retention in radiation-sterilized pork is better than in heat-sterilized pork [17]. Some of these results are presented in Figure 4. The data demonstrate again the improved retention of thiamin at lower temperatures during irradiation. Interestingly, they also show much better retention with electron irradiation than with gamma irradiation. At 60 kGy and −45°C, for instance, retention was 60% and 35%, respectively. The authors explain this as a consequence of the much higher dose rate delivered by the electron beam. There are not many examples in the literature of such a pronounced difference between effects caused by the two types of radiation.

Because an earlier study had suggested that irradiation might have caused the formation of antimetabolites to thiamin and pyridoxine in meats, a study of the possible occurrence of antithiamin and antipyridoxine factors in irradiated chicked and beef was carried out at the Letterman Army Institute of Research in San Francisco

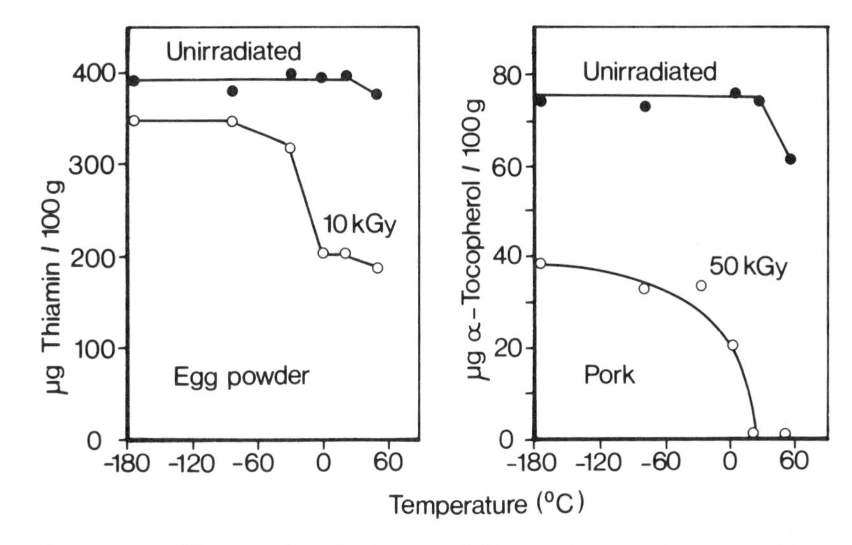

Figure 3 Effect of irradiation at different temperatures on the α-tocopherol (vitamin E) level of minced pork and the thiamin (vitamin B_1) level of egg powder (air not excluded). (From Ref. 14.)

[18]. This bioavailability study produced no evidence of antivitamin factors in any meat tested.

A rather complex situation emerges from the numerous studies on effects of irradiation on vitamin C levels in fruits and vegetables. The low radiation doses (<2 kGy) used to irradiate these products do not cause a significant radiolytic destruction of vitamin C. Only about 5% loss of vitamin C (ascorbic acid plus dehydroascorbic acid) was found when orange juice was irradiated with 2 kGy, increasing to about 60% at 10 kGy [19]. In black currant juice even a dose of 6 kGy caused no loss [20]. Vitamin C in dried vegetable products is quite resistant to radiation. No loss was observed in onion powder even when the extremely high dose of 270 kGy was applied to samples sealed in tin cans or 20 kGy to samples irradiated in commercial 50 lb boxes [21].

When losses of vitamin C are observed in response to irradiation with doses below 2 kGy in living tissue of fruits and vegetables, these losses are the result of metabolic changes and not of direct radiolytic action. This physiological response to irradiation is influenced by radiation dose, plant species and variety, time elapsed between irradiation and analysis, ripeness at the time of irradiation, temperature during irradiation and during postirradiation storage, and possibly by other factors.

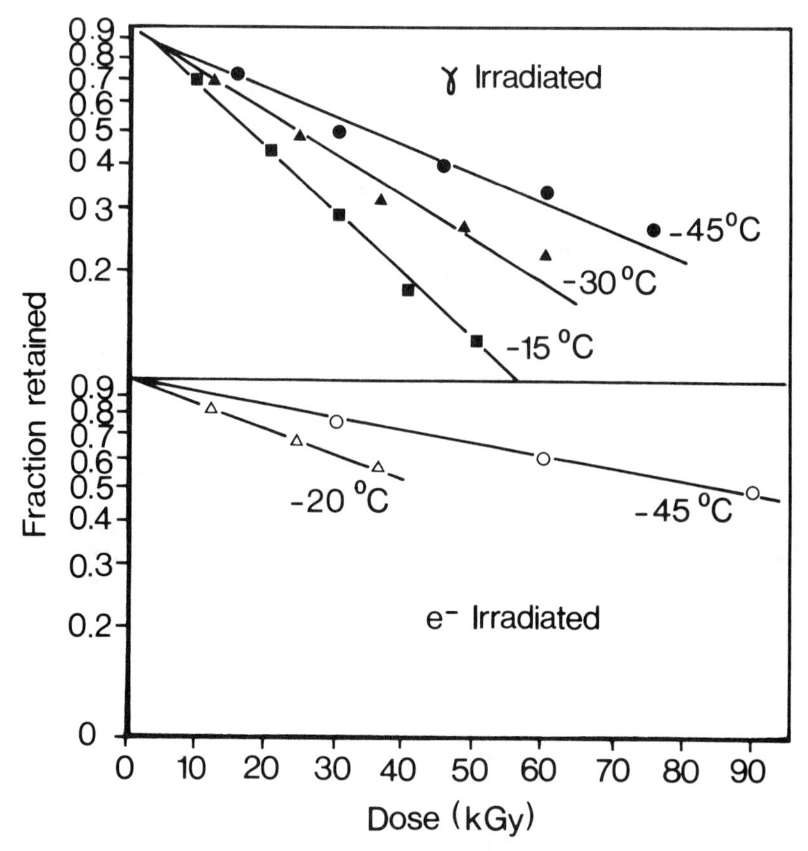

Figure 4 Effect of radiation dose and temperature during irradiation on thiamin retention in pork. Initial concentration of thiamin was 0.9 mg/100 g pork. (From Ref. 17, with permission.)

Many authors neglect to take into consideration that irradiation can cause a partial conversion of ascorbic acid to dehydroascorbic acid and that both have vitamin C activity in man. Reports on radiation effects in plant foods often mention only the ascorbic acid level, neglecting the dehydro form and reporting exaggerated losses. The data presented in Table 3 make it clear that an experimenter determining only ascorbic acid at the time of 2 h after irradiation might report a loss of 16% in response to irradiation with a dose of 4 kGy in lemons, while another investigator determining vitamin C as the sum of ascorbic acid and dehydroascorbic acid at 24 h after irradiation might report a 3.5% gain.

Table 3 Levels of Ascorbic Acid and Dehydroascorbic Acid in Juice from Lemons Exposed to 4 kGy of Ionizing Radiation

Time after irradiation (h)	Ascorbic acid (mg/100 ml juice)	% Change	Dehydroascorbic acid (mg/100 ml juice)	Sum of ascorbic plus dehydroascorbic acid (mg/100 ml)	% Change
Unirradiated controls	41.1 ± 0.8	—	1.5 ± 0.5	42.6	—
0	35.8	−13	4.5	40.3	−5.4
2	34.4	−16	4.7	39.1	−8.2
4	35.9	−13	5.2	41.1	−3.5
6	36.6	−11	4.9	41.9	−1.6
24	37.6	−9	6.5	44.1	+3.5

Source: Ref. 22.

Actually, the dose of 4 kGy is too high for any commercial application with fruits. In lemons it causes increased pitting during storage. In the dose range below 1 kGy, which is of practical interest, even those authors who have determined only ascorbic acid have found no loss or minimal losses, for instance, in oranges [23], grapefruit [24], mangoes and papayas [25], and bananas [26]. Strawberries irradiated at 2.5 kGy showed 15% loss of ascorbic acid immediately after irradiation. During refrigerated storage the ascorbic acid level decreased faster in unirradiated samples, so that after 1 week the irradiated samples had 8% and after 2 weeks only 3% less than the unirradiated controls [27]. The importance of determining both ascorbic and dehydroascorbic acid not only in the higher dose range where radiation chemical losses may occur but also in the low dose range where only the physiological response to radiation plays a role has been demonstrated by Wills in Australia in her studies of vitamin C levels in irradiated potatoes [28]. She found total vitamin C in Sequoia and Sebago potatoes unaffected by a dose of 50 Gy during 6 months of storage. When the dose was 100 Gy, Sequoia potatoes had 20% less total vitamin C compared to unirradiated controls after 2 and 4 months of storage, while after 6 months the levels tended to be somewhat higher in the irradiated samples.

Among the water-soluble vitamins thiamin is the most sensitive to radiation, followed by vitamin C, pyridoxine (B_6), riboflavin (B_2), and niacin. Such generalizations should not be taken too literally, however. The relative sensitivity of different vitamins depends on the irradiated material. For example, when riboflavin and thiamin were studied in different cereals under identical conditions, they exhibited approximately the same sensitivity in whole ground oats, whereas in whole ground wheat thiamin was clearly more sensitive to irradiation than riboflavin [29].

According to some rather early studies folic acid can be considered as quite resistant to radiation. However, analytical methods for folic acid were not reliable at that time, and the response of folic acid levels in foods to irradiation should again be studied, using the improved methods of analysis. Scant information exists on the radiation resistance of cobalamin (B_{12}).

Among the fat-soluble vitamins alpha-tocopherol is the most easily destroyed by radiation, followed by vitamin A and K, while vitamin D is relatively stable.

What counts, from a nutritional viewpoint, is the vitamin level in foods at the time of consumption. As indicated earlier for the case of thiamin in rolled oats, radiation-induced losses may become more evident after storage and cooking or, as indicated for B vitamins in dry legumes, they may be less evident in the irradiated cooked product. Some studies on combination effects of irradiation, storage, and heating have been carried out [4, 29a], but more would be desirable.

III. OTHER MICRONUTRIENTS

Other minor food constituents are the polyunsaturated fatty acids, sometimes called vitamin F. A report by British authors [30] who had irradiated a mixture of starch and herring oil and had found considerable destruction of highly unsaturated fatty acids during postirradiation storage in air, caused some concern that irradiation may generally act destructively on unsaturated fatty acids. However, there is no food that would correspond to this artificial mixture. When herring fillets were irradiated even a dose of 50 kGy did not affect the proportions of the polyunsaturated fatty acids [31]. When whole grain of rye, wheat and rice was irradiated, no loss of polyunsaturated fatty acids was observed in the dose range of 0.1 to 1 kGy, and only small losses at 63 kGy [32].

Minerals and trace elements are not affected by irradiation and there is no evidence that the bioavailability of these elements might be adversely affected by irradiation.

IV. CONCLUSIONS

In summary, the macronutrients—proteins, fats, and carbohydrates—are not significantly altered by irradiation in terms of nutrient value and digestibility. Among the micronutrients some of the vitamins are susceptible to irradiation, to an extent very much depending on the composition of the food, and on process and storage conditions. Whether or not the partial loss of a vitamin or of vitamins in an irradiated food is of importance depends on circumstances, such as the contribution that this food makes to the total diet. For instance, it would not be reasonable to oppose irradiation of spices with the argument that irradiation causes vitamin losses, because spices are not a significant source of vitamins. On the other hand, pork is an important source of thiamin for some populations. Regulatory bodies confronted with a request to permit high-dose irradiation of pork might insist that the meat be vacuum-packed and/or irradiated at −30°C to minimize losses of thiamin. If the request was for low-dose irradiation of pork, for instance the 0.3 kGy required to provide a trichina-safe product, concern about losses of thiamin would be unjustified.

As with regard to potential microbiological problems discussed in Chapter 6, it can be stated that potential nutritional losses in irradiated foods are not basically different from losses in foods treated by other processes. On the contrary, heating, drying, and some other traditional methods may cause higher nutritional losses than irradiation.

REFERENCES

1. Raica, N., Jr., J. Scott, and W. Nielsen, The nutritional quality of irradiated foods, *Radiation Res. Rev.*, 3:447 (1972).
2. Tobback, P. P., Radiation chemistry of vitamins, in *Radiation Chemistry of Major Food Components*, P. S. Elias and A. J. Cohen, eds. Elsevier, Amsterdam, 1977, p. 187.
3. Josephson, E. S., M. H. Thomas, and W. K. Calhoun, Nutritional aspects of food irradiation: an overview, *J. Fd. Proc. Preserv.*, 2:229 (1978).
4. Diehl, J. F., Effects of combination processes on the nutritive value of food, in *Combination Processes in Food Irradiation*, Proceedings of a Symposium held in Colombo, Sri Lanka, Nov. 1980. Int. Atomic Energy Agency, Vienna, 1981, p. 349.
5. Murray, T. K., Nutritional aspects of food irradiation, in *Recent Advances in Food Irradiation*, P. S. Elias and A. J. Cohen, eds. Elsevier, Amsterdam, 1983, p. 203.
6. Rhodes, D. N., The treatment of meats with ionizing radiation, *J. Sci. Food Agric.*, 17:180 (1966).
7. Underdal, B., J. Nordal, G. Lunde, and B. Eggum, The effect of ionizing radiation on the nutritional value of fish (cod) protein, *Lebensm. Wiss. Technol.*, 6:90 (1973).
8. Eggum, B. O., Effect of radiation treatment on protein quality and vitamin content of animal feeds, in *Decontamination of Animal Feeds by Irradiation*, Proceedings of a meeting held in Sofia, Oct. 1977. Int. Atomic Energy Agency, Vienna, 1979, p. 55.
9. Moran, E. T., Jr., J. D. Summers, and M. S. Blayley, Effect of cobalt-60 gamma irradiation on the utilization of energy, protein, and phosphorus from wheat bran by the chicken, *Cereal Chem.*, 45:469 (1968).
10. McGinnis, J., D. Honeyfield, M. B. Patel, and M. H. Pubols, Improvement in the nutritional value of rye by gamma irradiation, *Fed. Proc.*, 37:759 (1978).
11. Reddy, S. J., M. H. Pubols, and J. McGinnis, Effect of gamma irradiation on nutritional value of dry field beans (*Phaseolus vulgaris*) for chicks, *J. Nutrit.*, 109:1307 (1979).
12. Daghir, N. J., J. L. Sell, and G. G. Mateos, Effect of gamma irradiation on nutritional value of lentils (*Lens culinaris*) for chicks, *Nutrit. Rep. Internat.*, 27:1087 (1983).
13. Diehl, J. F., Thiamin in irradiated foods. 1. Influence of various conditions and of time after irradiation (in German), *Z. Lebensm. Unters. Forsch.*, 157:317 (1975).

13a. Diehl, J. F., Thiamin in irradiated foods. 2. Combined effects of irradiation, storage and cooling (in German), *Z. Lebensm. Unters. Fotsch.*, 158:83 (1975).
14. Diehl, J. F., Reduction of radiation-induced vitamin losses by irradiation of foodstuffs at low temperature and by exclusion of atmospheric oxygen (in German), *Z. Lebensm. Unters. Forsch.*, 169:276 (1979).
15. Kennedy, T. S., and F. J. Ley, Studies on the combined effect of gamma radiation and cooking on the nutritional value of food, *J. Sci. Fd. Agric.*, 22:146 (1971).
16. Sreenivasan, A., Compositional and quality changes in some irradiated foods, in *Improvement of Food Quality by Irradiation*, Int. Atomic Energy Agency, Vienna, 1974, p. 129.
17. Thomas, M. H., B. M. Atwood, E. Wierbicki, and I. A. Taub, Effect of radiation and conventional processing on thiamin content of pork, *J. Food Sci.*, 46:824 (1981).
18. Skala, J. H., E. L. McGown, and P. P. Waring, Wholesomeness of irradiated foods, *J. Food Protection*, 50:150 (1987).
19. Proctor, B. E., and J. P. O'Meara, Effect of high-voltage cathode rays on ascorbic acid—in vitro and in situ experiments, *Ind. Eng. Chem.*, 43:718 (1951).
20. M. Jensen, Irradiation of berry fruit with reference to their industrial utilization, Danish Atomic Energy Commission, *Risö Report*, 16:77 (1960).
21. Galetto, W., J. Kahan, M. Eiss, J. Welbourn, A. Bednarczyk, and O. Silberstein, Irradiation treatment of onion powder: effects on chemical constituents, *J. Food Sci.*, 44:591 (1979).
22. Romani, R. J., J. van Kooy, L. Lim, and B. Bowers, Radiation physiology of fruit—ascorbic acid, sulphydryl and soluble nitrogen content of irradiated citrus, *Radiation Botany*, 3:363 (1963).
23. Nagay, N. Y., and J. H. Moy, Quality of gamma irradiated California Valencia oranges, *J. Food Sci.*, 50:215 (1985).
24. Moshonas, M. G., and P. E. Shaw, Effects of low-dose gamma-irradiation on grapefruit products, *J. Agric. Food Chem.*, 32:1098 (1984).
25. Beyers, M., and A. C. Thomas, Gamma-irradiation of subtropical fruits. 4. Changes in certain nutrients present in mangoes, papayas and litchis during canning, freezing and irradiation, *J. Agric. Food Chem.*, 27:48 (1979).
26. Thomas, P., S. D. Sharkar, and A. Sreenivasan, Effect of gamma irradiation on the postharvest physiology of five banana varieties grown in India, *J. Food Sci.*, 36:243 (1971).
27. Zegota, H., Suitability of Dukat strawberries for studying effects on shelf life of irradiation combined with cold storage, *Z. Lebensm. Unters. Forsch.*, 187:111 (1988).

28. Wills, P. A., Some effects of gamma radiation on several varieties of Tasmanian potatoes, 2. Biochemical changes, *Austral. J. Exptl. Agr. Animal Husbandry*, 5:289 (1965).
29. Hanis, T., J. Mnukova, P. Jelen, P. Klir, B. Perez, and M. Pesek, Effect of gamma irradiation on survival of natural microflora and some nutrients in cereal meals, *Cereal Chem.*, 65:381 (1988).
29a. Fox, J. B., Jr., D. W. Thayer, R. K. Jenkins, J. G. Phillips, S. A. Ackerman, G. R. Beecher, J. M. Holden, F. D. Morrows, and D. M. Quirbach, Effect of gamma irradiation on the B vitamins of pork chops and chicken breasts, *Int. J. Radiat. Biol.*, 55:689 (1989).
30. Hammer, C. T., and F. D. Wills, The effect of ionizing radiation on the fatty acid composition of natural fats and on lipid peroxide formation, *Int. J. Radiat. Biol.*, 35:323 (1979).
31. Adam, S., G. Paul, and D. Ehlermann, Influence of ionizing radiation on the fatty acid composition of herring fillets, *Radiat. Phys. Chem.*, 20:289 (1982).
32. Vaca, C. E., and M. Harms-Ringdahl, Radiation-induced lipid peroxidation in whole grain of rye, wheat and rice: effects on linoleic and linolenic acid, *Radiat. Phys. Chem.*, 28:325 (1986).

8

Evaluation of the Wholesomeness of Irradiated Foods by Expert Groups and International Agencies

The terms wholesome or wholesomeness are often used in discussions on the safety of foods in general and of irradiated foods in particular. They require a few words of explanation because they do not occur in the Federal Food, Drug, and Cosmetic Act, which is the major legal instrument for ensuring safety of foods in the United States. In the 1950s some food companies in the United States had used food labels which suggested that their products were good for health or had some particular value in the control of disease. Any such suggestion was unacceptable under the misbranding sections of the Food, Drug, and Cosmetic Act. The term wholesome was therefore proposed to describe the qualities of food products in a legally acceptable way, and its use soon became quite general. The Poultry Products Inspection Act of 1957 introduced the following definition: "The term 'wholesome' means sound, healthful, clean, and otherwise fit for human food." Studies on the health safety of irradiated foods were begun on a large scale at that time, and these were generally described as wholesomeness studies.

It is difficult or impossible to find truly equivalent translations of this term in foreign languages. The expression "safety for consumption" is therefore usually preferred in documents for international distribution. With regard to irradiated foods, considerations of wholesomeness or safety for consumption involve the previously discussed aspects: radiological safety, toxicological safety, microbiological safety, and nutritional adequacy.

I. THE FAO/IAEA/WHO JOINT EXPERT COMMITTEE

The first international meeting exclusively devoted to a discussion of wholesomeness data and legislative aspects of irradiated foods was held in Brussels in October 1961. It was organized by FAO, WHO, and IAEA and attended by participants from 28 countries. Although a delegate from the United States reported that long-term toxicity studies had been conducted on 22 representative foods, and participants from many other countries presented the results of other such studies, the meeting decided that "general authorization of the commercial use of radiation for the treatment of food is premature." It was recommended that FAO, WHO, and IAEA consider the early establishment of a Joint Expert Committee to advise on the special requirements for the testing of the wholesomeness of irradiated foods [1].

The Joint FAO/IAEA/WHO Expert Committee on Irradiated Food had its first meeting in Rome in April 1964. The Committee stated that [2]

> extensive tests conducted by feeding to animals, and to a lesser extent to human volunteers, of irradiated food treated in accordance with procedures that should be followed in approved practice, have given no indication of adverse effects of any kind, and there has been no evidence that the nutritional value of irradiated food is affected in any important way.

The Committee recommended legal control of irradiated food "by the use of a list of permitted foods irradiated under specific conditions" and made recommendations as to which tests should be applied to an irradiated food to establish its safety for consumption; it suggested that these tests should be broadly similar to those used for testing the safety of food additives.

When the Committee met again in Geneva in April 1969, it pronounced "temporary acceptance" of irradiated potatoes (doses up to 0.15 kGy) and of wheat and wheat products (up to 0.75 kGy), while available data on irradiated onions were found not be be satisfactory for an evaluation [3]. The temporary nature of the acceptance for potatoes and wheat meant that the available data were insufficient to fully establish safety and that the Committee required additional evidence within a specified period of time.

At its third meeting, in Geneva in September 1976, the Committee gave "unconditional acceptance" to irradiated wheat (up to 1 kGy), potatoes (up to 0.15 kGy), papayas (up to 1 kGy), strawberries (up to 3 kGy) and chicken (up to 7 kGy), while onions (up to 0.15 kGy), rice (up to 1 kGy) and fresh cod and red-fish (up to 2.2 kGy) received "provisional acceptance." The latter category meant—as did the previously used term "temporary acceptance"—that some additional testing was required. The Committee also considered irradiated mushrooms, finding evaluation not possible with the available data [4].

The Committee gave much thought to the principles of testing the wholesomeness of irradiated foods and stressed the differences to the safety evaluation of food additives. It made clear that "irradiation is a physical process for treating foods and as such it is comparable to the heating or freezing of foods for preservation." This report also recognized the value of chemical studies as a basis for evaluating the wholesomeness of irradiated foods.

When the Committee had its fourth meeting, in Geneva in October 1980, it was provided with a wealth of additional data, mostly provided by the International Project. On this basis the Committee concluded [5] (emphasis added):

> (a) All the toxicological studies carried out on a large number of individual foods (from almost every type of food commodity) have produced no evidence of adverse effects as a result of irradiation.
> (b) Radiation chemistry studies have now shown that radiolytic products of major food components are identical, regardless of the food from which they are derived. Moreover, for major food components, most of these radiolytic products have also been identified in foods subjected to other, accepted types of food processing. Knowledge of the nature and concentration of these radiolytic products indicates that there is no evidence of a toxicological hazard.
> (c) Supporting evidence is provided by the absence of any adverse effects resulting from the feeding of irradiated diets to laboratory animals, the use of irradiated feeds in livestock production, and the practice of maintaining immunologically incompetent patients on irradiated diets. The Committee therefore concluded that the *irradiation of any food commodity up to an overall average dose of 10 kGy presents no toxicological hazard; hence, toxicological testing of foods so treated is no longer required.*
> The Committee considered that the irradiation of food up to an overall average dose of 10 kGy introduces *no special nutritional or microbiological problems.* However, the Committee emphasized that attention should be given to the significance of any changes in relation to each particular irradiated food and to its role in the diet.
> The Committee recognized that higher doses of radiation were needed for the treatment of certain foods but did not consider the toxicological evaluation and wholesomeness assessment of foods so treated because the available data are insufficient for this purpose. Further studies in this area are therefore needed.

Although the Joint Expert Committee had carefully considered microbiological aspects, FAO and WHO desired additional reassurance that nothing had been overlooked in this area. At their request, the Board of the International Committee on Food Microbiology and Hygiene (ICFMH) of the International Union of Microbiological Societies (IUMS) met in Copenhagen in December 1982 to reconsider the evidence concerning microbiological safety of the process. The Board found no cause for concern and concluded [6] that "food irradiation was an important addition to the methods of control of food-borne pathogens and did not present any additional hazards to health."

II. NATIONAL AND EUROPEAN ADVISORY GROUPS

The definitive conclusion of the 1980 Joint Expert Committee meeting that irradiation of any food with an average dose of up to 10 kGy presents no toxicological hazard and introduces no special nutritional or microbiological problems has been a turning point in the decades old debate over the safety of irradiated foods. Many national governments which had previously not granted any permission for irradiated foods did so after 1980. Few governments accepted the Committee's verdict outright as a basis for regulating irradiated foods. Usually they appointed their own expert committees which again went through the scientific evidence to see if one could agree with the conclusions reached in Geneva in 1980. Wherever the deliberations of these national expert committees were made public they fully support the judgment of 1980, to wit in Denmark [7], Sweden [8], and the United Kingdom [9]. Equally detailed committee proceedings are not available from France, but it has been reported that the French authorities do not require further toxicological testing in the dose range below 10 kGy [10]. In response to a critical report on food irradiation by the Standing Committee on Consumer and Corporate Affairs of the Canadian House of Commons, the Canadian Government in September 1987 submitted a detailed analysis of each point of the criticism [11]. This response confirms that the Government's Health Protection Branch "accepts in principle the lack of toxicological hazards for foods irradiated below 10 kGy."

The United States Food and Drug Administration established an Irradiated Foods Committee in 1979 and an Irradiated Foods Task Force in 1981. These groups concluded that studies with irradiated foods did not show adverse toxicological effects. On the basis of chemical studies they concluded that traditional toxicological testing of food irradiated at doses below 1 kGy could not be expected to provide meaningful answers to toxicity questions regarding such irradiated foods [12]. (Action taken by FDA on the basis of this advice will be described in Chapter 10.) A task force appointed by the

Council for Agricultural Science and Technology (CAST) went beyond the conclusions of the Joint Expert Committee when it confirmed the statement made by the Surgeon General in 1965 that irradiated foods are safe and nutritionally adequate [13]. The Surgeon General had spoken of foods irradiated with doses of up to 56 kGy.

Also in the United States, the Conference for Food Protection, sponsored by teh National Sanitation Foundation, asked a task force to report on the wholesomeness of irradiated foods. At its meeting in Ann Arbor, Michigan, in August 1986, the Conference accepted the recommendation of the task force to adopt the General Codex Standard on Irradiated Foods, fully confirming the decisions of the 1980 FAO/IAEA/WHO Joint Expert Committee. The Conference also accepted the recommendation: "Replace toxicologically unacceptable chemical sterilants by irradiation, which leaves no harmful residues in the treated foodstuffs" [14].

In the European Community, the Commission of the EC in Brussels has asked its Scientific Committee on Food for advice on the wholesomeness of foods irradiated by suitable procedures. In its report the Committee had expressly verified the conclusions of the Geneva meeting of 1980 and has confirmed that no further animal feeding studies need be carried out in order to assess the safety of a food irradiated up to a dose of 10 kGy [15].

III. CODEX ALIMENTARIUS COMMISSION

The Joint FAO/WHO Codex Alimentarius Commission in Rome was created in 1962 with the intention of facilitating worldwide food trade by harmonizing food laws. Close to 130 countries have joined and cooperate in this effort. The Commission generates the Codex Alimentarius, a collection of internationally recognized food standards. Achieving international agreement on such a sensitive issue as a food standard, which could seriously affect national economies and other national interests, is a delicate process. It involves a step-by-step acceptance procedure with a number of built-in phases of reconsideration and agreement-seeking operations, until a Codex Standard is finally accepted and published in the Codex Alimentarius. After several years of negotiations the Codex Alimentarius Commission adopted in 1983 a General Standard for Irradiated Foods and a Recommended International Code of Practice for the Operation of Radiation Facilities Used for the Treatment of Foods. Both documents are presented as Annexes to this book. The Joint Expert Committee's conclusions concerning the safety of any food irradiated with an overall average dose of up to 10 kGy have been fully adopted by the Codex Alimentarius Commission. The basis for worldwide legal acceptance of irradiated foods exists. In Chapter 10 we shall see to what extent this has led to legal approval in various countries.

IV. THE WORLD HEALTH ORGANIZATION

The World Health Organization has repeatedly expressed its full support for food irradiation. In a press release (In Point of Fact, No. 40/1987) WHO has again called attention to the conclusions of the FAO/IAEA/WHO Joint Expert Committee of 1980 and to the Codex Alimentarius Standard and continued:

> As a result of these internationally agreed documents, all countries—regardless of their stage of development—are encouraged to apply food irradiation. The process not only allows for a larger supply of safe food but also has the advantage of reducing dependence on food treated with chemical substances . . . . The World Health Organization sees food irradiation as a process which has the potential to increase safe food supplies, thus contributing to primary health care.
> Widespread information campaigns are still required for food irradiation to be fully accepted. WHO is concerned that rejection of the process, essentially based on emotional or ideological influences, may hamper its use in those countries which may benefit the most.

A more extensive presentation of WHO's views on food irradiation has been published as a book [16].

An International Conference on the Acceptance, Control of, and Trade in Irradiated Food, jointly held by WHO and several other organizations of the United Nations, met in Geneva, Switzerland, December 12–16, 1988. Some 60 governments sent delegations and numerous international organizations participated. A Conference Document was adopted which reemphasized the validity of the 1980–decision on the wholesomeness of foods irradiated at dose levels of up to 10 kGy, and of the Codex General Standard for Irradiated Foods and the Recommended International Code of Practice for the Operation of Radiation Facilities for the Treatment of Food. Regulatory control by competent authorities was considered as a necessary prerequisite for introduction of the process in accordance with the principles of the Codex Standards. "International trade in irradiated foods would be facilitated by harmonization of national procedures based on internationally recognized standards for the control of food irradiation." The Conference Document further concluded that the acceptance of irradiated food by the consumer is a vital factor in the successful commercialization of the irradiation process, and that the provision of clear and adequate information about food irradiation to the public can contribute to this acceptance. WHO has announced that the proceedings of the Conference, including the sometimes heated debates between proponents and opponents of the process, will be published in late 1989.

REFERENCES

1. FAO, Report of the Meeting on the Wholesomeness of Irradiated Foods. Brussels, Oct. 1961, FAO, Rome, 1962.
2. WHO, The Technical Basis for Legislation on Irradiated Food, WHO Technical Report Series 316, Geneva, 1965.
3. WHO, Wholesomeness of Irradiated Food with Special Reference to Wheat, Potatoes and Onion, WHO Technical Report Series 451, Geneva, 1970.
4. WHO, Wholesomeness of Irradiated Food, WHO Technical Report Series, 604, Geneva, 1977.
5. WHO, Wholesomeness of Irradiated Food, WHO Technical Report Series 659, Geneva, 1981.
6. Codex Alimentarius Commission, The Microbiological Safety of Irradiated Food, Report CX/FH 83/9, Rome, 1983.
7. LST, Irradiation of Food, Report of a Danish Working Group, National Food Agency Publication no. 120, Soborg, 1986.
8. SOU, Irradiation of foods? Report of an Expert Committee (in Swedish), Statens Offentliga utredningar no. 26, Ministry of Agriculture, Stockholm, 1983.
9. Department of Health and Social Security, Report on the Safety and Wholesomeness of Irradiated Foods Advisory Committee on Irradiated and Novel Foods, HMSO Publications Centre, London, 1986.
10. Henon, Y., The present stage of development of ionizing radiation treatment in France, in *Food Irradiation Processing*, Proceedings of a Symposium held in Washington, D.C., March 1985. Int. Atomic Energy Agency, Vienna, 1985, p. 311.
11. Health and Welfare Canada, Comprehensive Federal Government Response to Report of the Standing Committee on Consumer and Corporate Affairs on the Question of Food Irradiation and the Labelling of Irradiated Foods, Ottawa, Sept. 1987.
12. Pauli, G. H., and C. A. Takeguchi, Irradiation of foods—an FDA perspective, *Food Revs. Internat.*, 2:79 (1986).
13. CAST, Ionizing Energy in Food Processing and Pest Control: 1. Wholesomeness of Food Treated with Ionizing Radiation, Council for Agricultural Science and Technology, Report no. 109, Ames, Iowa, 1986.
14. Elias, P. S., Task force on irradiation processing wholesomeness studies, in *Food Protection Technology*, C. W. Felix, ed. Lewis Publishers, Inc., Chelsea, Michigan, 1987, p. 349.
15. Scientific Committee for Food, Report on Irradiated Foods, EUR Series Food-Science and Techniques No. 10840, Brussels, 1987.
16. World Health Organization and Food and Agriculture Organization of the United Nations, *Food Irradiation—A Technique for Preserving and Improving the Safety of Food*, Geneva, 1988.

9

Potential and Actual Applications of Food Irradiation

I. AN OVERVIEW

Irradiation of foods can have different purposes, as shown in the overview presented in Table 1. The classification into low-, medium-, and high-dose applications follows a proposal of the FAO/IAEA/WHO Joint Expert Committee on Food Irradiation at its meeting in 1980. The separation of low-dose from medium-dose treatments, and of medium-dose from high-dose treatments is rather arbitrary, and the existence of a sharp dividing line between these treatments should not be assumed.

The dose levels indicated in Table 1 can serve only as a rough guideline. As explained in Chapter 2, it is not possible to irradiate a food package in such a way that each food particle in that package receives exactly the same dose. It should be understood that the dose distribution ($D_{max}:D_{min}$) depends on the radiation facility actually used. If, for instance, a minimum dose of 8 kGy is required to achieve a particular purpose, a laboratory irradiator with a uniformity ratio of 1.8 would give a maximum dose of 14.4 kGy and an average dose of about 11 kGy, while an industrial irradiator with a uniformity ratio of 3 would give a maximum dose of 24 kGy and an average dose of about 16 kGy.

In practical situations, the purpose of irradiation can sometimes not be separated as clearly as in Table 1. For instance, a treatment of meat for prevention of food poisoning (destruction of salmonella and other pathogens) may automatically lead to increased refrigerated shelf life, or irradiation of fruits for reduction of populations of molds and yeasts may at the same time lead to delayed maturation of the fruits.

Table 1 Purposes of Food Irradiation

Food	Main objective	Means of attaining objective	Dose (kGy)
1. *Low-dose treatments* (up to about 1 kGy)			
Potatoes, yams, onions garlic, shallots	Extension of storage life	Inhibition of sprouting	0.05–0.15
Certain fruits and vegetables	Improved keeping properties	Delay in maturation and senescence	0.25–1
Cereal grain, flour, dried fruits, nuts, pulses	Prevention of losses caused by insects	Killing or sexual sterilization of insects	0.2–0.7
Fruits	Prevention of spreading of pests. Quarantine treatment	Killing or sexual sterilization of insects	0.2–0.7
Meat	Prevention of parasitic disease transmitted through food	Destruction of parasites such as *Trichinella spiralis* and *Taenia saginata*	0.3–0.5
2. *Medium dose treatments* (about 1–10 kGy)			
Certain fruits and vegetables, sliced bread	Improved keeping properties	Reduction of populations of bacteria, molds and yeasts	1–3

Meat, poultry, fish	Improved refrigerated storage	Reduction of population of microorganisms capable of growth at refrig. temperature	1–5
Meat, poultry, eggs, egg powder, froglegs, frozen seafood, and other foods carrying pathogenic microorganisms	Prevention of food poisoning	Destruction of *Salmonella*, *Shigella*, *Campylobacter*, *Vibrio*, *Yersinia*, and other non–spore-forming pathogens	3–10
Spices, dried vegetables, and other food ingredients	Prevention of contamination of food to which the ingredients are added	Reduction of population of microorganisms in the ingredient	3–10
3. *High dose treatments* (about 10–45 kGy)			
Meat, poultry	Safe long-term storage without refrigeration	Destruction of spoilage organisms and pathogens, including spore-formers	25–45
Complete hospital meals or constituents of such meals	To supply patients with sterile meals	Same as above	25–45

This overview of potential applications of food irradiation is not in any way comprehensive. Several thousand journal articles and many patents describe applications of food irradiation for various purposes. More extensive descriptions of various practical applications can be found in the books of Josephson and Peterson [1] and Urbain [2].

A thorough discussion of the economic aspects of food irradiation is beyond the scope of this book. Some comments and literature references concerning this topic were presented in Chapter 2, Sec. IV.

II. LOW-DOSE APPLICATIONS

A. Inhibition of Sprouting of Tubers, Bulbs, and Root Crops

In order to provide consumers with a year-round supply of potato tubers, onion bulbs, and other sprouting plant foods, storage over periods of many months is necessary—unless imports from other climatic zones, usually at a much higher price, can replace local production during off-season. Such long-term storage is possible with the aid of refrigeration, which is costly, particularly in subtropical and tropical regions. Chemical sprout inhibitors, such as maleic hydrazide as a preharvest treatment, and propham (isopropyl carbamate, IPC) or chloropropham (isopropyl chlorocarbamate, CIPC) as a postharvest treatment, are effective for most of these crops and relatively cheap, but they leave residues in the produce, and for health reasons the use of some or all of the chemical sprout inhibitors has been banned in many countries.

Under the climatic conditions prevailing in temperate zones it is possible to store untreated *potatoes* without sprouting until about March (in the Northern Hemisphere) if a storage temperature of about 5°C is maintained by blowing cold air into the storehouse at night. Such potatoes are suitable for household use, but not for chips manufacture because a low storage temperature results in elevated levels of reducing sugars, and these cause undesirable browning when the chips are fried. Potatoes destined for processing into chips are therefore usually stored at 9—10°C. Under these conditions glucose/fructose levels remain low until about May, but sprouting begins as early as January, unless chemical sprout inhibitors are employed. Irradiation is an alternative to the use of these chemicals. Numerous studies on the inhibition of germination of white potatoes by irradiation have been carried out since Sparrow and Christensen reported in 1954 on the work they carried out at Brookhaven National Laboratory [3]. The required radiation dose is in the range of 40—150 Gy and depends on the potato variety used, on postirradiation storage conditions, intended use of the tubers, and time elapsed between harvest and irradiation [4, 5]. A dose of 150 Gy completely prevents sprouting, due to inhibition of cell division. If such tubers are stored in a sufficiently humid atmosphere to prevent water loss and shrivelling, they can be kept in a cool basement (15°C) for a year without much loss of quality. However, this dose also reduces the ability of the tubers to heal cuts and bruises received during harvest and transport. (This also applies to chemical sprout inhibitors, which also act by affecting cell division.) Rot-causing organisms can enter through these cuts and cause decay. Unless the tubers have been handled extremely carefully, irradiation at this dose level may cause increased losses due to rotting.

One way of preventing this is to allow a curing time of about 2 weeks between harvest and irradiation, to let any damage heal that was caused during harvesting. Preferably, the tubers should be cured, irradiated, and stored after irradiation in the same pallet boxes or bulk containers, so as to prevent skin damage due to repeated loading and unloading. Pallet boxes can be stacked one on top of another without increasing the pressure on tubers in the bottom layers. Another approach is to irradiate with an average dose of only 100 Gy. With some varieties even 40 Gy will suffice. This will permit small buds to develop, which will not interfere with most uses, and it will maintain a sufficient wound-healing capacity.

The required dose depends on preirradiation storage time. Irradiation inhibits sprouting most effectively when applied soon after harvest when the tubers are dormant. Higher radiation doses are needed after the dormant period. It is necessary to determine the proper dose for the particular variety and the intended storage temperature and storage time. Results obtained with potato variety Maritta are shown in Figure 1.

Some varieties have a tendency to turn dark after boiling, and this is somewhat aggravated by irradiation. This effect is not seen in other varieties, and even in the susceptible varieties it has not been observed in all studies. Soil or climatic conditions seem to play a role. After-cooking darkening—in unirradiated and in irradiated potatoes—is thought to be due to formation of iron-phenolic complexes. Irradiation causes increased polyphenol and decreased citric acid levels, and both conditions favor the formation of these colored complexes [6]. While this type of greyish darkening is not related to sugar levels in the tubers, irradiated potatoes stored at cool temperatures for 9 months or longer do have elevated levels of reducing sugars and this causes increased browning if these potatoes are used for industrial production of chips [7], a problem also seen with chemically treated potatoes kept at low temperature. This can be overcome by allowing a 2-week period of storage at about 20°C before processing. This lowers the sugar levels, but increased handling also means increased cost.

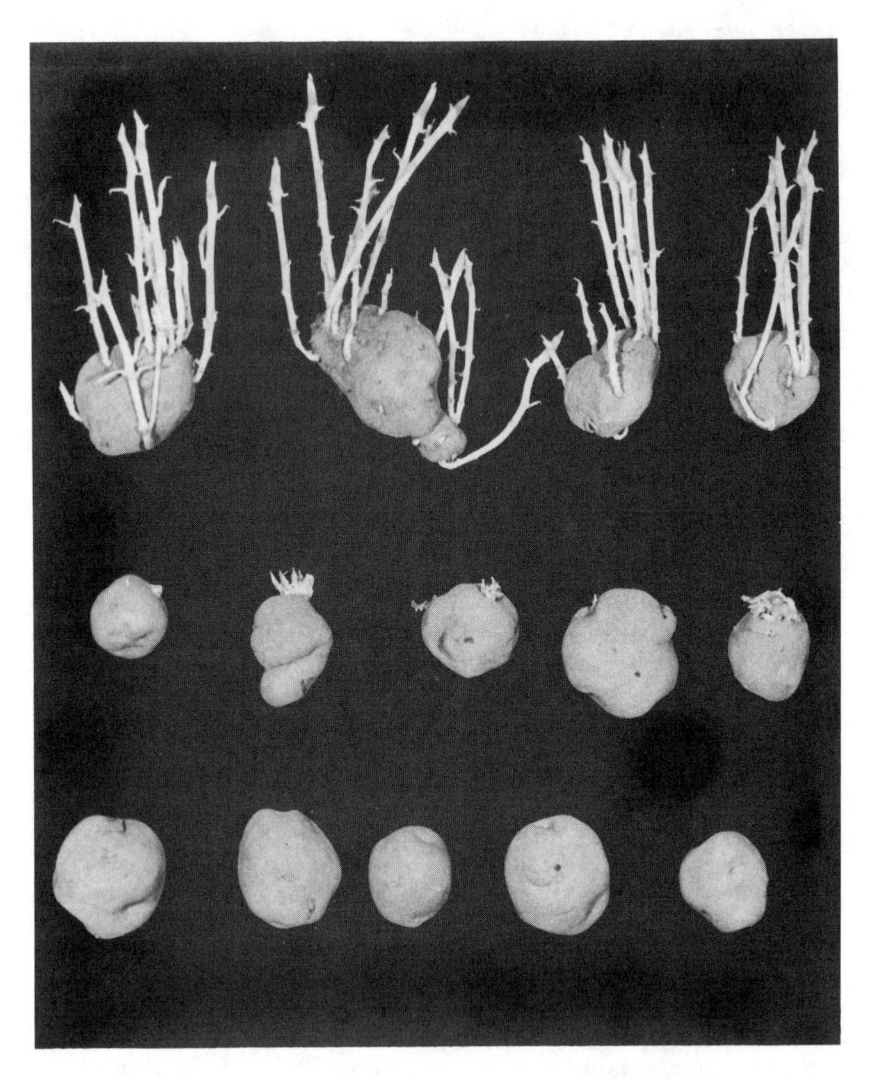

Figure 1 Potato tubers, variety Maritta, harvested in October and stored at 10°C until June. Top row, unirradiated; center row, irradiated at 60 Gy; bottom row, irradiated at 120 Gy. Irradiation was carried out with X-rays about 4 weeks after harvest. (From Federal Research Center for Nutrition, Karlsruhe, unpublished.)

Test marketing trials conducted in Chile, Hungary, Israel, Italy, South Africa, and Uruguay with many hundred tons of irradiated potatoes have all shown positive consumer acceptance for the irradiated tubers [5]. In these tests the potatoes were sold through supermarkets, greengrocers, or open markets, and the consumer response was solicited with questionnaires. In another type of test, carried out in the Federal Republic of Germany [8], each participating family received three coded batches of potatoes, variety Hansa. They knew that one or two of these were irradiated with a dose of 120 Gy but they were not told which. The potatoes were stored in the homes, and during a period of 5 months the housewives were asked every month to judge these tubers for their appearance, sprouting, rotting, suitability for cooking, and sensory quality of the cooked meal. As indicated in Figure 2, the percentage of participants preferring the irradiated tubers increased after December, and in March 75% preferred the irradiated, 18% the unirradiated potatoes. The extent of the difference noted between the unirradiated and the irradiated tubers depended on the way in which they were prepared (Fig. 3). The comparison could not be continued beyond March because the unirradiated potatoes were no longer usable. Some families continued with the evaluation of the irradiated potatoes and found them acceptable until May or June, when shrivelling due to water loss began to affect the quality.

In countries where chemical sprout inhibitors are permitted, the use of irradiation is quite unlikely, because it is more expensive. Chemical sprout inhibition is not allowed in Japan, the only country where irradiation of potatoes is carried out on a commercial scale. The potato irradiator of the Shihoro Agricultural Cooperative Association commenced operation in 1973 and has processed about 30,000 t of potatoes annually since then. The irradiated tubers are sold for industrial processing, not for household usage. Cost of irradiation has been estimated at $7–10/ton [4]. Chemical treatment would cost less than half as much. The Japanese experience has shown that irradiation is a viable alternative when chemical inhibition of sprouting is not allowed.

Inhibition of sprouting of *onions* can be achieved with a dose of 20–40 Gy [9]. In general, onions should not be kept very long after harvest and before irradiation. Maximum sprout inhibition is achieved if the bulbs are irradiated within 2 weeks after harvest [10]. Long-term storage of irradiated onions is accompanied by some browning of the growth center (inner buds) of the bulb. Most authors conclude that this discoloration does not make the onions objectionable for industrial use or for marketing [11, 12]. Alternatives to irradiation are refrigeration (usually 0°C and relative humidity of 70%) or spraying with maleic hydrazide, which is not permitted in many countries for health reasons. The chemical is usually applied to the crop

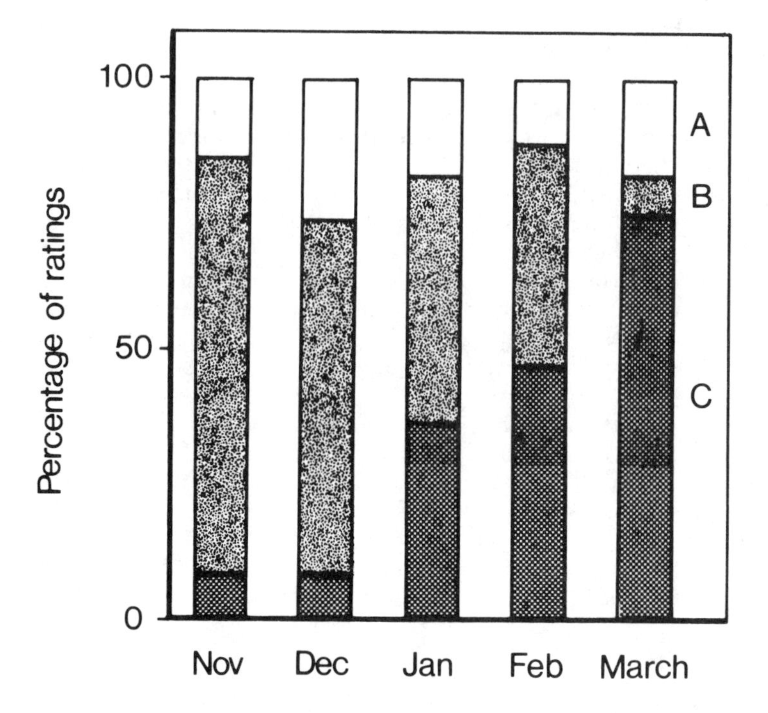

Figure 2 Consumer preference for untreated or irradiated potatoes. Tubers of variety Hansa were harvested in mid-September and distributed to participating families, either unirradiated or after irradiation on October 10 with a dose of 120 Gy. Quality ratings were collected once a month. (From Ref. 8.)

2 or 3 weeks before harvest when enough green foliage is present to facilitate its absorption and translocation to the bulb.

Marketing trials with irradiated onions have been successfully carried out in several countries, particularly Argentina, Hungary, and Israel. However, the only country where the process is carried out on a commercial scale is the German Democratic Republic. Some 4000 t/year of onions were reportedly irradiated there in 1986 and 1987.

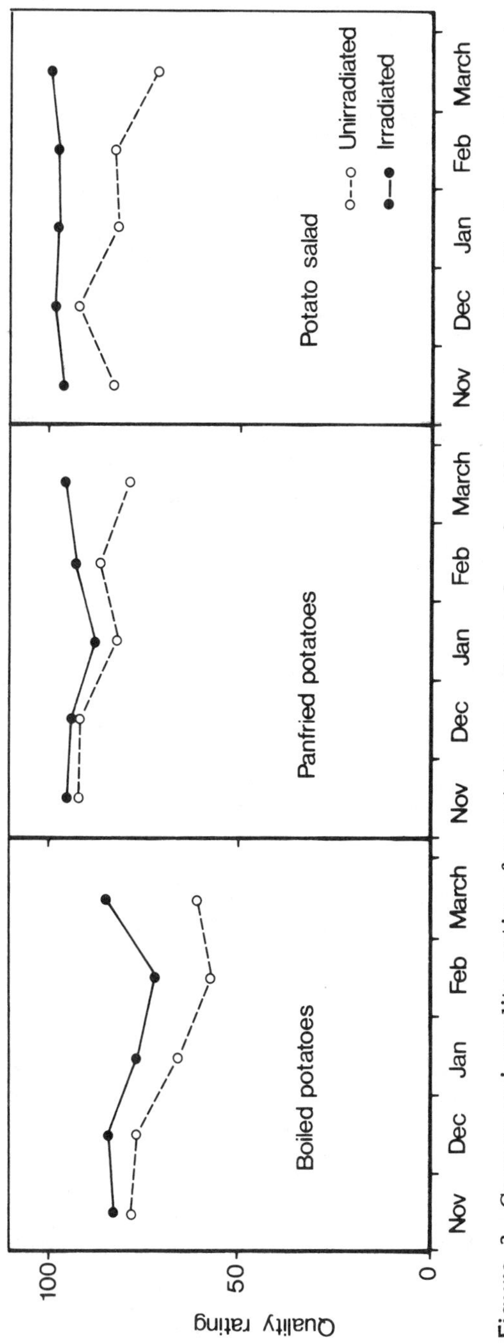

Figure 3 Consumers' quality rating for potatoes, untreated or irradiated, stored in the homes for up to 5 months and prepared as boiled or panfried potatoes or as potato salad. 100 = all participants gave the highest quality rating, 0 = all participants gave the lowest quality rating. (From Ref. 8.)

Table 2 Effect of Irradiation on Percentage of Marketable Garlic and Onion Bulbs Stored at Ambient Temperature[a]

Days postharvest	Garlic		Onion	
	0 Gy	50 Gy	0 Gy	50 Gy
180	91.6	93.6	89.1	95.3
210	52.5	85.5	85.4	95.1
240	18.7	96.3	76.6	92.2
270	11.4	80.1	48.2	81.5
300	1.1	55.6	3.0	65.5

[a]Average values from eight consecutive harvests.
Source: Ref. 14.

As in the case of onions, the dose required for sprout inhibition in *garlic* and *shallots* (*Allium ascalonicum* L.) depends on the time of irradiation after harvest. Doses in the range of 20–60 Gy, if applied during the dormancy period, result in complete inhibition of sprouting [10]. Studies carried out in Korea indicated that the time of irradiation after harvest was not very critical if a dose of 75 Gy was used for garlic [13]. A summary of results obtained with onions and garlic from 8 consecutive years (1979–1986) in Argentina is presented in Table 2. After irradiation the bulbs were stored in a local warehouse under ambient conditions, which are described by the authors [14] as ranging from −7°C to +38°C and 35–87% R.H. After 9 months, only 48% of control onions and 11% of control garlic were still in marketable condition. In contrast, after irradiation with 50 Gy, 81% of the onions and 80% of the garlic were still marketable at that time.

Sprout inhibition in *sweet potatoes* (*Ipomea batatus* L.) can be achieved by irradiation with a dose of 100 Gy. Sensory quality, as evaluated with baked sweet potatoes, was not affected at this level [15]. With varieties common in Taiwan a dose of 50 Gy was reported sufficient to extend storage life from 1 month (unirradiated) to 5 months under ambient storage conditions [16].

Yams, the tubers of *Dioscorea* spp., are of great importance as a staple food in many tropical regions. Usually only one crop of yams is grown per year, and this is harvested over a limited period of 3–4 months. Long-term storage is therefore important. It is limited by rotting, sprouting, and dehydration of the tubers. Refrigerated storage is not possible because yams, like some other tropical products quickly deteriorate at temperatures below 12°C.

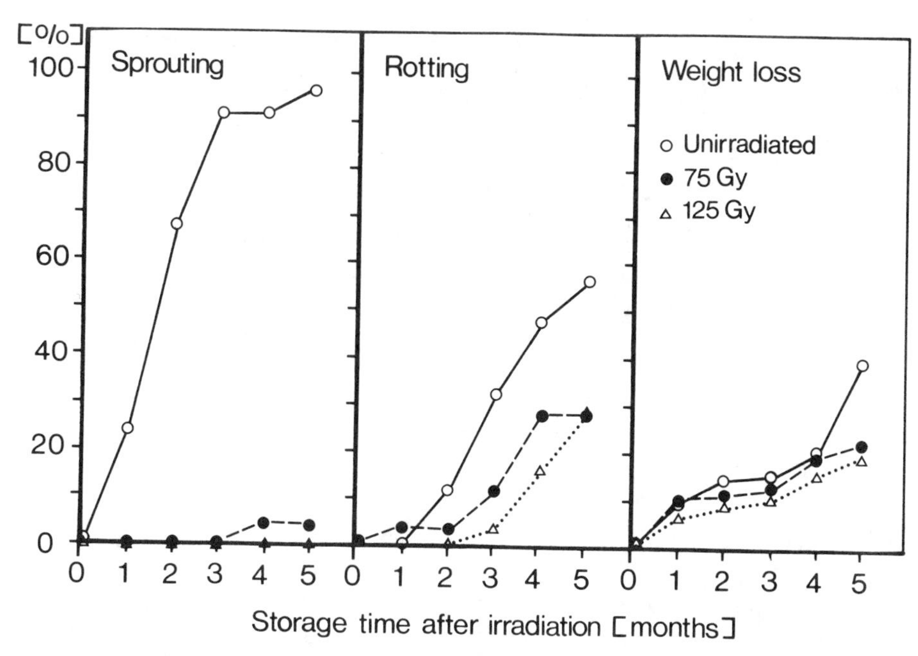

Figure 4 Effect of irradiation on cumulative percentage of sprouting and rotten yam tubers and on weight loss of the remaining tubers during storage at ambient temperature in Nigeria. (Adapted from Ref. 7.)

Chemical sprout inhibitors are ineffective. Traditional storage is in thatched shelters, huts, yam barns, or in small earthen silos in trenches. Losses are high, often reaching the 100% level after 5 or 6 months. Better results are obtained through controlled temperature storage at 16°C and 70% R.H., but storage facilities of this type are not available in most tropical countries. Excellent results with irradiation were obtained in Nigeria in studies reported by Adesuyi and Mackenzie [17], using the species *Dioscorea rotunda* poir, the most important in Africa. In storage tests employing traditional yam barns, doses of 75–125 Gy applied before the tubers began to sprout prevented sprouting and reduced rotting and weight loss during a period of 5 months (Fig. 4). Sensory evaluation resulted in highest rating for the tubers irradiated with 125 Gy, the lowest for the uniradiated control (Table 3).

It should be noted that the prospects in tropical regions for irradiation of potatoes, onions, and garlic are not as favorable as they are for yams. In tropical climates the storage life of potatoes, onions, and garlic is primarily limited by rotting, and irradiation cannot stop the rotting. For long-term storage in tropical countries

Table 3 Quality Assessment after 5 Months' Storage of Irradiated and Nonirradiated Yams, Expressed as Percentage of Tasters Giving a Grading of A (very good) to D (poor)

	Dose (Gy)						
Grading	150	125	100	75	50	25	Control
A (very good)	85	95	45	55	—	—	—
B (good)	15	5	55	45	40	20	—
C (fair)	—	—	—	—	60	80	15
D (poor)	—	—	—	—	—	—	85

Source: Ref. 17.

these commodities must be kept at temperatures of around 14–16°C after irradiation. The cost of irradiation and 5 to 6 months' storage at 16°C has been estimated at $12.50 to $16.00 per ton for the situation in Pakistan, while storage at refrigeration temperature without irradiation was reported to cost twice as much [18]. Irradiation of these commodities in tropical countries is thus still of interest, but the advantages are not as great as they are in the case of yams, which can be stored at ambient temperature.

Studies on *carrots*, *gingerroots*, *Jerusalem artichoke tubers*, *red beets*, and *turnips* have shown that sprouting in these commodities can also be inhibited by irradiation [4].

B. Delay of Maturation and Senescence in Fruits and Vegetables

The softening and browning associated with ripening of *bananas* can be delayed by irradiating the fruits while they are still in the preclimacteric stage. The optimal dose to achieve this and the maximum dose the fruits can tolerate seem to differ among the varieties and even for the same variety grown in different geographical areas [19]. In South Africa, where bananas are irradiated commercially, the dose applied is 600 Gy [20].

Irradiated *mangoes* also remain edible for longer periods before they pass into the senescent phase. Doses between 250 and 1000 Gy have been reported as optimal for different varieties, although some authors found skin spotting and other indications of phytotoxicity at doses of 750 Gy or higher [19], while others insist that doses above 250 Gy should be avoided in order to prevent radiation injury in mangoes [20a].

The *avocado* is one of the most sensitive fruits to radiation. Depending on variety and geographic origin, doses of 20–100 Gy have been reported as optimal for delay of ripening, while doses higher than 100–200 Gy have caused skin blemishes and internal discoloration [21].

The extension of marketable shelf life in these and some other fruits is in the range of 3–5 days, and it is difficult to achieve this consistently because the success of the treatment depends much on the fruit's stage of maturity at the time of irradiation. Considering the cost of irradiation and the time and expense required for transporting the fruits to the irradiator, it appears questionable that delay of maturation by irradiation of these fruits will be applied on a large scale.

More promising is the delay of maturation in cultured *mushrooms*. Most studies have been done on *Agaricus bisporus*, which is preferred in an immature (button) stage, with the cap unopened, the gills not visible and the stem plump rather than elongated. Within a day at 10°C, two days at about 4°C, and 5 days at 0°C, quality deteriorates, with opening of the cap, elongation of the stem and darkening of the gills, which now become visible [22]. A dose of about 1 kGy applied soon after picking at the closed button stage prolongs shelf life at 10°C from 1 day to 5 or 6 days. Most authors agree that irradiation has no adverse effects on sensory quality. Thus, in a study reported by Kovacs and Vas in Hungary [23], a panel of judges could not detect any change in the odor, flavor, and texture of irradiated mushrooms, either immediately after treatment or during storage.

Commercial irradiation of fresh mushrooms was carried out for some time in the Netherlands in the early 1970s. The irradiated mushrooms sold well until the supermarket chain selling them began to label the bins with irradiated mushrooms as "irradiated," those with unirradiated mushrooms as "fresh." From then on consumers preferred to buy the unirradiated product and the sale of the irradiated mushrooms soon came to a halt.

C. Insect Disinfestation

Insect control in stored *grain* and *grain products*, in stored *fruits*, *nuts*, *dried fruits*, and many other foodstuffs is usually achieved by fumigation. The use of some fumigants such as ethylene dibromide and ethylene oxide has been banned or severely restricted in most countries because of health concerns. Other fumigants such as hydrogen phosphide and methyl bromide are still in use although they have serious limitations. They require that the grain or other material remain undisturbed for several days before shipment (shorter treatment periods suffice for fruits), they leave undesirable residues on the food, and they cannot penetrate some commodities in sufficient concentration to control all pests. Accidental exposure of workers has occurred, sometimes with lethal consequences.

Alternative methods of insect disinfestation are sought, and treatment with ionizing radiation has long been considered the most promising. As mentioned in Chapter 1, effects of X-rays on the tobacco beetle were studied over 70 years ago. Much has been learned since then about the radiation sensitivity of different insect species and about the factors which modify sensitivity such as sex, age, temperature, dose rate, and host food. (This topic has been discussed in Chapter 4, Sec. VII.)

Much practical experience with radiation disinfestation has been gathered in many countries, and it appears that this is one of the more promising applications of radiation processing. Grain irradiation on a commercial scale became reality in the Soviet Union in 1981 when a radiation facility equipped with two electron accelerators was installed in the port of Odessa, which is reportedly processing 400,000 tons of wheat a year.

The purpose of insect disinfestation can be either the prevention of losses caused by insects in stored grain, flour, cereals, coffee beans, pulses, dried fruits, dried nuts, and other dehydrated food products or the fulfillment of quarantine requirements for fresh fruits. In both cases the necessary radiation dose is in the range of 200–700 Gy. At the lower end of this range emergence of viable adults from eggs or larvae and sterility of adults can be achieved, while doses in the upper end of this range are required for killing off adult insects. As described in Chapter 4, Sec. VII, killing of adult moths requires an even higher dose.

In dehydrated products the dose applied for insect disinfestation generally do not cause deleterious effects to the irradiated food. Literature concerning studies on cereal grains has been reviewed by Lorenz [24], on dried fruits and nuts by Thomas [22]. It is often assumed that radiation treatment is unsuitable for nuts because it might stimulate development of rancidity in the lipids. However, practical experience has not confirmed this assumption. For instance, almonds irradiated with a dose of 1 kGy and stored for up to 12 months could not be distinguished from nonirradiated controls in flavor tests with raw and roasted almonds [25]. Similar conclusions were reached with regard to shelled walnuts stored under different temperature and packaging conditions [26]. No deleterious effects of irradiation even with a dose of 10 kGy could be detected in pistachio kernels [27].

To prevent the spreading of insect pests, many countries require quarantine treatment of imported fruits. Especially the occurrence of fruit flies, such as the Mediterranean, Oriental, Mexican, or Caribbean fruit flies, has repeatedly disrupted trade among countries and among states within the United States. A number of quarantine treatments permitted in the past have recently been banned, fumigation with ethylene dibromide (EDB) being the most prominent example. The remaining chemical fumigants, such as methyl bromide or hydrogen cyanide, and cold treatments or heat treatments are either not allowed for some purposes or are damaging to some commodities. Radiation is effective as a quarantine treatment, and many fruits tolerate the required dose well. Some other fruits are particularly sensitive to radiation (avocado has already been mentioned) and require a careful balancing between a dose that is high enough to fulfill the quarantine requirement and yet low enough to prevent deterioration of quality.

An International Conference on Radiation Disinfestation of Food and Agricultural Products was held in Honolulu, Hawaii in 1983, and the proceedings of this conference provide a good source of information on worldwide activities in this field [28]. Some shorter reviews are also available [29–31].

A. Parasite Disinfection in Foods of Animal Origin

Fresh *pork* in the United States is purchased with the knowledge that it must be heated to a core temperature of at least 77°C (or 170°F), because it may be infected with *Trichinella spiralis*. In spite of this knowledge, trichinosis occurs. Its severity varies from asymptomatic illness to death, depending largely on the numbers of larvae ingested which have not been killed by cooking. Some consumers do not eat pork for fear that microwave heating—or other heat treatments—might be insufficient. Irradiation with a dose of 300 Gy would inactivate the larvae of the parasite and irradiated pork could be sold as "trichina-safe." A program supported by the National Pork Producers Council, the USDA Animal Parasitology Institute and the U.S. Department of Energy has demonstrated the technical and economic feasibility of pork irradiation for this purpose [32]. The cost of the treatment was estimated at $12.30/t or $0.95/hog for a facility processing 1000 hogs/day; this would come down to $4.20/t or $0.32/hog for a facility processing 6000 hogs/day [33].

Other diseases arising from consumption of undercooked or raw pork are cysticercosis, caused by the pork tapeworm, and toxoplasmosis, caused by *Toxoplasma gondii*. The latter organism can also be transmitted through other species, but it has been estimated that pork causes half to three quarters of the cases occurring in the United States [34]. As described in Chapter 4, Secs. V and VI, low-dose irradiation could eliminate these risks.

Infections caused by consumption of undercooked *beef* containing cysts of the beef tapeworm are rare in the United States, but very common in other parts of the world. In certain areas of Africa, 20–30% of the human population are infected. A dose of 400 Gy renders the cysts incapable of development in man.

A number of other parasitic protozoa and helminths, some of which are of enormous public health significance in tropical countries, have also been studied with regard to their radiation sensitivity. It appears that all of them could be inactivated by low-dose irradiation without affecting the sensory quality of the treated meat.

III. MEDIUM-DOSE APPLICATIONS

Populations of bacteria, molds, and yeasts on fresh *fruits* and *vegetables* can be eliminated or reduced by irradiation with doses in the range of 1–3 kGy to retard spoilage caused by these organisms. However, many commodities will show radiation-induced damage at this dose level, such as surface pitting, mottling, darkening, softening of texture, or altered flavor. Maxie and co-workers [36], after many years of work in this area at the University of California in Davis, came to the conclusion that strawberries were the only domestic commodity with even a remote potential for commercial irradiation because other fruits and vegetables could not tolerate the radiation dose needed for disease control. This pessimistic evaluation was perhaps necessary to counterbalance overly optimistic conclusions which had come from other authors. In the meantime commercial irradiation of strawberries with a dose of 2 kGy has become a reality in South Africa. Shelf life at refrigeration temperature can be extended by about 5 days, and in situations where long distances between production area and city centers have to be overcome this may be a decisive benefit. With the Dukat variety of strawberries a shelf life extension of 9 days was achieved with a dose of 2.5 kGy in experiments reported from Poland [36a].

It has also been shown that the radiation dose needed for disease control can be considerably lowered by combining the radiation treatment with hot water dipping. The two treatments act synergistically on rot-causing fungi, and particularly for some subtropical fruits such combination treatments seem to be advantageous. Papayas and mangoes irradiated with a dose of 750 Gy and hot water dipped (50°C for 10 min) kept much better than control fruits or fruits receiving only the radiation or only the hot water treatment [37]. At this dose level the radiation treatment is actually in the low-dose range.

Radiation treatment with a dose of 1 kGy, just at the border line between the low-dose and the medium-dose ranges, was used by Langerak in the Netherlands [38] for extending shelf life of prepackaged leaf vegetables stored at 10°C. As demonstrated in Figure 5, irradiation of endive (*Chichorium endiva* L.) in unperforated packages assured good quality (8 on a 10-point scale) even after 5 days, while the quality of the unirradiated control fell below that level after 1 day in perforated and after 2 days in unperforated packages. Irradiation also improved hygienic quality by eliminating potentially pathogenic microorganisms (*Enterobacteriaceae*). The favorable results obtained in this study are clearly due to the combination of ir-

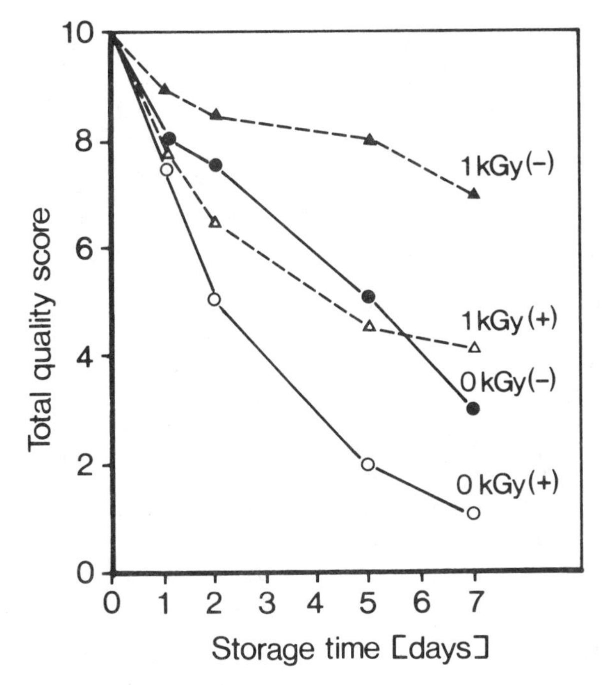

Figure 5 Effect of irradiation (1 kGy gamma rays) and packaging on the total quality of endive packed in perforated (+) or nonperforated (−) polyethylene bags and stored at 10°C. Total quality is based on sensory evaluation of visible decay, odor, and color. (Adapted from Ref. 38.)

radiation and airtight packaging. As Langerak points out, the low oxygen and high carbon dioxide atmosphere, caused by the respiratory activity of the vegetable, prevented discoloration, diminshed loss of vitamin C, and helped to retard senescense.

Dehydrated vegetables, spices, herbs, and other dry ingredients can be irradiated with doses in the range of 3−10 kGy in order to improve their hygienic quality [39, 40, 40a]. Radiation sensitivity of these dry products is much less than that of plant tissues having a high moisture content, and radiation-induced quality losses are not observed in most dry ingredients even at the 10 kGy level. The fresh plants from which these dry products are prepared inevitably contain microorganisms from the soil. During the drying process these organisms can multiply, and the end product often contains 10^6 microorganisms ("total viable counts") per gram. Black pepper and some other spices such as allspice, caraway, cumin, celery seed, paprika, and onion powder may have 10^8 counts/g [39]. When such products are used as ingredients in the manufacture of processed foods, and the manufacturing process does not include a sterilizing step, foods or meals prepared with such ingredients can spoil quickly. In many cases contaminated seasoning is responsible for the spoilage of canned meats or may cause defective sausage products [41]. The contaminating microorganisms may include pathogens, i.e., organisms capable of causing disease in man. For instance, examination of 357 samples of 47 different dried food products in Germany in the period 1980–1983 revealed contamination with salmonella species in 29 samples, or 8.1% [42]. Epidemics of salmonella infection in Canada and Norway have been traced to the use of contaminated white pepper and black pepper [43]. A recent outbreak of salmonellosis in Sweden was also caused by contaminated white pepper [43a].

Decontamination of dry ingredients by heat treatments is usually not psosible because of the resulting quality loss. Fumigation with ethylene oxide was routinely practiced for many years, but has recently been banned in most countries since ethylene oxide has been recognized as a potential carcinogen. Irradiation remains as the only satisfactory treatment. As far as sensory quality of the product is concerned, some spices actually rate better after irradiation than after ethylene oxide treatment [44, 45]. In granulated or powdered paprika, irradiation at 6.5 kGy was found to be more effective than ethylene oxide treatment over 48 h at 25–30°C in a sterilizing chamber [46].

Irradiation of spices on a commercial scale is a reality in about a dozen countries, the United States among them. Irradiation of dehydrated onions, asparagus, mushrooms, and some other dried vegetables is carried out on an increasing scale in some of those countries. Irradiation of gum arabic and some other gums and enzyme preparations and of herbal teas is also applied with the aim of improving the hygienic quality of such products.

Radiation processing of *meat and poultry* with doses in the range of up to 10 kGy can have two principal aims. One, called radurization, is the improvement of shelf life at refrigeration temperature, the other, called radicidation, is the prevention of food poisoning by destruction of pathogenic microorganisms present in the meat, such as salmonella. (More on the origin and use of these terms in Chapter 4, Sec. I.) Successful radurization of meat has been achieved [47], but there seems to be little interest in applying irradiation for this purpose in industrialized countries, where efficient distribution chains exist and where deep-freezing is available for extended storage of meat. The situation may be quite different in developing countries, where reliable transportation systems and freezing facilities are often not available.

Radicidation of meat and poultry is of potential interest in all countries because meat and poultry have been recognized as major sources of foodborne infections in all countries where good health statistics exist [48]. Various species of *Salmonella* and of *Campylobacter* are the most important but not the only organisms responsible [43]. Both are rather easily destroyed by radiation, and various public health experts have advocated irradiation as a means of eliminating these and other pathogens from foods of animal origin [48, 49]. For example, it has been suggested that the salmonellosis problem in Canada be attacked by irradiating essentially all of the poultry produced in that country at a dose of 3 kGy and at an estimated cost of 3–5 cents/kg [50]. Costs of foodborne diseases and possible public health benefits from radicidation of chicken, pork, and beef in the United States have been outlined by Morrison and Roberts [34].

There seems to have been only one instance up until now where proposals of this nature have become commercial reality. A poultry processor in Vannes, France, has installed an electron accelerator for the irradiation of frozen deboned chicken meat. The facility went into production at the beginning of 1987 and is processing 7000 t of chicken meat per year at a dose of 3 kGy [50a].

The situation with regard to *seafood* is similar to that described for meat and poultry: not much interest in radurization, at least not in industrialized countries, but considerable interest and some actual use of radicidation. Especially shellfish is often contaminated with pathogenic organisms such as *Salmonella, Vibrio parahaemolyticus*, and *Shigella*. A number of disease outbreaks with high numbers of persons infected and some with a relatively high fatility rate have been reported in recent years as having been caused by consumption of contaminated crustaceans or molluscs. For instance, 14 deaths were attributed to *Shigella* in a shrimp cocktail in the Netherlands in December 1983 [43]. Irradiation of imported frozen shrimps is now carried out commercially in the Netherlands. Logistic/economic aspects of radiation processing of fishery products in the United States have been presented by Giddings [51].

The shrimp industry in northern Europe has shown considerable interest in radiation processing as a means of extending refrigerated shelf life of shrimps without chemical preservatives (radurization) and as a means of reducing hygienic risks (radicidation). This interest is related to the small size of European brown shrimp (*Crangon vulgaris*) and to the difficulties encountered when attempts are made to peel brown shrimp by machine. Only some prototypes of peeling machines are available, and they do not meet practical requirements. Almost all peeling is done by hand in private homes. The time required for home peeling and for transportation to and from the homes means reduced shelf life in trade. Exposure to pathogenic microorganisms in the homes is inevitable. In order to avoid quick spoilage and hygienic risks, preservative mixtures containing benzoic acid and citric acid are used. A coordinated research program in

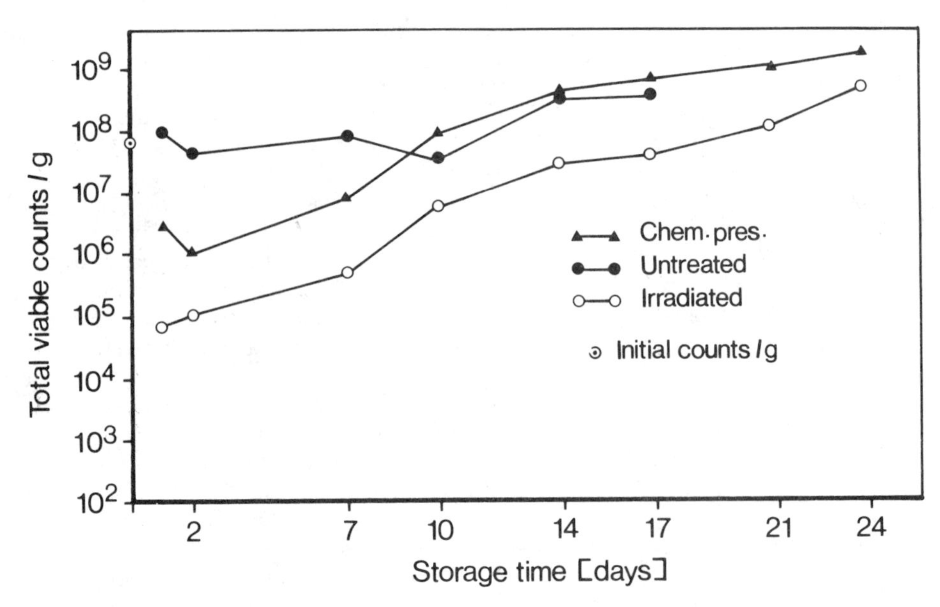

Figure 6 Effect of irradiation (1.3 kGy gamma rays) and of chemical preservation (1% benzoic acid) on growth of microorganisms in brown shrimp packed in polyethylene bags and stored at 0°C. (Adapted from Ref. 52.)

several countries of the European Community has demonstrated radiation processing to be a superior alternative. Some of the results are shown in Figure 6. The untreated raw material had almost 10^8 total viable counts/g. Chemical preservation as commercially practiced reduced this to 10^6 but within about 10 days of refrigerated storage the level of 10^8 was reached again. Irradiation with a dose of 1.3 kGy after manual peeling reduced the total counts to about 10^5/g, and the level of 10^8/g was reached after about 21 days. The advantage of irradiation was even more convincing with regard to the presence of pathogenic staphylococci (not shown in Fig. 6). After 17 days none were found in the irradiated sample, 1.9×10^4 in the chemically preserved sample, and 6.9×10^4 in the untreated sample [52]. Logistic/economic considerations indicated that irradiation of brown shrimp was commercially feasible [53], but the process is not yet used in practice.

Promising results with radicidation treatments of frozen whole *eggs*, *egg powder*, *egg albumen*, *blood plasma*, frozen *froglegs*, and some other potentially contaminated products have been described by various authors, but no information is available as to what extent any of these items are actually irradiated in countries where this is permitted.

Radurization and radicidation treatments may produce additional benefits, as when *soy sauce* is treated with a dose of 5 kGy. According to Chinese researchers this not only eliminates pathogenic *Enterobacteriaceae* and reduces the total bacterial count to below the national standard of 5×10^4/ml but also significantly improves the sensory quality [54]. A process developed in France uses a dose of 3 kGy to achieve preservation of *garlic paste* without addition of chemical preservatives [55].

One aspect that could become important where the cost of electrical energy is high or in tropical countries, where the energy consumption for freezing is high, is the low energy consumption for radurization and refrigerated storage as compared to freezing and frozen storage. The following case example may demonstrate this.

Both in home cooking and restaurant cooking, *french fried potatoes* are now usually made not from raw potatoes but from a convenience product offered by the potato processing industry. The steps used to prepare the prefabricated product involve washing, peeling, and slicing, followed by a heat treatment (blanching) to inactivate enzymes which otherwise would cause browning. The blanched slices are then packaged, frozen in a blast freezer, and placed in cold storage at −26°C until consumption. If the blanched slices are irradiated they can be kept at refrigeration temperature (0°C) in lieu of freezing and frozen storage. The potato processors of the Pacific Northwest produced 3.11 billion pounds of frozen potatoes for french fried potato preparation in 1980, and on that basis a study carried out by CH2M-HILL Company estimated potential energy savings of 300 million kWh if radurization and refrigerated storage were used instead of freezing (Table 4). In tropical regions the energy consumption for freezing and frozen storage is much higher than the values assumed in Table 4, and the benefit of irradiation plus refrigerated storage is correspondingly higher.

Instead of blanching the potato slices before freezing one can also carry out the deep-frying step directly and freeze the ready-prepared product. This obviates the deep frying in the kitchen. All that is required is heating in a microwave oven. Here again radurization would make freezing and frozen storage unnecessary. As can be seen from Figure 7, the unirradiated french fried potatoes can be stored at 4°C for about 10 days, while irradiation with a dose of 5 kGy allows about 25 days of storage at this temperature. Off-flavors were noted when doses higher than 5 kGy were used. No increased peroxide values were found in this dose range, even after 25 days of storage [57].

Table 4 Potential Energy Savings from Use of Irradiation in the Frozen French Fried Potato Industry

Energy requirement (kWh/lb)	Irradiation process	Freezing process
for radurization	0.0025	—
for heat treatment (blanching)	0.1238	0.1238
for freezing to −26°C	—	0.0459
for storage (42 days at 0°C)	—	0.1031
for storage (42 days at −26°C)	0.0495	—
	0.1761	0.2728
Energy cost at 6 cents/kWh	0.0106 $/lb	0.0164 $/lb
Total energy consumed for 3.11 billion pounds of product	547.7 million kWh	848.5 million kWh
Total energy conserved	300.8 million kWh	
Total energy cost savings	$18,050,000.—	

Source: Ref. 56.

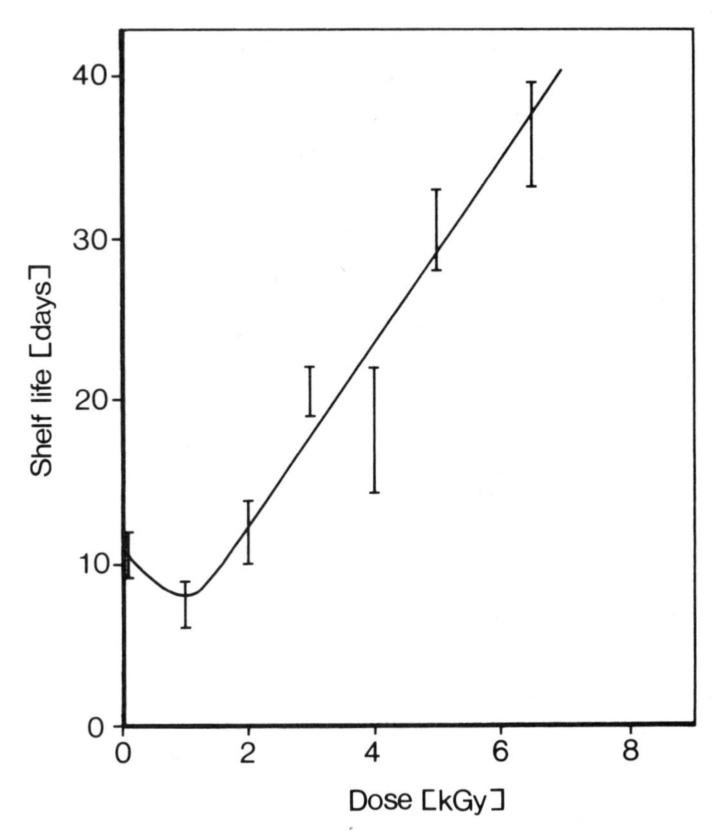

Figure 7 Effect of electron irradiation on refrigerated shelf life (4°C) of prefried french potatoes sealed in polyethylene bags. Shelf life, based on sensory evaluation by a trained panel, was defined as the time when a quality score of below 6 on a 9-point scale (0 = lowest, 9 = highest quality) was reached. (Adapted from Ref. 57.)

IV. HIGH-DOSE APPLICATIONS

As indicated in the previous section, many foods, especially fresh fruit and vegetables, tolerate only low-dose or medium-dose applications without loss of quality. However, some foods, such as meat, poultry, and some seafood items, tolerate high-dose applications if certain precautions are taken. Commercial sterility, that is, the practically complete absence of viable bacteria, yeasts, and molds, can be achieved by doses in the range of 25–45 kGy. Autolytic enzymes present in all raw foods are resistant to these radiation doses and must be inactivated by a mild heat treatment (blanching

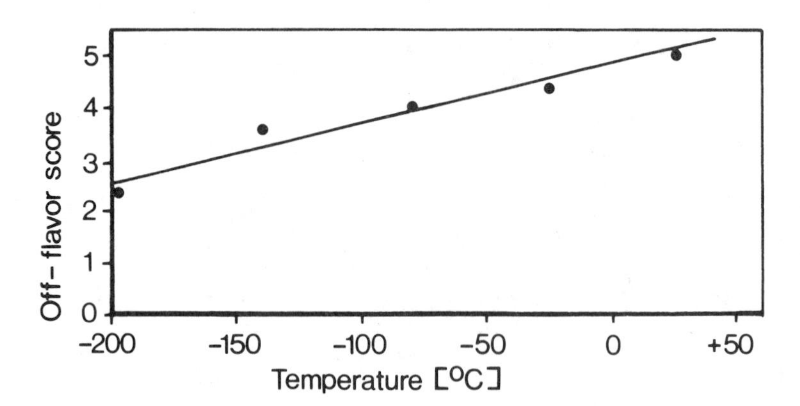

Figure 8 Modifying effect of temperature during irradiation on radiation-induced off-flavor in beefsteak. Radiation dose 60 kGy. Six-point scale: 1 = no irradiation off-flavor, 6 = very pronounced off-flavor. (Adapted from Ref. 58.)

at about 70°C) if long-term shelf stability without refrigeration is to be attained. In order to prevent off-flavors, oxygen must be excluded by vacuum packaging in metal cans or in laminated flexible pouches. Furthermore, it is usually necessary to irradiate the product while frozen at temperatures of −20 to −40°C. The beneficial effect of low temperature during irradiation is demonstrated in Figure 8 for beefsteak irradiated with the very high dose of 60 kGy.

Practically all of the work on radiation-sterilization was carried out by the U.S. Army's laboratories, first at the Quartermaster Food and Container Institute in Chicago, Illinois (1953−1962), then at Natick, Massachusetts (1962−1980). Some of the successfully developed radiation-sterilized products were beef, pork and lamb roasts, beef patties, ham, bacon, corned beef, pork sausage, cooked salami, chicken, turkey slices, shrimp, and codfish cake. These products received good ratings in panel tests, and some of them were part of the meals consumed by American astronauts on the Apollo 17 flight to the moon in December 1972 and on the joint flight in space with Soviet cosmonauts in 1975 (Apollo-Soyuz Test Program). All irradiated foods eaten on these space flights were considered to be highly acceptable [59]. Many visitors to the Natick Laboratories had the opportunity to taste these foods and can attest to the rating of "highly acceptable." Sensory evaluation was carried out not only immediately after radiation processing but also after 2−5 years of storage at ambient temperature, and practically no loss of quality was noted.

The interest of the Army in development of shelf-stable meat products was at its peak during the Vietnam war. The prospect of being able to ship meat to troops in a tropical region without the need for freezing or even refrigeration must have been very appealing to the quartermaster. The big stumbling block to actual large-scale use of radiation-sterilized foods was the lack of approval by the Food and Drug Administration. The Army financed massive test programs to evaluate the wholesomeness of radiation-sterilized beef and chicken, but usable results did not become available until 1985 (see Chapter 5). By that time the Army had lost interest. What remained of the research activities in this field has been taken over by the Department of Agriculture's Eastern Regional Research Center in Philadelphia in 1980.

What chances radiation-sterilized foods will have in the civilian market once FDA approval has been granted is difficult to say. The cost of a process which requires not only a high radiation dose but also vacuum packaging, enzyme inactivation by heating, and freezing to a temperature of −20 to −40°C during irradiation must be considerable. For use on space flights, and by hikers, mountain climbers, or around-the-world sailors even a relatively high price may be acceptable in view of the great advantages: better quality than heat-sterilized meat products, no need for temperature control as in frozen meat. Chances that radiation-sterilized food items may become part of the normal grocery-line depend partly on the further development of energy costs. Display cases for frozen food use almost one third of a typical supermarket's entire power requirement. Faced with steadily rising utility bills, supermarket owners may eventually see a financial advantage in replacing some of the frozen meats by shelf-stable meats.

For one important purpose, although in small quantities, radiation-sterilized foods have been used for many years in several countries to feed hospital patients whose immune system is suppressed. The immune response of cancer patients is weakened and is additionally compromised by anticancer therapies. Patients receiving bone marrow transplants for treatment of leukemia and aplastic anemia are given whole-body irradiation and immunosuppressive drugs. These patients are highly vulnerable even to otherwise harmless organisms present in everyday foodstuffs, which are not sterile. As described by Aker [60], the Fred Hutchinson Cancer Research Center in Seattle has used radiation-sterilized meals or components of meals to protect such vulnerable patients since 1969. Foods sterilized with a dose of 35 kGy include bread and noodles, pastry products, cereals, dry beverages (instant coffee, tea, lemonade, bouillon cubes, etc.), snacks, candies, nutritional supplements and condiments (salt, sugar, pepper, jam, catsup, etc.). The irradiated items are not offered instead of heat-sterilized items but in addition to them. The intention is to provide the patients with a more varied, more balanced, more appealing choice of foods. In Saudi Arabia, Grecz and co-workers have focused their efforts on irradiation of local ethnic foods such as Arabic bread [61] and dates [62] which would make their patients more comfortable than a monotonous diet of standard canned foods. According to Grecz [61] "industrial sterility," defined as <10 cells/g, is sufficient for this purpose. The required radiation dose is then considerably lower than the 35 kGy used in Seattle.

V. MISCELLANEOUS APPLICATIONS

Many advantageous uses of food irradiation have been reported which are not related to any of the effects mentioned in Table 1 or in the previous sections of this chapter. Only a few examples can be mentioned here in order to illustrate the variety of different applications. Whether any of the reported beneficial effects of irradiation are substantial enough to justify the expense of irradiation remains to be seen.

Low doses of radiation were reported to improve the malting properties of barley and some other grains. For instance, a dose of 18 Gy raised diastatic power, beta-amylase, and alpha-amylase in Nigerian sorghum [63].

A number of studies have revealed shortened cooking time and improved rehydration in pulses and dried vegetables irradiated with a dose of about 10 kGy [64, 65].

Juice yield from grapes and other fruits was found to be increased by irradiating the fruits prior to pressing with a dose of about 10 kGy [65].

Flatulence-causing oligosaccharides (stachiose, raffinose) in certain legumes—soybeans, navy beans, red gram, green gram—can be effectively eliminated by irradiating the germinated legume seeds with doses in the range of 2.5–10 kGy [66, 67]. Irradiation of alginate and carrageenan with a dose of 50 kGy accelerated the clarification rate of turbid sake to which these were added [68].

Frozen shovel-nosed lobsters (*Ibacus peronii*) often exhibit a garliclike off-odor which is caused by traces of bis-(methylthio)-methane. This compound can be destroyed and the off-odor can be removed by irradiation with a dose of 25 kGy [69].

Irradiation not only allows manufacture of cured meat products with greatly reduced levels of added nitrite but also actively decreases levels of nitrite and nitrosamines present in these meat products [70, 71].

Such a list of beneficial effects of irradiation should by no means create the impression that irradiation is a micracle method. Many attempted radiation treatments have failed. Even very low doses of radiation cause repulsive off-flavors in some foods, for instance in

Table 5 Commercial Uses of Food Irradiation

Nation	Company (city)	Food item (tonnage)	Starting date	Reference
Argentina	Nat. Atomic Energy Commission (Buenos Aires)	Spices, cocoa powder (50 t/y)	1986	72
Belgium	IRE (Fleurus)	Spices, dehydrated vegetables, deep-frozen foods, gum arabic (10,000 t/y)	1981	74
Brazil	EMBRARAD (Sao Paulo)	Spices, dehydraged vegetables (120 t/y)	1985	75
Canada	Canadian Irradiation Centre (Laval, Quebec)	Spices	1987 (?)	73
	Various companies primarily irradiating medical disposables			
Chile	CCHEN (Santiago)	Onions, potatoes, spices, and dehydrated vegetables (500 t/y)	1983	72
China, People's Rep. of	Nine multipurpose irradiators (Beijing, Chengdu, Jinan, Lanzhou, Nanjing, Shanghai, Shenzhen, Tianjin, Zhengzhou)	Various foods	1985	—
China, Rep. of	Inst. of Nuclear Energy Research (Lung Tan)	Spices, dehydrated mushrooms	1986 (?)	—
Cuba	Food Industries Research Inst. (El Guatao, Havana)	Potatoes (1,000 t/y), onions (100 t/y)	1987	76

Table 5 Continued

Nation	Company (city)	Food Food item (tonnage)	Starting date	Reference
Denmark	Riso National Laboratory	Spices	1986	72
Finland	KOLMI-SET Oy (Ilomantsi)	Spices	1986	72
France	Conservatome (Lyon)	Spices (500–600 t/y)	1982	77
	CARIC (Paris)	Spices, poultry (300 t/y)	1986	77
	SPI (Vannes)	Deboned chicken meat (7,000 t/y)	1987	50a
	ORIS (Nice)	Spices (200 t/y)	1988	77
German Democratic Republic	Centr. Inst. Isotope Rad. Res. (Weideroda)	Onions (1,000 t/y), garlic (10 t/y)	1982 1984	77a
	Queis Agric. Coop. (Spickendorf)	Onions (5,000 t/y), egg powder (70 t/y)	1986 1987	
	VEB Prowiko (Schoenebeck)	Enzyme solutions (300 t/y)	1986	
Hungary	AGROSTER (Budapest)	Spices (400 t/y)	1982	78
Israel	Sorvan Radiation Ltd. (Yavne)	Spices (120 t/y)	1986	72
Japan	Shihoro Agric. Coop. (Hokkaido)	Potatoes (30,000 t/y)	1973	79
Korea	Korea Advanced Energy Res. Inst. (Seoul)	Garlic powder	1985	72
Mexico	Centro Nuclear de Mexico (ININ)	Spices, onion and garlic powder, peanuts, cocoa powder, cocoa nut, starch, rice flour, dehydrated mushrooms, pancreatic enzyme	1984 (?)	80
	Instituto de Ciencias Nucleares (ICN-UNAM)			80a

Netherlands	Gammaster (Ede)	Spices, dehydrated vegetables, dehydrated blood, egg powder, starch products; frozen poultry, game, froglegs, seafood	1978	81
	Pilot Plant for Food Irradiation (Wageningen)	Spices (together 18,000 t/y)	1978	
Norway	Institute for Energy Technology (Kjeller)	Spices	1982	72
South Africa	Nuclear Development Corporation (Pelindaba)	Fresh fruits, spices, dehydrated vegetables, herbal tea, mango pickle (achar), sorghum malt, cane yeast, chicken, fish (together 30,000 t/y)	1981	82
	ISO-STER (Kempton Park)		1982	
	HEPRO (Tzaneen)		1981	
	HEPRO (Cape Town)		1986	
Soviet Union	Odessa Port Elevator RDU (Odessa)	Cereal grains (400,000 t/y)	1983	72
Thailand	OAEP (Bangkok)	Onions, fermented sausages (600 t/y)	1987	72
United States	Several companies primarily irradiating medical disposables	Spices (3,000 t/y)	1984	72
Yugoslavia	Ruder Boskovic Institute (Zagreb)	Spices	1985	72
	Boris Kidric Institute (Belgrade)	Spices	1987 (?)	

beer and milk. In other cases irradiation caused undesirable effects on texture or color. This does not mean that irradiation should be condemned. It simply means that the usefulness of irradiation, like that of canning, freezing, drying, extrusion cooking or any other process, is limited to certain products and certain purposes.

VI. ACTUAL COMMERCIAL USES OF THE PROCESS

The following overview (Table 5) is based primarily on information provided by the International Atomic Energy Agency [72], supplemented by more recently published papers mentioned in the table, and by communications obtained directly from persons who are involved in radiation processing of food in various companies or institutes. Countries not listed either have no commercial food irradiation or have not made information available on existing activities.

Considering the variety of commodities and the tonnage irradiated, the Netherlands and South Africa must be considered the leaders in food irradiation processing, with Belgium following closely behind. The Soviet Union and Japan show a sizeable tonnage—but this is limited to wheat and potatoes, respectively, and there are no indications that either of these countries will progress to other irradiated commodities in the foreseeable future. The situation of food irradiation in the People's Republic of China is not clear. Nine industrial irradiators have been constructed in the last few years, but it is not known to what extent they are utilized for irradiation of medical disposables and to what extent for food, nor is it known which foods are actually irradiated on a commercial scale. During a personal visit at the radiation facility in Chengdu in 1988 the author was given the information that ton quantities of pork sausages and of brandy made from sweet potatoes are irradiated there for sale to the public. According to the operators of the Chengdu plant the purpose of sausage irradiation is threefold: elimination of pathogenic organisms, extension of shelf-life, and reduced use of nitrite in manufacture. The purpose of irradiation of sweet-potato brandy is improvement of sensory quality. The untreated brandy has an unpleasant bitter taste which disappears after irradiation.

REFERENCES

1. Josephson, E. S., and M. S. Peterson, eds., *Preservation of Food by Ionizing Radiation*, 3 volumes. CRC Press, Boca Raton, Florida, 1982/3.
2. Urbain, W. M., *Food Irradiation*. Academic Press, Orlando, Florida, 1986.

3. Sparrow, A. H., and E. Christensen, Improved storage quality of potato tubers after exposure to Co^{60} gammas, *Nucleonics*, 12(8):16 (1954).
4. Matsuyama, A., and K. Umeda, Sprout inhibition in tubers and bulbs, in *Preservation of Food by Ionizing Radiation*, vol. 3, E. S. Josephson and M. S. Peterson, eds. CRC Press, Boca Raton, Florida, 1983, p. 159.
5. Thomas, P., Radiation preservation of foods of plant origin. I. Potatoes and other tuber crops, *Crit. Revs. Food Sci. Nutrit.*, 19:327 (1984).
6. Thomas, P., S. Adam, and J. F. Diehl, Role of citric acid in the after-cooking darkening of gamma-irradiated potato tubers, *J. Agric. Food Chem.*, 27:519 (1979).
7. Diehl, J. F., Experiences with irradiation of potatoes and onions, *Lebensmittel-Wiss. Technol.*, *10*:178 (1977).
8. Grünewald, T., Consumer Test with Irradiated Potatoes (in German), Reports of the Federal Research Center for Food Preservation, BFL 1/1970.
9. Dallyn, S. L., R. L. Sawyer, and A. H. Sparrow, Extending onion storage life by gamma irradiation, *Nucleonics*, 13(4):48 (1955).
10. Thomas, P., Radiation preservation of foods of plant origin. II. Onions and other bulb crops, *Crit. Revs. Food Sci. Nutrit.*, 21:95 (1984).
11. Grünewald, T., Studies on sprout inhibition of onions by irradiation in the Federal Republic of Germany, in *Food Preservation by Irradiation*, Proceedings of a Symposium held in Wageningen, November 1977. Internat. Atomic Energy Agency, Vienna, 1978, p. 123.
12. Curzio, O. A., and C. A. Croci, Extending onion storage life by gamma-irradiation, *J. Food Processing Preserv.*, 7:19 (1983).
13. Kwon, J. H., M. W. Byun, and H. O. Cho, Effects of gamma irradiation dose and timing of treatment after harvest on the storeability of garlic bulbs, *J. Food Sci.*, 50:379 (1985).
14. Curzio, O. A., and C. A. Croci, Radioinhibition process in Argentinian garlic and onion bulbs, *Radiat. Phys. Chem.*, *31*: 203 (1988).
15. Lu, J. Y., S. White, P. Yakubu, and P. A. Loretan, Effects of gamma radiation on nutritive and sensory qualities of sweet potato storage roots, *J. Food Quality*, 9:425 (1986).
16. Wang, U. P., C. Y. Lee, J. Y. Chang, and C. L. Yet, Gamma-radiation effects on Taiwan sweet potatoes (in Chinese), *J. Chin. Agric. Chem. Soc.*, 20:133 (1982) (as quoted in Food Sci. Techn. Abstr. No. 11J1718 (1983)).
17. Adesuyi, S. A., and J. A. Mackenzie, The inhibition of sprouting in stored yams, *Dioscorea rotunda* poir, by gamma radiation and chemicals, in *Radiation Preservation of Food*, Proceedings

of a symposium held in Bombay, Nov. 1972. Intern. Atomic Energy Agency, Vienna, 1973, p. 127.
18. Khan, I., and M. Wahid, Feasibility of radiation preservation of potatoes, onions and garlic in Pakistan, in *Food Preservation by Irradiation*, Proceedings of a symposium held in Wageningen, November 1977. Internat. Atomic Energy Agency, Vienna, 1978, p. 63.
19. Thomas, P., Radiation preservation of foods of plant origin. III. Tropical fruits: bananas, mangoes, and papayas, *Crit. Revs. Food Sci. Nutrit.*, 23:147 (1986).
20. Brodrick, H. T., R. S. Thord-Gray, and G. J. Strydom, The radurisation of bananas under commercial conditions. II. Shelf- or market-life extension, *Citrus and Sub-Tropical Fruit Journal* (603):4 (1984).
20a. Spalding, D. H., and D. L. von Windeguth, Quality and decay of irradiated mangoes, *Hort. Science*, 23:187 (1988).
21. Thomas, P., Radiation preservation of foods of plant origin. IV. Subtropical fruits: citrus, grapes, and avocados, *Crit. Revs. Food Sci. Nutrit.*, 24:53 (1986).
22. Thomas, P., Radiation preservation of foods of plant origin. VI. Mushrooms, tomatoes, minor fruits and vegetables, dried fruits, and nuts, *Crit. Revs. Food Sci. Nutrit.*, 26:313 (1988).
23. Kovacs, E., and K. Vas, Effect of ionizing radiations on some organoleptic characteristics of edible mushrooms, *Acta Aliment.*, 3:11 (1974).
24. Lorenz, K., Irradiation of cereal grains and cereal grain products, *Crit. Revs. Food Sci. Nutrit.*, 6:317 (1975).
25. O'Mahony, M., S. Y. Wong, and N. Odbert, An initial sensory examination of the effect of postharvest irradiation on almonds, *J. Ind. Irrad. Technol.*, 3:135 (1985).
26. Jan, M., D. Langerak, T. G. Wolters, J. Farkas, H. J. V. D. Kamp, and B. G. Muuse, The effect of packaging and storage conditions on the keeping quality of walnuts treated with disinfestation doses of gamma rays, *Acta Aliment.*, 17:13 (1988).
27. Kashani, G., and L. R. G. Valadon, Effects of gamma irradiation on the lipids, carbohydrates and proteins of Iranian pistachio kernels, *J. Food Technol.*, 19:631 (1984).
28. Moy, J. H., ed., *Radiation Disinfestation of Food and Agricultural Products*, Proceedings of an International Conference held in Honolulu, Hawaii, Nov. 1983, Hawaii Inst. of Tropical Agric. and Human Resources. Univ. of Hawaii at Manoa, Honolulu, Hawaii, 1985.
29. Burditt, Jr., A. K., Food irradiation as a quarantine treatment of fruits, *Food Technology* (Chicago), 36(11):51 (1982).
30. Kader, A. A., Potential applications of ionizing radiation in postharvest handling of fresh fruits and vegetables, *Food Technolgy* (Chicago), 40(6):117 (1986).

31. Tilton, E. W., and A. K. Burditt, Jr., Insect disinfestation of grain and fruit, in *Preservation of Food by Ionizing Radiation*, vol. 3, E. S. Josephson and M. S. Peterson, eds. CRC Press, Boca Raton, Florida, 1983, p. 215.
32. Brake, R. J., K. D. Murrell, E. E. Ray, J. D. Thomas, B. A. Muggenburg, and J. S. Sivinski, Destruction of *Trichinella spiralis* by low-dose irradiation of infected pork, *J. Food Safety*, 7:127 (1985).
33. Sivinski, J. S., and R. K. Switzer, Low-dose irradiation: a promising option for trichina-safe pork certificaiton, in *Radiation Disinfestation of Food and Agricultural Products*, J. H. Moy, ed. Hawaii Inst. of Tropical Agric. and Human Resources, Univ. of Hawaii at Manoa, Honolulu, Hawaii, 1985, p. 181.
34. Morrison, R. M., and T. Roberts, *Food Irradiation: New Perspectives on a Controversial Technology*, U.S. Dept. of Commerce, National Technical Information Service, Springfield, VA, 1985.
35. King, B. L., and E. S. Josephson, Action of radiation on protozoa and helminths, in *Preservation of Food by Ionizing Radiation*, vol. 2, E. S. Josephson and M. S. Peterson, eds. CRC-Press, Boca Raton, Florida, 1983, p. 245.
36. Maxie, E. C., N. F. Sommer, and F. G. Mitchell, Infeasibility of irradiating fresh fruits and vegetables, *Hort. Science*, 6(3): 202 (1971).
36a. Zegota, H., Suitability of Dukat strawberries for studying effects on shelf life of irradiation combined with cold storage, *Z. Lebensm. Unters. Forsch.*, 187:111 (1988).
37. Brodrick, H. T., and H. J. van der Linde, Technological feasibility studies on combination treatments for subtropical fruits, in *Combination Processes in Food Irradiation*, Proceedings of a symposium held in Colombo, November 1980. Internat. Atomic Energy Agency, Vienna, 1981, p. 141.
38. Langerak, D. I., The influence of irradiation and packaging on the quality of prepacked vegetables, *Ann. Nutr. Alim.*, 32: 569 (1978).
39. Farkas, J., *Irradiation of Dry Food Ingredients*. CRC Press, Boca Raton, Florida, 1988.
40. Kiss, I., and J. Farkas, Irradiation as a method for decontamination of spices, *Food Revs. Internat.*, 4:77 (1988).
40a. Katusin-Razem, B., S. Matic, D. Razem, and V. Mihokovic, Radiation decontamination of tea herbs, *J. Food Sci.*, 53:1120 (1988).
41. Eiss, M. I., Irradiation of spices and herbs, *Food Technology in Australia*, 36:362 (1984).
42. Bockemühl, J., and B. Wohlers, Concerning the problem of contamination with salmonella of unprocessed dry products in

the food industry (in German), *Zbl. Bakteriol. Parasitenk. Infektionskr. Hyg. Orig.*, B178:535 (1984).
43. Farkas, J., Decontamination, including parasite control, of dried, chilled and frozen foods by irradiation, *Acta Alimentaria*, 16:351 (1987).
43a. Persson, L., Warning against unirradiated spices! (in Swedish), *Läkartidningen*, 85:2641 (1988).
44. Vajdi, M., and N. N. Pereira, Comparative effect of ethylene oxide, gamma irradiation and microwave treatment on selected spices, *J. Food Sci.*, 38:839 (1973).
45. Farkas, J., and E. Andrássy, Comparative analysis of spices decontaminated by ethylene oxide or gamma radiation, *Acta Alimentaria*, 17:17 (1988).
46. Llorente Franco, S., J. L. Giménez, F. Martinez Sánchez, and F. Romojaro, Effectiveness of ethylene oxide and gamma irradiation on the microbiological population of three types of paprika, *J. Food Sci.*, 51:1571 (1986).
47. Urbain, W. M., *Food Irradiation*. Academic Press, Orlando, FL, p. 124.
48. Kampelmacher, E. H., Prospects of eliminating pathogens by the process of food irradiation, in *Combination Processes in Food Irradiation*, Proceedings of a symposium held in Colombo, November 1980. Internat. Atomic Energy Agency, Vienna, 1981, p. 265.
49. Mossel, D. A. A., and H. Stegeman, Irradiation: an effective mode of processing food for safety, in *Food Irradiation Processing*, Proceedings of a symposium held in Washington, D.C., March 1985. Internat. Atomic Energy Agency, Vienna, 1985, p. 251.
50. Ouwerkerk, T., Salmonella control in poultry through the use of gamma irradiation, in *Combination Processes in Food Irradiation*, Proceedings of a symposium held in Colombo, November 1980. Internat. Atomic Energy Agency, Vienna, 1981, p. 335.
50a. Wagner, J., and M. Sillard, Pasteurizing with electrons, *Food Engineerg. Internat.*, 14(2):38 (1989).
51. Giddings, G. G., Radiation processing of fishery porducts, *Food Technology* (Chicago), 38(4):61 (1984).
52. Ehlermann, D., and R. Münzner, Radiation preservation of North Sea shrimps (in German), *Arch. Lebensmittelhyg.*, 27: 50 (1976).
53. Ehlermann, D., and J. F. Diehl, Economic aspects of the introduction of radiation preservation of brown shrimp in the Federal Republic of Germany, *Radiat. Phys. Chem.*, 9:875 (1977).
54. Jingtian, Y., J. Xinhua, G. Guoxing, and Y. Guichun, Studies of soy sauce sterilization and its special flavour improvement by gamma-ray irradiation, *Radiat. Phys. Chem.*, 32:209 (1988).

55. Irigaray, J. L., and J. C. Capelani, Comparison of garlic paste preservation by gamma-irradiation and by chemical treatment (in French), *Ind. Alim. Agric.*, 99:979 (1982).
56. CH2M-HILL Company, *An update on Food Irradiation Technology in the United States.* DOE/USDA/AIBS Workshop on Low-dose Irradiation Treatment of Agricultural Commodities, Washington, D.C., April 1982.
57. Grünewald, T., and G. Rumpf, On treatment of prefried potatoes with ionizing radiation (in German), *Lebensm.-Wiss. Technol.*, 2:86 (1969).
58. Kauffman, F. L., J. W. Harlan, and C. E. Rasmussen, Final report, U.S. Army Natick Laboratory Contract No. DA-19-129-QM-2000 Swift Company, 1964.
59. Josephson, E. S., Radappertization of meat, poultry, finfish, shellfish and special diets, in *Preservation of Food by Ionizing Radiation*, vol. 3, E. S. Josephson and M. S. Peterson, eds. CRC Press, Boca Raton, Florida, 1983, p. 231.
60. Aker, S. N., On the cutting edge of dietetic science, *Nutrition Today*, 19(4):24 (1984).
61. Grecz, N., R. Brannon, R. Jaw, R. Al-Harithy, and E. W. Hahn, Gamma processing of Arabic bread for immune system-compromised cancer patients, *Appl. Environm. Microbiol.*, 50: 1531 (1985).
62. Grecz, N., R. Al-Harithy, R. Jaw, R. Al-Suwaine, M. A. El-Mojaddidi, and S. Rahma, Radiation processing of dates grown in Saudi Arabia, *Radiat. Phys. Chem.*, 31:161 (1988).
63. Uwaifo, A. O., Biological and biochemical changes in two Nigerian species of sorghum (SK 5912 and HP3) following pre-malting gamma-irradiation treatment, *J. Agr. Food Chem.*, 31: 1292 (1983).
64. Rao, V. S., and U. K. Vakil, Effects of gamma-radiation on cooking quality and sensory attributes of four legumes, *J. Food Sci.*, 50:376 (1985).
65. Kiss, I., J. Farkas, S. Ferenczi, B. Kalman, and J. Beczner, Effects of irradiation on the technological and hygienic qualities of several food products, in *Improvement of Food Quality by Irradiation*, Proceedings of a Panel, International Atomic Energy Agency, Vienna 1974, p. 157.
66. Hasegawa, Y., and J. H. Moy, Reducing oligosaccharides in soybeans by gamma-radiation controlled germination, in *Radiation Preservation of Food*, Proceedings of a Symposium held in Bombay, November 1972. International Atomic Energy Agency, Vienna, 1973, p. 89.
67. Rao, V. S., and U. K. Vakil, Effect of gamma-irradiation on flatulence-causing oligosaccharides in green gram (*Phaseolus aureus*), *J. Food Sci.*, 48:1791 (1983).

68. Kume, T., and M. Takehisa, Effect of gamma-irradiation on sodium alginate and carrageenan, *Agric. Biol. Chem.*, 47:889 (1983).
69. Freeman, D. J., M. W. Izzard, and F. B. Whitfield, Removal of garlic-like off-odours from crustacea by gamma-irradiation, *Australian Fisheries*, 44:(4) 35 (1985).
70. Fiddler, W., J. W. Pensabene, R. A. Gates, R. K. Jenkins, and E. Wierbicki, Observations on the use of gamma irradiation to control nitrosamine formation in bacon, in *Food Irradiation Processing*, Proceedings of a Symposium held in Washington, D.C., March 1985. International Atomic Energy Agency, Vienna, 1985, p. 238.
71. Singh, H., Radiation preservation of low nitrite bacon, *Radiat. Phys. Chem.*, 31:165 (1988).
72. IAEA, Commercialisation of food irradiation, *Food Irradiation Newsletter* 10(2):48 (1986) and 12(2):42 (1988).
73. Kunstadt, P., Canadian perspectives on food irradiation, *Radiat. Phys. Chem.*, 31:239 (1988).
74. Constant, R., and J. P. Lacroix, Irradiation in the service of mankind (in French), *Revue IRE Tijdschrift*, 11(1):2 (1987).
75. Hutzler, R. V., and D. M. Vizeu, Multiple uses of JS 7400, *Radiat. Phys. Chem.*, *31*:357 (1988).
76. IAEA, New food irradiation facilities—Cuba, *Food Irradiation Newsletter*, 11(2):52 (1987).
77. Laizier, J., Status report on food irradiation—France, *Food Irradiation Newsletter* 11(2):47 (1987).
77a. Hübner, G., Status of food irradiation in the German Democratic Republic, *AIII-Newsletter*, 19:169 (1988).
78. Kalman, B., Conclusions from the last five years of experiments in the field of food irradiation in Hungary, *Radiat. Phys. Chem.*, 31:225 (1988).
79. Takehisa, M., and H. Ito, Experiences of food irradiation in Japan, *Food Revs. Internat.*, 2:19 (1986).
80. Bustos, E., P. C. Luna, and J. Reyes, Status report on food irradiation—Mexico, *Food Irradiation Newsletter*, 11(2):48 (1987).
80a. Piña, G., Radiation research and processing in Mexico, *AIII-Newsletter*, 19:159 (1988).
81. Leemhorst, J. G., Food irradiation: Status and prospects in Europe, *Radiat. Phys. Chem.*, 31:235 (1988).
82. van der Linde, H. J., and R. A. Basson, Factors determining the viability of radiation processing in developing countries, *Radiat. Phys. Chem.*, 31:257 (1988).
83. Giddings, G. G., Radiation preservation of foods: substerilizing applications, *J. Food Safety*, 5:191 (1983).

84. Wetzel, K., G. Huebner, and M. Baer, Irradiation of onions, spices and enzyme solutions in the German Democratic Republic, in *Food Irradiation Processing*, Proceedings of a Symposium held in Washington, D.C., March 1985. Internat. Atomic Energy Agency, Vienna, 1985, p. 35.
85. van Kooij, J. G., International trends in and uses of food irradiation, *Food Revs. Internat.* 2:1 (1986).

10
Government Regulation of Irradiated Foods

I. UNITED STATES OF AMERICA

The basis for regulating food irradiation in the United States is the Food Additives Amendment to the Food, Drugs and Cosmetic Act of September 6, 1958. In Section 201(s), the Act includes sources of radiation in its definition of a food additive (emphasis added):

> The term "food additive" means any substance the intended use of which results or may reasonably be expected to result, directly or indirectly, in its becoming a component or otherwise affecting the characteristics of any food (including any substance intended for use in producing, manufacturing, packing, processing, preparing, treating, packaging, transporting, or holding food; and *including any source of radiation* intended for any such use), if such substance is not generally recognized, among experts qualified by scientific training and experience to evaluate its safety, as having been adequately shown through scientific procedures (or, in the case of a substance used in food prior to January 1, 1958, through either scientific procedures or experience based on common use in food) to be safe under the conditions of its intended use.

This concept of considering radiation sources as food additives has caused much consternation, especially in foreign countries, where it was often mistakenly stated that FDA defined food irradiation as a food additive although any sensible person must recognize that it is a process rather than an additive. Pauli and Takeguchi [1] of the Food and Drug Administration have presented an enlightening overview of how this concept developed historically. They explain that the inclusion of radiation sources under the food additive definition is not any more illogical than the inclusion of a filter or other equipment that may affect the characteristics of a food. The Agency does not claim that irradiation is a food additive—but it regulates food irradiation in the same way as if it were an additive.

According to Section 402(a)(7), a food shall be deemed to be adulterated if it has been intentionally subjected to radiation, unless the use of the radiation was in conformity with a regulation or exemption in effect pursuant to Section 409.

Section 409(a) states that

> A food additive shall, with respect to any particular use or intended use of such additives, be deemed to be unsafe . . . unless (2) there is in effect, and it and its use or intended use are in conformity with, a regulation issued under this section prescribing the conditions under which such additive may be safely used.

In Section 409(c)(5), the Act describes some of the factors that should be taken into account when the safety of a food additive is considered:

> In determining, for the purpose of this section, whether a proposed use of a food additive is safe, the Secretary shall consider among other relevant factors—
>
> (A) the probable consumption of the additive and of any substance formed in or on food because of the use of the additive;
>
> (B) the cumulative effect of such additive in the diet of man or animals, taking into account any chemically or pharmacologically related substance or substances in such diet;
>
> and
>
> (C) safety factors which in the opinion of experts qualified by scientific training and experience to evaluate the safety of food additives are generally recognized as appropriate for the use of animal experimentation data.

In Chapter 5 some of the difficulties of trying to apply these principles to the safety testing of irradiated foods were mentioned. How could one (A) estimate probable consumption or (B) cumulative effects of any substance formed in or on food because of the use of a radiation source, as long as the knowledge of radiation chemical effects was very limited? How could one (C) consider safety factors without such knowledge? Reading these requirements laid down in Section 409, originally intended for the safety evaluation of intentional food additives and later extended to the safety evaluation of the incidental food additive "any substance formed" in and on irradiated food, one understands that animal feeding studies alone could never satisfy critical interpreters of the Food Additives Amendment. Fortunately, the progress made in the radiation chemistry of foods has changed this situation.

The Act foresees two procedures for obtaining premarket approval for an additive's safe use. In the principal procedure a proponent of such use files a food additive petition, accompanied by sufficient data to demonstrate safety. A second, far less frequently used procedure allows the Secretary of the Department of Health and Human Services to establish regulations on his or her own initiative. Such a government proposal must meet the same standard of demonstrating safety as a petition from industry. The public is given at least 30 days to comment on the proposal, and all substantive comments must be considered.

As mentioned in Chapter 5, FDA has published a number of food irradiation regulations in the 1960s in response to petitions. These regulations provided for the safe use of gamma radiation or electron beam radiation for the processing of wheat, potatoes, and bacon. Petitioners were the Army, the Atomic Energy Commission, industrial firms, or, in the case of wheat irradiation, private citizens. These regulations have been superseded by new regulations published in 1986, this time not based on a petition but on a government proposal [2].

The decision to make this proposal was obviously not taken lightly. FDA had decided in 1979 to reevaluate its approach to regulating food irradiation. It established the Bureau of Foods Irradiated Foods Committee on October 23, 1979, "to provide a total reassessment of all relevant issues applicable to irradiated foods." On March 27, 1981, FDA published an Advance Notice of Proposed Rule-making in the Federal Register. In response the Agency received some 80 comments from consumers, companies, and various organizations. On February 14, 1984, FDA published a Proposed Rule. This time it received over 5000 comments. Only after these had been considered was the Final Rule announced on April 18, 1986. These new regulations specify as permitted energy sources:

Gamma rays from sealed units of cobalt-60 or cesium-137.

Electrons generated from machine sources at energies not to exceed 10 MeV.

X-rays generated from machine sources at energies not to exceed 5 MeV.

They permit manufacturers to irradiate:

Fresh foods for inhibition of growth and maturation at doses not to exceed 1 kGy.

Any food for disinfestation of arthropod pests at doses not to exceed 1 kGy.

Dry or dehydrated enzyme preparations, including immobilized enzymes, for microbial disinfection at doses not to exceed 10 kGy.

Dry or dehydrated aromatic vegetable substances for microbial disinfection at doses not exceeding 30 kGy: culinary herbs, seeds, spices, teas, vegetable seasonings, and blends of these aromatic vegetable substances. Turmeric and paprika may also be irradiated when they are to be used as color additives. The blends may contain sodium chloride and minor amounts of dry food ingredients ordinarily used in such blends.

Pork carcasses or fresh, non–heat-processed cuts of pork carcasses for control of *Trichinella spiralis* at doses of at least 0.3 kGy, and not to exceed 1 kGy.

The regulations require that irradiated foods be labeled to show this fact, both at the wholesale and retail level. The labeling shall bear the following logo

along with either the statment "Treated with radiation" or the statement "Treated by radiation" in addition to information required by other regulations. The logo shall be placed prominently and conspicuously in conjunction with the required statement. For unpackaged foods, the required logo and statement shall be displayed either with the labeling of the bulk container plainly in view or with a counter sign, card, or other appropriate device bearing the information that the product has been treated with radiation. As an alternative, each item, for instance each fruit, amy be individually labeled.

FDA permits manufacturers to include the labeling any phrase such as "treated with radiation to control spoilage" or "treated with radiation to extend shelf life." Additional labeling statements as part of a consumer education effort are also acceptable if they are truthful and not misleading to the consumer. For example, the manufacturer may wish to state that "this treatment does not induce radioactivity."

Similarly, the type of radiation used in the treatment may be indicated by stating "treated with gamma radiation," "treated with ionizing radiation," or "treated with X-radiation."

In the explanatory part of the regulations, FDA repeatedly emphasizes that the labeling requirement is not based on any concern about the safety of the uses of radiation. The Agency points out that it has historically required the disclosure of processing when the process is not obvious to the consumer, as it is in the case of canned and frozen foods. Pasteurized milk, for example, or pasteurized orange juice, or bleached flour must be labeled to indicate that these processes have been applied. Analogously, irradiation of food is a material fact that must be disclosed to the consumer to prevent deception. Whether the labeling information is material under the Act depends not on the abstract worth of the information, "but on whether consumers view such information as important and whether the omission of label information may mislead a consumer" [2].

The Agency's verdict that not labeling irradiated foods would deceive or mislead the consumer is not easy to understand, considering the fact that the Agency did not demand retail labeling when it approved irradiation of wheat and wheat flour in 1963 and of potatoes in 1964. Nor did the Proposed Rule of February 14, 1984, include the requirement of retail labeling. On the contrary, the 1984 Proposed Rule had expressly stated: "The agency has concluded that the information about radiation processing is not material in this sense and therefore need not be provided on the label of retail foods." If the marketing of unlabeled irradiated foods was not considered to deceive or mislead the consumer in 1963 or 1984, why should it be considered deceptive or misleading in 1986? The reason for the Agency's change of mind was apparently the protest expressed by consumer groups after the Proposed Rule was announced in 1984.

Importantly, the labeling requirement applies only to a food that has been irradiated (first-generation food), not to a food that merely contains an irradiated ingredient but that has not itself been irradiated (second-generation food). Some consumer groups had demanded labeling of second- and even third-generation foods, and the Agency has had the wisdom not to follow this request. In contrast, the Danish Government has linked its permission for spice irradiation with an unlimited labeling requirement. This means, for instance, that a salami containing irradiated pepper must be so labeled (second generation), and even if a pizza bakery uses a few slices of such salami (third generation), it must inform its customers accordingly. This places an unreasonable burden on the food trade and also on the authorities charged with control of compliance. Displaying a label or a counter sign is only one part of fulfilling such a requirement. Since there is no method available for recognizing whether or not the pepper in a salami has been irradiated, control of compliance would only be possible if the manufacturer of the salami kept records of how much irradiated pepper he bought and in which products he used it. He would have to keep the irradiated pepper separate from nonirradiated pepper, probably locked, to avoid accidental use in unlabeled products. The whole procedure would be so cumbersome that the Danish food industry has preferred not to make any use of the permission granted in 1985. Spices are being irradiated in Denmark, but only for export to countries with more realistic regulations.

In response to concerns raised by the New York State Consumer Protection Board, FDA reconfirmed in June 1988 its opinion that no need exists for labeling of irradiated ingredients in food products. Saying that deliberations on the proper labeling for irradiated ingredients had considered various options, FDA concluded:

> The Agency did not find an approach, however, that would supply useful information to the consumer sufficient to justify the cost, difficulty, and potential confusion such a requirement would cause.

At the same time FDA reconfirmed its evaluation of the safety of irradiated foods. Addressing the New York Board's concerns about the validity of research studies on food irradiation, the Agency said it would not have considered approving food irradiation "if there were not a reliable data base on which to make such an important decision" [3].

The FDA regulations of April 18, 1986, also contain general provisions for food irradiation regarding current good manufacturing practices (CGMP):

Any firm that treats food with ionizing radiation must comply with other regulations applicable to food handling. These other regulations specify in particular the sanitation requirements and controls for the raw materials, the facilities, the equipment, and the process.

A food treated with ionizing radiation shall receive only the minimum dose reasonably required to accomplish the purpose and not more than the maximum dose specified for that use.

Only such packaging materials may be subjected to irradiation of prepackaged foods which have been specifically authorized for such use (see Table 1).

Radiation treatment of food shall conform to a scheduled process, that means a written procedure specifying the radiation dose and other processing conditions, such as atmosphere and temperature. The scheduled process shall be established by qualified persons having expert knowledge in radiation processing, specifically for that food and for that manufacturer's irradiation facility.

A food irradiation processor shall maintain records, as specified, for a period of time that exceeds the shelf life of the irradiated food by 1 year, up to a maximum of 3 years. Such

Table 1 Packaging Materials Approved by the U.S. Food and Drug Administration for Use During Irradiation of Prepackaged Foods

Material	Maximum dose (kGy)
Paper, kraft (flour only)	0.5
Paper, glassine	10
Paperboard, wax-coated	10
Cellophane, either nitrocellulose-coated or vinylidene chloride copolymer-coated	10
Polyolefin film	10
Polystyrene film	10
Rubber hydrochloride film	10
Vinylidene chloride-vinyl chloride copolymer film	10
Nylon-11	10
Vegetable parchment	60
Polyethylene film	60
Polyethylene terephthalate film	60
Nylon-6	60
Vinyl chloride-vinyl acetate copolymer film	60
Acrylonitrile copolymers	60

Source: U.S. Code of Federal Regulations, CFR Title 21, Part 179, Sub-part C. 179.45 (1985).

> records describe the food treatment, lot identification, scheduled process, evidence of compliance with the scheduled process, ionizing energy sources, source calibration, dosimetry, dose distribution in the product, and the date of irradiation. The records shall be available for inspection and copying by authorized employees of FDA.

At the present time FDA apparently has no intention of proposing a general regulation allowing doses of up to 10 kGy for treatment of all foodstuffs. It will continue to evaluate and respond on a case-by-case basis to petitions involving irradiation of foodstuffs [1].

Irradiation of meat and poultry requires, in addition to FDA approval, permission from the United States Department of Agriculture's Food Safety and Inspection Service (FSIS). In January 1986,

FSIS approved the use of gamma irradiation for trichina control in pork carcasses and fresh (unfrozen), non–heat-processed cuts of pork at the minimum absorbed dose of 0.3 kGy and a maximum absorbed dose of 1 kGy. FSIS finalized this rule in the December 5, 1986, *Federal Register*. Specifics on labeling, facilities, equipment, quality control, and trichina certification are being developed by FSIS for notice and comment rulemaking.

A petition for irradiation of poultry products to reduce the potential of foodborne illness was developed by FSIS and filed with FDA on November 19, 1986. The petition proposes the use of ionizing radiation on packaged, fresh, or frozen poultry at an absorbed dose of 1.5–3.0 kGy to reduce some food borne pathogens, such as salmonella, campylobacter, and yersinia.

In a lengthy document published in the Federal Register of December 30, 1988, the Food and Drug Administration denied requests for a hearing on its Final Rule of April 18, 1986, and responded to various written objections to this rule. Based on a comprehensive evaluation of all relevant evidence the agency concluded once more that food irradiation under the conditions of the Final Rule was safe. The agency said many of the objections expressed general opposition to food irradiation but identified no substantive question to which the agency could respond.

II. CANADA

Like the use of food additives, irradiation of foods is generally forbidden in Canada, but permissions can be granted for particular foods. Canada was one of the first countries to clear an irradiated food when it allowed irradiation of potatoes for inhibition of sprouting in 1960. Other clearances are listed in Table 2.

For many years food irradiation in Canada was regulated under the Food Additive Tables, Food and Drug Regulations. Health and Welfare Canada's Health Protection Branch issued in July 1983 a document known as the "Information Letter No. 651," entitled "Proposed Revised Regulations for the Control of Food Irradiation." It proposed to remove food irradiation from the Food Additive Tables and to consider it as a process. The document confirmed that there was no evidence of adverse effects of the consumption of foods irradiated in the dose range up to 10 kGy and required no new toxicological testing of foods irradiated up to this dose level. Nevertheless, the Branch proposed to examine food irradiation petitions on a case-by-case basis, demanding proof of technical efficacy of the process under commercial conditions.

Health and Welfare Canada does not concern itself with labeling of food products. Labeling is the authority of Consumer and Corporate Affairs Canada. In their Communique No. 50 of November 1985, Consumer and Corporate Affairs Canada published the "Recom-

Table 2 List of Exemptions for Irradiated Foods in Countries Outside the United States

Country (Organization)	Product	Purpose of irradiation	Dose permitted (kGy)	Date of approval
Argentina	Strawberries	Shelf-life extension	2.5 max.	30 April 1987
	Potatoes	Sprout inhibition	0.03 to 0.15	30 April 1987
	Onions	Sprout inhibition	0.02 to 0.15	30 April 1987
	Garlic	Sprout inhibition	0.02 to 0.15	30 April 1987
Bangladesh	Chicken	Shelf-life extension/ decontamination	Up to 8	28 December 1983
	Papaya	Insect disinfestation/ control of ripening	Up to 1	28 December 1983
	Potatoes	Sprout inhibition	Up to 0.15	28 December 1983
	Wheat and ground wheat products	Insect disinfestation	Up to 1	28 December 1983
	Fish	Shelf-life extension/ decontamination/ insect disinfestation	Up to 2.2	28 December 1983
	Onions	Sprout inhibition	Up to 0.15	28 December 1983
	Rice	Insect disinfestation	Up to 1	28 December 1983
	Froglegs	Decontamination		
	Shrimp	Shelf-life extension/ decontamination		

Table 2 Continued

Country (Organization)	Product	Purpose of irradiation	Dose permitted (kGy)	Date of approval
Bangladesh (continued)	Mangoes	Shelf-life extension/ insect disinfestation/ control of ripening	Up to 1	28 December 1983
	Pulses	Insect disinfestation	Up to 1	28 December 1983
	Spices	Decontamination/insect disinfestation	Up to 10	28 December 1983
Belgium	Potatoes	Sprout inhibition	Up to 0.15	16 July 1980
	Strawberry	Shelf-life extension	Up to 3	16 July 1980
	Onions	Sprout inhibition	Up to 0.15	16 October 1980
	Garlic	Sprout inhibition	Up to 0.15	16 October 1980
	Shallots	Sprout inhibition	Up to 0.15	16 October 1980
	Black/white pepper	Decontamination	Up to 10	16 October 1980
	Paprika powder	Decontamination	Up to 10	16 October 1980
	Arabic gum	Decontamination	Up to 7	29 September 1983
	Spices (78 different products)	Decontamination	Up to 10	29 September 1983
	Dried vegetables (7 different products)	Decontamination	Up to 10	29 September 1983

Brazil	Rice	Insect disinfestation	Up to 1	7 March 1985
	Potatoes	Sprout inhibition	Up to 0.15	7 March 1985
	Onions	Sprout inhibition	Up to 0.15	7 March 1985
	Beans	Insect disinfestation	Up to 1	7 March 1985
	Maize	Insect disinfestation	Up to 0.5	7 March 1985
	Wheat	Insect disinfestation	Up to 1	7 March 1985
	Wheat flour	Insect disinfestation	Up to 1	7 March 1985
	Spices (13 different products)	Decontamination/insect disinfestation	Up to 10	7 March 1985
	Papayas	Insect disinfestation/ control of ripening	Up to 1	7 March 1985
	Strawberries	Shelf-life extension	Up to 3	7 March 1985
	Fish and fish-products (fillets, salted, smoked, dried, dehydrated)	Shelf-life extension/ decontamination/ insect disinfestation	Up to 2.2	8 March 1985
	Poultry	Shelf-life extension/ decontamination	Up to 7	8 March 1985
Canada	Potatoes	Sprout inhibition	Up to 0.1 up to 0.15	9 November 1960 14 June 1965
	Onions	Sprout inhibition	Up to 0.15	25 March 1965
	Wheat, flour, wholewheat flour	Insect disinfestation	Up to 0.75	25 February 1969

Table 2 Continued

Country (Organization)	Product	Purpose of irradiation	Dose permitted (kGy)	Date of approval
Canada (continued)	Spices and certain dried vegetables seasonings	Decontamination	Up to 10 aver.	3 October 1984
	Onion powder	Decontamination	Up to 10 aver.	12 December 1983
Chile	Potatoes	Sprout inhibition	Up to 0.15	29 December 1982
	Papaya	Insect disinfestation	Up to 1	29 December 1982
	Wheat and ground wheat products	Insect disinfestation	Up to 1	29 December 1982
	Strawberry	Shelf-life extension	Up to 3	29 December 1982
	Chicken	Decontamination	Up to 7	29 December 1982
	Onions	Sprout inhibition	Up to 0.15	29 December 1982
	Rice	Insect disinfestation	Up to 1	29 December 1982
	Teleost fist and fish products	Shelf-life extension/ decontamination insect disinfestation	Up to 2.2	29 December 1982
	Cocoa beans	Decontamination/insect disinfestation	Up to 5	29 December 1982
	Dates	Insect disinfestation	Up to 1	29 December 1982

	Mangoes	Shelf-life extension/ insect disinfestation/ control of ripening	Up to 1	29 December 1982
	Pulses	Insect disinfestation	Up to 1	29 December 1982
	Spices and condiments	Decontamination/insect disinfestation	Up to 10	29 December 1982
China, People's Rep.	Potatoes	Sprout inhibition	Up to 0.20	30 November 1984
	Onions	Sprout inhibition	Up to 0.15	30 November 1984
	Garlic	Sprout inhibition	Up to 0.10	30 November 1984
	Peanut	Insect disinfestation	Up to 0.40	30 November 1984
	Grain	Insect disinfestation	Up to 0.45	30 November 1984
	Mushroom	Growth inhibition	Up to 1	30 November 1984
	Sausage	Decontamination	Up to 8	30 November 1984
	Apples	Shelf-life extension	Up to 0.4	30 September 1988
Denmark	Spices and herbs	Decontamination	Up to 15 max. up to 10 aver.	23 December 1985
Finland	Dry and dehydrated spices and herbs	Decontamination	Up to 10 aver.	13 November 1987
	All foods for patients requiring sterile diet	Sterilization		13 November 1987

Table 2 Continued

Country (Organization)	Product	Purpose of irradiation	Dose permitted (kGy)	Date of approval
France	Potatoes	Sprout inhibition	0.075–0.15	8 November 1972
	Onions	Sprout inhibition	0.075–0.15	9 August 1977
	Garlic	Sprout inhibition	0.075–0.15	9 August 1977
	Shallot	Sprout inhibition	0.075–0.15	9 August 1977
	Spices and aromatic substances (72 products inclusive powdered onions and garlic)	Decontamination	Up to 11	10 February 1983
	Gum Arabic	Decontamination	Up to 9	16 June 1985
	Muesli-like cereal	Decontamination	Up to 10	16 June 1985
	Dehydrated vegetables	Decontamination	Up to 10	16 June 1985
	Mechanically deboned poultry meat	Decontamination	Up to 5	16 February 1985
	Dried fruits	Insect disinfestation	1 max.	6 January 1988
	Dried vegetables	Insect disinfestation	1 max.	6 January 1988
	Frozen froglegs	Decontamination	4 to 8	3 May 1988
	Strawberries	Shelf-life extension	3 max.	29 December 1988

German Democratic Republic	Onions	Sprout inhibition	0.05	16 October 1985
	Enzyme solutions	Decontamination	10	16 October 1985
	Spices and dried herbs	Decontamination	4 to 10	10 July 1987
	Garlic	Sprout inhibition	0.05	2 January 1987
	Potato powder	Decontamination	4	25 November 1987
	Egg powder	Decontamination	2	15 January 1988
	Strawberries	Shelf-life extension	2	27 June 1988
Hungary	Onions	Sprout-inhibition	0.05 max.	23 June 1982
	Spices	Decontamination	8 average	19 August 1986
India	Potatoes	Sprout inhibition	Codex Standard	January 1986
	Onions	Sprout inhibition	Codex Standard	January 1986
	Spices	Disinfection (for export only)	Codex Standard	January 1986
	Frozen shrimps and froglegs	Disinfection (for export only)	Codex Standard	January 1986
Indonesia	Dried spices	Decontamination	10 max.	29 December 1987
	Tuber and root crops (potatoes, shallots, garlic and rhizomes)	Sprout inhibition	0.15 max.	29 December 1987

Table 2 Continued

Country (Organization)	Product	Purpose of irradiation	Dose permitted (kGy)	Date of approval
Indonesia (continued)	Cereals	Disinfestation	1 max.	29 December 1987
Israel	Potatoes	Sprout inhibition	0.15 max.	5 July 1967
	Onions	Sprout inhibition	0.10 max.	25 July 1968
	Poultry and poultry sections	Shelf-life extension/ decontamination	7 max.	23 April 1982
	Onions	Sprout inhibition	0.15	6 March 1985
	Garlic	Sprout inhibition	0.15	6 March 1985
	Shallots	Sprout inhibition	0.15	6 March 1985
	Spices (36 different products	Decontamination	10	6 March 1985
	Fresh fruits and vegetables	Disinfestation	1 average	January 1987
	Grains, cereals, pulses, cocoa and coffee beans, nuts, edible seeds	Disinfestation	1 average	January 1987
	Mushrooms, straw-berries	Shelf-life extension	3 average	January 1987

	Poultry and poultry sections	Decontamination	7 average	January 1987
	Spices and condiments, dehydrated and dried vegetables, edible herbs	Decontamination	10 average	January 1987
	Poultry feeds	Decontamination	15 average	January 1987
Italy	Potatoes	Sprout inhibition	0.075–0.15	30 August 1973
	Onions	Sprout inhibition	0.075–0.15	30 August 1973
	Garlic	Sprout inhibition	0.075–0.15	30 August 1973
Japan	Potatoes	Sprout inhibition	0.15 max.	30 August 1972
Korea Republic of	Potatoes	Sprout inhibition	0.15 max.	28 September 1987
	Onions	Sprout inhibition	0.15 max.	28 September 1987
	Garlic	Sprout inhibition	0.15 max.	28 September 1987
	Chestnut	Sprout inhibition	0.25 max.	28 September 1987
	Fresh and dried mushrooms	Growth inhibition/ insect disinfestation	1.00 max.	28 September 1987
Netherlands	Mushrooms	Growth inhibition	2.5 max.	23 October 1969
	Deep-frozen meals	Sterilization (hospital patients)	25 min.	27 November 1969

Table 2 Continued

Country (Organization)	Product	Purpose of irradiation	Dose permitted (kGy)	Date of approval
Netherlands (continued)	Potatoes	Sprout inhibition	0.15 max.	23 March 1970
	Fresh, tinned and liquid foodstuffs	Sterilization (hospital patients)	25 min.	8 March 1972
	Onions	Sprout inhibition	0.05 max.	9 June 1975
	Chicken	Shelf-life extension/ decontamination	3 max.	10 May 1976
	Spices	Decontamination	10	4 April 1978
	Frozen froglegs	Decontamination	5	25 September 1978
	Rice and ground rice products	Insect disinfestation	1	15 March 1979
	Rye bread	Shelf-life extension	5 max.	12 February 1980
	Spices	Decontamination	7 max.	15 April 1980
	Frozen shrimp	Decontamination	7 max.	9 May 1980
	Malt	Decontamination	10 max.	8 February 1983
	Boiled and refrigerated shrimp	Shelf-life extension	1 max.	8 February 1983
	Frozen shrimp	Decontamination	7 max.	8 February 1983
	Frozen fish	Decontamination	6 max.	25 August 1983

	Egg powder	Decontamination	6 max.	25 August 1983
	Dry blood protein	Decontamination	7 max.	25 August 1983
	Dehydrated vegetables	Decontamination	10 max.	27 October 1983
Norway	Spices	Decontamination	Up to 10	
Philippines	Potatoes	Sprout inhibition	0.15 max.	13 September 1972
	Onions	Sprout inhibition	0.07	1981
	Garlic	Sprout inhibition	0.07	1981
Poland	Potatoes	Sprout inhibition	Up to 0.15	1982
	Onions	Sprout inhibition		March 1983
Pakistan	Potatoes	Sprout inhibition	Up to 0.15	13 June 1988
	Onions	Sprout inhibition	Up to 0.15	13 June 1988
	Garlic	Sprout inhibition	Up to 0.15	13 June 1988
	Spices	Decontamination/ disinfestation	Up to 10	13 June 1988
South Africa	Potatoes	Sprout inhibition	0.12–0.24	19 January 1977
	Dried bananas	Insect disinfestation	0.5 max.	28 July 1977
	Avocados	Insect disinfestation	0.1 max.	28 July 1977
	Onions	Sprout inhibition	0.5–0.15	25 August 1978
	Garlic	Sprout inhibition	0.1–0.20	25 August 1978

Table 2 Continued

Country (Organization)	Product	Purpose of irradiation	Dose permitted (kGy)	Date of approval
South Africa (continued)	Chicken	Shelf-life extension/ decontamination	2–7	25 August 1978
	Papaya	Shelf-life extension	0.5–1.5	25 August 1978
	Mango	Shelf-life extension	0.5–1.5	25 August 1978
	Strawberries	Shelf-life extension	1–4	25 August 1978
	Bananas	Shelf-life extension	"	
	Litchis	Shelf-life extension	"	
	Pickled mango (achar)	Shelf-life extension	"	
	Avocados	Shelf-life extension	"	
	Almonds	Insect disinfestation	"	
	Cheese powder	Insect disinfestation	"	
	Yeast powder	Decontamination	"	
	Herbal tea	Decontamination	"	
	Various spices	Decontamination	"	
	Various dehydrated vegetables	Decontamination	"	

Spain	Potatoes	Sprout inhibition	0.05–0.15	4 November 1969
	Onions	Sprout inhibition	0.08 max.	1971
Thailand	Onions	Sprout inhibition	0.1 max.	20 March 1973
	Potatoes, onions, garlic	Sprout inhibition	0.15	4 December 1986
	Dates	Disinfestation	1	4 December 1986
	Mangos, papaya	Disinfestation/delay of ripening	1	4 December 1986
	Wheat, rice, pulses	Disinfestation	1	4 December 1986
	Cocoa beans	Disinfestation	1	4 December 1986
	Fish and fishery products	Disinfestation	1	4 December 1986
	Fish and fishery products	Reduce microbial load	2.2	4 December 1986
	Strawberries	Shelf-life extension	3	4 December 1986
	Nham	Decontamination	4	4 December 1986
	Moo yor	Decontamination	5	4 December 1986
	Sausage	Decontamination	5	4 December 1986

Table 2 Continued

Country (Organization)	Product	Purpose of irradiation	Dose permitted (kGy)	Date of approval
Thailand (continued)	Frozen shrimps	Decontamination	5	4 December 1986
	Coca beans	Reduce microbial load	5	4 December 1986
	Chicken	Decontamination/ shelf-life extension	7	4 December 1986
	Spices and condiments, dehydrated onions, and onion powder	Insect disinfestation/ decontamination	1 10	4 December 1986 4 December 1986
Union of Soviet Socialist Republics	Potatoes	Sprout inhibition	0.1 max.	14 March 1958
	Potatoes	Sprout inhibition	0.3 (1 MeV-electrons)	17 July 1973
	Grain	Insect disinfestation	0.3	1959
	Dried fruits	Insect disinfestation	1	15 February 1966
	Dry food concentrates (buckwheat mush, gruel, rice pudding)	Insect disinfestation	0.7	6 June 1966
	Onions	Sprout inhibition	0.06	17 July 1973

United Kingdom	Any food for consumption by patients who require a sterile diet as essential factor of their treatment	Sterilization		1 December 1969
Uruguay	Potatoes	Sprout inhibition		23 June 1970
Yugoslavia	Cereals	Insect disinfestation	Up to 10	17 December 1984
	Legumes	Insect disinfestation	Up to 10	17 December 1984
	Onions	Sprout inhibition	Up to 10	17 December 1984
	Garlic	Sprout inhibition	Up to 10	17 December 1984
	Potatoes	Sprout inhibition	Up to 10	17 December 1984
	Dehydrated fruits and vegetables	Sprout inhibition	Up to 10	17 December 1984
	Dried mushrooms		Up to 10	17 December 1984
	Egg powder	Decontamination	Up to 10	17 December 1984
	Herbal teas, tea extracts	Decontamination	Up to 10	17 December 1984
	Fresh poultry	Shelf-life extension/ decontamination	Up to 10	17 December 1984

Source: *IAEA Food Irradiation Newsletter*, April 1988.

mended Labeling Provisions for Irradiated Foods." This proposal foresaw the use of the same logo that in the meantime has been introduced in the United States by the FDA's final rule of April 18, 1986.

Both proposals have been attacked by activist groups opposed to food irradiation. Letter-writing campaigns to Members of Parliament have aimed at preventing the new legislation and have even requested that already existing clearances be withdrawn. Politicians responded to this pressure, and the Standing Committee on Consumer and Corporate Affairs of the Canadian House of Commons published a report "On the Question of Food Irradiation and the Labelling of Irradiated Foods" in May 1987, which was very critical of the two Government proposals. The Government responded in September 1987 and defended its position.

On June 4, 1988, the Health Protection Branch issued "Information Letter No. 746," which reconfirmed the intention to regulate food irradiation as a process, rather than under the Food Additive Tables. The Branch also reconfirmed that it accepted in principle the lack of toxicological hazards for food irradiated below 10 kGy. Nevertheless it insisted on case-by-case examination of petitions and specified that petitions shall be accompanied by a submission to the Director containing the following information:

(a) the purpose and details of the proposed irradiation, including the source of ionizing radiation and the proposed frequency of and minimum and maximum dose of ionizing radiation;

(b) data indicating that the minimum dose of ionizing radiation proposed to be used accomplishes the intended purpose of the irradiation and the maximum dose of ionizing radiation proposed does not exceed the amount required to accomplish the purpose of the irradiation;

(c) information on the nature of the dosimeter used, the frequency of the dosimetry on the food and data pertaining to the dosimetry and phantoms used to assure that the dosimetry readings reflect the dose absorbed by the food during irradiation;

(d) data indicating the effects, if any, on the nutritional quality of the food, raw and ready-to-serve, under the proposed conditions of irradiation and any other processes that are combined with the irradiation;

(e) data establishing that the irradiated food has not been significantly altered in chemical, physical or microbiological characteristics to render the food unfit for human consumption;

(f) where the Director so requests, data establishing that the proposed irradiation is safe under the conditions proposed for the irradiation;

(g) the recommended conditions of storage and shipment of the irradiated food including the time, temperature and packaging and a comparison of the recommended conditions for the same food that has not been irradiated;

(h) details of any other processes to be applied to the food prior to or after the proposed irradiation; and

(i) such other data as the Director may require to establish that consumers and purchasers of the irradiated food will not be deceived or misled as to the character, value, composition, merit or safety of the irradiated food.

Comparison of the proposals of 1983 and 1988 shows that the Canadian Government has considerably tightened its proposed pre-market clearance and compliance rules for irradiated food. Where the 1983 proposal considered any food treated with up to 10 kGy as toxicologically safe, the 1988 proposal requires data indicating effects on the nutritional quality and on chemical, physical, and microbiological characteristics of the treated food. In addition, "where the Director so requests," toxicological data must be submitted regardless of the radiation dose applied. This does not appear to be entirely logical when the proposal emphasizes at the same time that the Branch still accepts in principal the lack of toxicological hazards for food irradiated at levels below 10 kGy. Obviously, far-reaching concessions have been made to the anti–food irradiation movement.

Also on June 4, 1988, the Department of Consumer and Corporate Affairs issued a proposed labeling regulation requiring that, along with the international irradiation symbol, prepackaged irradiated food be further identified with the words "treated with radiation," "treated by irradiation," "irradiated," or "a written statement that has the same meaning." An ingredient or component of a prepackaged food that has been irradiated would have to be included in the list of ingredients and preceded by the statement "irradiated" if it constituted 10% or more of the prepackaged product.

III. UNITED KINGDOM

Irradiation of food and food products intended for human consumption is controlled in the United Kingdom by the "Food (Control of Irradiation) Regulations," which came into force on June 1, 1967. The control is achieved by means of prohibition, with exemption to be granted under conditions to be specified.

Amendments dates December 1, 1969 and February 16, 1972 modified these Regulations by granting exemption for:

a) inspection of food and control of food processing (e.g., detection of empty packages on a filling line) with the aid of ionizing radiation, provided the absorbed dose is not more than 0.5 Gy, and

b) irradiation of food intended for consumption by patients who require a sterile diet as an essential factor in their treatment.

In 1982 the British government established the Advisory Committee on Irradiated and Novel Foods (ACINF) under the chairmanship of Sir Arnold Burgen of the University of Cambridge. The membership was drawn from experts independent of industry and government. The Committee published its Report on the Safety and Wholesomeness of Irradiated Foods in 1986, fully confirming the conclusions of the 1980 FAO/IAEA/WHO Joint Expert Committee, namely, that the irradiation of food up to an overall average dose of 10 kGy presents no toxicological hazard and introduces no special nutritional or microbiological problems [4]. The Committee further concluded: "We are satisfied that there is no justification on public health grounds for the present United Kingdom Regulations not to be amended to permit the ionizing irradiation of food up to an overall average dose of 10 kGy." With regard to labeling, ACINF declared itself satisfied "that there are no scientific or public health reasons which would require an indication at the point of retail sale that a food had been irradiated." In spite of this, the Committee recommended, for the purpose of informing the consumer, that all irradiated foods and compound foods containing irradiated ingredients should bear an indication of the treatment in specified terms. This recommendation was based on the view that failure to declare the fact of irradiation would be seen by the public not as a neutral act, but as a positive concealment, and that as such it would be likely to exacerbate public fears and obscure the real benefits of irradiation.

If the members of the Committee had hoped that the proposed labeling requirement would mollify the opponents of food irradiation, they must have been surprised by the reception their report received. As in the United States and Canada, the proposal to facilitate the introduction of irradiated foods by modifying existing regulations immediately elicited massive criticism. The publication of the ACINF report in April 1986 almost coincided with the news about radioactivity in foods from the reactor accident at Chernobyl. Not a few commentators apparently confused irradiation of food with radioactivity in food and registered their vehement opposition. The movement against food irradiation was (and still is) spearheaded by the London Food Commission (LFC) which, in spite of its official-sounding name, is a private organization led by two self-appointed consumer advocates, Tony Webb and Tim Lang. In their publications they have expressed the view that the safety of irradiated foods has not been proven (on the contrary, "a cancer connection" is implied), that the major impetus for irradiation has come from the nuclear industry, that food irradiation is not needed, that it creates undesirable changes in the food, that it is too costly, that "those with vested interests" are playing on public ignorance and using their influence within the machinery of government in order to achieve a rapid change in the law, that workers in irradiation plants are not sufficiently protected against radiation exposure, and that the introduction of food irradiation could lead to widespread concealment of contamination of food, a lowering of food hygiene standards, and an increased risk to public health. LFC also attacked the personal integrity of members and advisors of ACINF, hinting at possible conflict of interest [5].

The Parliamentary Secretary of Health announced to the press on February 4, 1988, that 6000 letters had been received in response to the ACINF report of 1986. In light of these, the Government had sought further advice from the ACINF. In a report dated November 1987, the Committee had reaffirmed its earlier advice that, under the conditions prescribed, irradiated food was safe and wholesome. ACINF did not consider that any of the comments they had received caused them to change the advice given in their report of 1986. Nevertheless, the government had decided to maintain the present general prohibition on irradiated foods until they were satisfied that effective regulatory controls could be drawn up for the irradiation of specific foods in order to bring about good industrial and marketing practices and ensure informed consumer choice. Such controls, continued the press release, should include assurances on the quality of the particular foods to be irradiated, requirements as to the maintenance of documentary records, the licensing of premises, and inspection of operations. The availability of detection tests would be one of the factors to be taken into consideration, and there would also be statutory provision for the labeling of irradiated foods and food ingredients [6].

The new ACINF response makes clear that, as in Canada, the organized public protests have not revealed any previously overlooked facts or have in any other way modified the experts' advice that foods irradiated in the dose range of up to 10 kGy are safe to eat. However, political considerations have weighed more heavily. With the exception of food intended for consumption by a small group of hospital patients, food irradiation remains prohibited in the United Kingdom. If those suspected by LFC of having "vested interests" have tried to use their "influence within the machinery of government," they have failed miserably. When the intended controls, assurances, requirements, licenses, and inspection procedures are finally in force, permissions for irradiation of a few food items will probably be as overburdened by regulations as they already are or are becoming in the United States and Canada.

IV. FEDERAL REPUBLIC OF GERMANY

Other than the government of the United States, probably no other government has spent as much on food irradiation research as has the Federal Republic of Germany. The first projects were started around 1955. An Institute of Radiation Technology specially designed for food irradiation research and equipped with a 6 kW electron linear accelerator was opened in 1966 as a unit of the Federal Research Center for Food Preservation in Karlsruhe. From 1970 until 1982 the FRG was the host country of the International Project in the Field of Food Irradiation and its principal financial supporter.

However, this enthusiam for research was never matched by a willingness to take political risks when legal permissions for food irradiation were requested. A new Food Law came into force in 1959, which controlled food irradiation by means of prohibition with provision for exemptions. The only exemption granted until now is specified in the Food Irradiation Regulation of December 19, 1959. It permits irradiation of foods with electron, gamma, and X-rays for control purposes only, with an absorbed dose of not more than 0.1 Gy. It also permits treatment of certain foods with ultraviolet radiation.

A group of representatives of the German fishing industry presented a petition for an exemption for fish irradiation to the federal authorities on December 28, 1971. The purpose of the request was on-board irradiation of iced ocean fish for the extension of shelf life. Permission was never granted or refused. When nothing had happened after 3 years, the fishing industry lost interest. In the meantime, most countries bordering the oceans had established the 200-mile Exclusive Economic Zone, and this had deprived German fishermen of most of their traditional fishing grounds. The size of the German fishing fleet had to be drastically reduced, and the fishing industry had too many other problems to worry about irradiation.

In 1980 several spice manufacturers applied for permission to irradiate spices. At that time it could be foreseen that ethylene oxide, which was then used for improving the hygienic quality of spices, would be forbidden because it had been recognized as a potential carcinogen. The request was given support by all relevant expert committees and advisory boards, and the regulation permitting spice irradiation was only awaiting the signature of the Minister of Health. It has rested at this stage for years. In view of the hostile attitude of the media and consumer organizations, the clearance for spice irradiation is considered as "politically not feasible."

In May 1984 the Green Party in the Federal Parliament moved to change the Food Law in such a way that no exemptions for food irradiation could have been granted. After 2 years of debates in parliamentary committees and in plenary session, the move was finally defeated.

The prolonged debates and delayed decisions on these issues have strengthened the widespread impression that something must be wrong with food irradiation. Until the harmonization of food laws in the member countries of the European Community introduces a European guideline on food irradiation, which would commit the German Government to comply, it is quite unlikely that any clearances for food irradiation will be granted in the Federal Republic of Germany.

The methods used by some opponents of food irradiation to bring about this situation may be illustrated by a few examples. The Arbeitsgemeinschaft der Verbraucher (AgV), the national head organization of the German consumer unions, a politically influential group financed from public funds, submitted a statement to the Federal Parliament at the time when the Green Party's move to change the Food Law was under debate. In this statement, AgV flatly opposed any permissions for food irradiation, claiming that such treatment favored the production of highly toxic and carcinogenic compounds in foods. The key reference supporting this claim was designated as "Document of the Food and Drug Administration (FDA) Nr. 81N/0004 of May 16, 1984." Scrutiny of this source later revealed that it was a comment on FDA's proposed food irradiation regulation, submitted to FDA by Kathleen M. Tucker and Robert Alvarez, both well-known consumer advocates in the United States. This comment did carry the FDA docket number 81N/0004, but to describe it as a document of the FDA is grossly misleading. Its claims of harmful effects of irradiation were based on studies such as the Indian reports on increased aflatoxin production which, as discussed in Chapter 6, are irrelevant for practical food irradiation.

The second example is an information brochure published by Stiftung Verbraucherinstitut in Berlin, another consumer-interest organization supported form public funds [7]. The brochure mixes many anti–food irradiation comments, like that of Tucker and Alvarez, with a few neutral or pro–food irradiation comments and illustrates the text with cartoons such as those shown in Figure 1. An objective informing of the readers is clearly not intended.

With the obvious aim of causing mistrust and prejudice in the population, the organized opponents of food irradiation have consistently used and continue to use the term "radioactive irradiation" (*radioaktive Bestrahlung*) to describe the process. This has been generally adopted by the media, and even the government speakers used this term when the topic was discussed in Parliament. It can be assumed that the majority of consumers consider "radioactive irradiation" synonymous with "radioactive contamination." At any rate, politicians or government officials who denouce "radioactive irradiation of our foods" can be sure of general applause. As long as this situation prevails, it is rather unlikely that the food industry will try to put labeled irradiated foods on the market, even if the

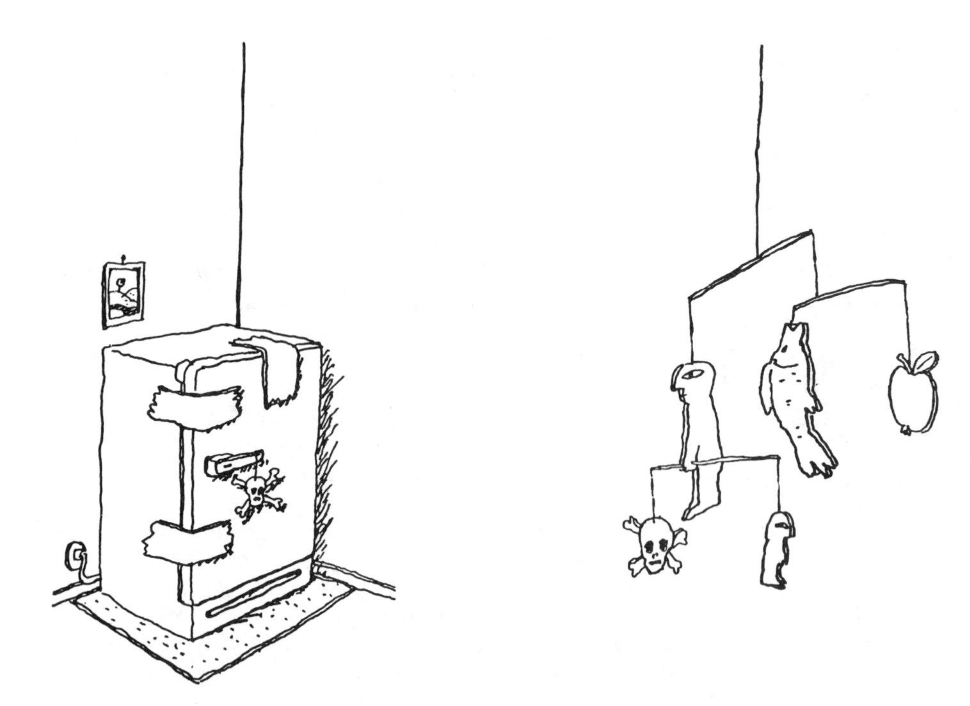

Figure 1 Cartoons from an information brochure on food irradiation widely circulated in the Federal Republic of Germany. (From Ref. 7 with permission.)

harmonization of European food laws should make this possible in the future.

V. OTHER COUNTRIES

It is not possible to discuss in detail the regulatory situation in other parts of the world. Most other countries also have legislation that provides for a general interdiction on irradiated foods, with authority to grant exemption for the production, importation, and sale of individual irradiated foods in accord with designated limitations. So far, 32 countries have approved over 40 irradiated food items for consumption, either on an unconditional or a restricted basis.

The International Atomic Energy Agency periodically publishes in its Food Irradiation Newsletter a list of such exemptions granted by the governments of various countries. That list is very long, because it contains many exemptions for experimental batches and for test marketing of limited quantities of irradiated foods. Table 2 is a shortened version, containing only those exemptions characterized in the IAEA listing as provisional or unconditional, and which permits the processing of unlimited quantities of the particular food. Some exemptions granted since the publication of IAEA's latest listing in April 1988 have been added.

The existence of regulations permitting irradiation of certain foods in a country does not, of course, mean that the food industry in that country actually makes use of this exemption. Information on where food irradiation actually is carried out on a commercial scale was presented in Chapter 9.

Information on the situation in the United States was left out of Table 2 as it was presented in Sec. I. Other countries not appearing in this list may either not have granted any exemptions other than for experimental batches or may not have an interdiction on irradiated food in their food law. It is noteworthy that no government has given blanket approval for irradiation of all foods up to a dose of 10 kGy. Until now, all national authorities, like FDA, have proceeded on a case-by-case basis. Opponents of food irradiation often refer to this and claim that these authorities do not agree with the JECFI conclusions of 1980. This is a misunderstanding. JECFI, as a committee of experts, had no regulatory authority. The experts studied the available evidence and concluded that the safety of foods irradiated up to a dose of 10 kGy had been established. In contrast, the national authorities have to make regulations, and in doing so they have to take into account safety data *and* other considerations, for instance, the view that irradiation should be permitted of only those commodities that show a clear advantage of the radiation treatment. If government agencies decide that the permitted radiation dose should be limited to the technically required level, even if that level is below JECFI's 10 kGy, that does not mean that there is disagreement with the conclusions of JECFI.

The governments of the member countries of the European Community (EC) have recently confirmed their intention to create an unrestricted common market by the year 1992. This means unrestricted trade of goods, including foodstuffs, between the EC countries. A precondition would be the complete harmonization of food laws and regulations. The present situation, where some EC countries such as the Netherlands, Belgium, and France permit irradiation of certain commodities, while other member countries such as Germany have not granted any exemptions, cannot persist. Since commercial application of food irradiation is a reality in several EC countries, and since the practical experience of many years in these countries has not revealed any technical or regulatory problems, one may expect the issuance of a European regulation that opens the common market, at least for all those irradiated foods now allowed in only some of the EC countries. However, if the anti–food irradiation campaign, organized so successfully in the Federal Republic of Germany and in the United Kingdom, spreads to other European countries, the outcome will be quite uncertain. (Public attitudes to food irradiation will be further discussed in Chapter 11.)

Attention is called to Appendixes I and II of this book, which present the Codex Alimentarius General Standard for Irradiated Foods and the Recommended International Code of Practice for the Operation of Irradiation Facilities Used for the Treatment of Foods—two important documents that have had considerable influence on worldwide developments in the regulation of irradiated foods. A look at the approval dates in the last column of Table 2 will show that most of these approvals have been granted since these Codex documents were adopted in 1983.

REFERENCES

1. Pauli, G. H., and C. A. Takeguchi, Irradiation of foods—an FDA perspective, *Food Revs. Internat.*, 2:79 (1986).
2. FDA, Irradiation in the production, processing, and handling of food, *Federal Register,* 51:13376 (1986).
3. Anonymous, FDA sees "no need" for irradiated ingredient labeling, *Food Chem. News,* June 20, p. 46 (1988).
4. HMSO, *Report on the Safety and Wholesomeness of Irradiated Foods,* Advisory Committee on Irradiated and Novel Foods, Her Majesty's Stationery Office, London, 1986.
5. Webb, T., and T. Lang, *Food Irradiation: The Facts.* Thorsons Publ. Group, Wellingborough, Northamptonshire, 1987.
6. Department of Health and Social Security, The Response of the Advisory Committee on Irradiated and Novel Foods (ACINF) to Comments Received on the "Report on the Safety and Wholesomeness of Irradiated Foods," November 1987. Available from DHSS Store, Health Publication Unit, Manchester Rd., Heywood, Lancs. OL10 2PZ.
7. Kursawa-Stucke, H.-J., ed., *STRAHLENKONSERVIERUNG.* Stiftung Verbraucherinstitut, Berlin, 1985.

SUGGESTED READINGS

Engel, R. E., The regulatory involvement of the food safety and inspection service in food irradiation, in *Food Irradiation Processing,* Proceedings of a Symposium held in Washington, D.C., March 1985. Int. Atomic Energy Agency, Vienna, 1985, p. 297.

Giddings, G. G., Food irradiation: the "reluctant" food additive for all agendas, *Food Revs. Internat.*, 2:109 (1986).

IAEA, Legislations in the Field of Food Irradiation, IAEA-TECDOC-422, Vienna, 1987.

11

Consumer Attitudes

I. EXPERIENCE IN VARIOUS COUNTRIES

Food manufacturers and retailers take a risk with every new product they offer. Consumer acceptance is difficult to predict. Of the many new products introduced to the U.S. market each year, less than 25% make it into the following year. Over a longer period of time nine out of ten product launchers are destined to failure. The situation is probably not very different in other countries where consumers are offered a plentiful choice of products.

The uncertainty is even greater when it comes to irradiated foods. Initial reaction of many consumers to foods openly labeled as "irradiated," "treated with famma radiation," or with any other term related to the word "radiation" is likely to be negative. The word evokes associations like radioactivity, danger, cancer, and nuclear waste.

Experience in some countries has shown that consumers provided with sufficient information can be convinced of the advantages offered by irradiated foods. In some of these countries consumer organizations have been closely involved in the introduction of irradiated foods and have taken a positive stand. The Consumers' Association of Canada (CAC), for instance, at its 1982 Annual Meeting passed two resolutions supporting irradiated food. As a result, CAC accepts and supports the Codex International General Standard for Irradiated Food and the Codex Recommended Code of Practice for Operation of Radiation Facilities. CAC has asked the Government of Canada to do likewise and to demand labeling of irradiated foods with the internationally accepted logo; a CAC representative predicted at that time that the benefits of food irradiation were going to be welcomed by the public once they were properly informed [1].

The need for early involvement of consumer representatives has also been recognized in the Republic of South Africa. Van der Linde and Brodrick [2] have described activities in that country:

> In order to attain our ultimate goal of establishing radurisation as a commercial process, we recognised at a very early stage that it was necessary to formulate and execute a well-planned strategy involving all partners in the food production and distribution chain, from farmers and producers, to food handlers, retailers, consumers and government departments involved in regulating food production and distribution country-wide.

Consequently, a National Steering Committee on Food Irradiation was constituted, with representatives from the consumer organization, various government departments, the trade, the radiation processors and other interested organizations. It coordinates all aspects relating to the marketing of irradiated food and serves as a link between research activities and practical realization.

Radiation processing of foods has reached a level of 30,000 t a year in South Africa. No adverse reaction from consumers has been reported. It should be noted, however, that the South African authorities do not demand labeling at the retail level. Irradiated products must be labeled with the irradiation logo at the wholesale level only (crates, shipping boxes, etc.). The retailer knows which shipments have been irradiated, but he is under no obligation to advertise openly whether his product has been irradiated or not. "As probably could be expected, no retailer initially commited himself to this but as the process is becoming increasingly accepted, this fact is starting to become advertised" [2]. Whether or not South African consumers would react as favorably if the irradiated products were labeled with words like "irradiated" instead of the logo is open to speculation.

South Africa is one of the few countries where "first-generation" irradiated foodstuffs reach the consumer. Another country where fruits and other first-generation irradiated products are sold is the People's Republic of China. Here, also, labeling is not required. The food items irradiated in countries like the Netherlands and Belgium are spices, dried vegetables, enzyme preparations, egg powder, and similar products, which are sold to food manufacturers and become ingredients of "second-generation" irradiated foods. Labeling of second-generation irradiated items is not mandatory in these countries. Consumers therefore do not know when they are buying or eating foods containing irradiated ingredients. Similarly, in Japan potatoes irradiated for inhibition of sprouting are sold only to chips manufacturers and other potato processors, and the chips and other potato products are not labeled.

As mentioned in Chapter 9, a Dutch food chain began to offer irradiated mushrooms around 1970. The purpose of irradiation was improved keeping quality. Although signs at the sales counters clearly marked these mushrooms as irradiated, they sold well. When some customers complained about not having a choice between irradiated and unirradiated mushrooms the food chain began to offer two types of mushrooms, one labeled "irradiated mushrooms," the other labeled "fresh mushrooms." That, of course, made the customers think that irradiated samples were less desirable, and they stopped buying them. That was the end of mushroom irradiation in the Netherlands.

Marketing trials with openly labeled irradiated potatoes and onions were carried out in several countries and were always described as successful. Most of these trials were carried out some 10 to 15 years ago at a time when the attitude toward nuclear energy was generally more positive than it is today and when fewer consumers were concerned about overprocessing of foods.

Typical for such trials was the sale of 15 tons of potatoes in Italy in 1976. The tubers had been irradiated at the Casaccia Nuclear Center near Rome in the fall of 1975. In March 1976 they were offered for sale in four cities. An illustrated pamphlet was included in each sack of potatoes in order to provide the consumer with information about this preservation process, together with a tear-out questionnaire. Wide press coverage both preceding and during the sales was given to promote the marketing campaign; newspaper articles, radio and television interviews all served to inform the public about the general application of irradiation for food conservation purposes. It was reported at the time that all irradiated potatoes were sold on the first day and that the response of the buyers was positive.

A marketing test with irradiated onions was carried out in Bahia Blanca, Argentina, in October 1985. Signs explained the benefits offered by the irradiation process, and trained personnel were at hand to answer questions from the public. The sale price was 20% below the sale price of nonirradiated onions. Questionnaires were handed to the buyers. Of those who returned the questionnaires, 94% rated the irradiated onions as "very good."

The first test marketing of an irradiated food in France was carried out in May and June 1987. During a 3-week period, irradiated strawberries were offered in several stores and markets in the city of Lyon. The display boxes were labeled "Protégée par ionisation" (protected by ionization). The irradiated fruits carried a freshness guarantee and were 30% more expensive than unirradiated strawberries. They reportedly sold well.

In Thailand a consumer survey carried out in 1986/87 showed good acceptance of irradiated nham, a fermented pork sausage. Nham, which is normally consumed without cooking, is often contaminated by salmonella species and occasionally by *Trichinella spiralis*. A dose of 2 kGy eliminates these risks. Although the irradiated sausage was offered at a price of 4 US cents per piece (150 g) higher than the nonirradiated nham offered at the same market, ten times as much was sold; 95% of the surveyed consumers indicated that they would buy again irradiated nham at this price; 71% were willing to buy the irradiated sausage even at a price of 8 US cents per piece above that of the unirradiated product.

II. TEST MARKETING AND CONSUMER SURVEYS IN THE UNITED STATES

A few market tests with irradiated fruits were carried out in the United States in recent years, but the quantities sold were probably too small and the sales outlets too limited to draw far-reaching conclusions from the results.

An in-store study of consumer acceptance of irradiated papayas was conducted in two California groceries, one in Irvine, the other in Anaheim on March 28, 1987. Consumers were offered papayas that had received two different types of pest-control treatment—irradiation at a dose of 450 Gy or hot-water double-dipping. It was reported that the customers found the irradiated products more appealing in appearance than the double-dipped papaya and that 81% of the customers in Irvine and 66% in Anaheim indicated a willingness to purchase the irradiated papayas in the future; 15% in Irvine and 32% in Anaheim said "no," the rest were uncertain. The irradiated papayas outsold the double-dipped papayas at a ratio of better than 10:1 [3]. While the double-dip method requires picking the fruits at a mature-green stage, papayas to be irradiated are tree-ripened. Papaya experts consider irradiation as the better quarantine treatment as compared to the double-dip method, and the California consumers obviously agreed. Since the ban of ethylene dibromide (EDB) in 1984, the Hawaiian papaya producers have utilized the double-dip method as an interim disinfestation procedure. Although fruit quality is not fully satisfactory, this is the only way Hawaiian papayas can meet mainland quarantine regulations as long as irradiation is not accepted. The untreated (neither dipped nor irradiated) fruits for this test were shipped from Hawaii under a special permit from USDA-APHIS (Animal and Plant Health Inspection Service) and irradiated in California. Press coverage of the event was uneven. An account in the *Wall Street Journal* was headlined "You would think these papayas would be easy to find at night," indicating the somewhat tired joke that irradiated foods glow in the dark. The *New York Times* reported that the irradiated papayas "were sneaked into Los Angeles last week and unobtrusively delivered to supermarkets." The *Associated Press* quoted a member of the California-based National Coalition to Stop Food Irradiation (NCSFI): "When we find the test site, then we are going down to the stores, picket them and hand out literature about what food irradiation is and what our concerns are." Apparently the irradiated papayas were sold out before CSFI members could find the test site.

Irradiated Puerto Rican mangoes went on sale in a supermarket in North Miami Beach on September 11, 1986, marking the first time in history that irradiated food was made commercially available in the United States. The two tons of mangoes, at $1.49 a pound, were sold out within a week, although the three-by-three-foot signs around the mango bins left no doubt that the product was irradiated. The event was well covered by the media. No protesting pickets appeared.

This test must be seen in the context of the banning of EDB by the Environmental Protection Agency in 1984. Due to this ban Puerto Rico has difficulties to ship its large and growing mango crop to the U.S. mainland. The test confirmed that irradiation in Puerto Rico at a dose of 500 to 1000 Gy could solve this problem. John Plana of Miami's Sun Fruit Company, who had imported the test fruits, was quoted in a press report:

> The produce industry is losing business opportunities because of bans on some pesticides. We want to give the American consumer a greater selection of fruits from around the world. Irradiation will allow us to do so in the future by meeting USDA quarantine restrictions. Without irradiation, we cannot meet the growing demands of the American public for imported speciality and tropical fruits and vegetables.

Sponsored by the National Marine Fisheries Service, Brand Group, Inc. conducted a study on consumer acceptance of irradiated seafood, the result of which was published in March 1986 [4]. According to this study, consumer attitudes toward irradiation reflect three groups: rejectors, estimated to represent 5–10% of the population and objecting because of personal beliefs on "natural" diets or antinuclear concerns; undecideds, estimated at 55–65% of the population, who have adopted a wait-and-see attitude because of lack of knowledge about the process; and acceptors, estimated to be 25–30% of the population and accepting irradiation processing because they believe they understand its advantages and consider it safe. The study further concluded that the rejectors were so firm in their beliefs and values that no educational campaign could change their attitudes. In contrast, such educational efforts could be effective in reducing concern and uncertainties and fostering more positive attitudes toward irradiation in the undecideds. The potential was seen for gaining accpetance for irradiated foods among 90% of the population.

More concern for preservatives and chemical sprays than for food irradiation was expressed by consumers interviewed in California in 1985 [5]. Using a group discussion technique, this study focused on attitude changes as a result of educational efforts. While conventional consumers' attitudes toward food irradiation could be positively influenced by printed information and by interaction with experts, alternative consumers remained unwilling to buy irradiated foods, and their concerns even increased after educational efforts. This group apparently corresponded to the rejectors in the Brand Group study. The same group of authors from the University of California, Davis [6], found leaflets and posters as effective as group discussions in generating change in attitude toward food irradiation among conventional consumers.

At the request of the U.S. Department of Energy and the National Pork Producers Council, a study on consumer reaction to food irradiation was carried out by Wiese Research Associates, Inc. and published in March 1984 [7]. The study, based on nationwide telephone interviews, concluded that only about one in four adults had heard of this method of food preservation. Among those who were aware of food irradiation, 30% had no concern about food irradiation, 30% had minor concern, 37% had major concern, and 3% were undecided. Among those who were unaware of the process, 16% had no concern, 29% had minor concern, 39% had major concern, and 16% were undecided. The cause of concern most frequently mentioned was the word "radiation" itself. Other causes were "might be harmful to people," "may have side-effects," "may be cancer-causing." Only 1% of the concerns mentioned were related to cost or cost efficiency. As in the California study [5], the level of concern for chemical sprays and preservatives was higher than for irradiation.

III. ARGUMENTS AGAINST FOOD IRRADIATION

Most of the fears of those who are opposed to food irradiation are related to health issues. Irradiated foods are thought to be radioactive or to be nutritionally inferior or microbiologically dangerous or to pose some unforeseeable threat upon long-term consumption. Results of research relevant to these questions were presented in Chapter 5 and subsequent chapters.

To those who claim that the safety of irradiated foods has not yet been proven and that more research should be done before the process is permitted, one can only suggest that they take cognizance of the enormous amount of research that has been done over a period of almost 40 years, practically all of it supported from public funds.

Those who claim that research has demonstrated adverse health effects of the consumption of irradiated foods must be asked to document this claim. For example, the California-based National Coalition to Stop Food Irradiation (NCSFI) states in its "Food Irradiation Fact Sheet" without any documentation:

> Studies from around the world have shown adverse effects when animals have been fed irradiated food. Some of these are: tumors, cataracts, kidney damage, chromosome breakage, fewer offspring, and higher mortality.

Various committees of experts who have scanned the scientific literature for studies on the wholesomeness of irradiated foods and have evaluated these studies have not found any evidence that would justify such a claim. Until NCSFI documents this statement, it must be considered a result of misunderstanding, if no fabrication. According to the "Fact Sheet," "New chemicals called unique radiolytic products (URPS) are formed in the foods. Most URPS are unknown and untested." As discussed in Chapter 3, many attempts were made to find URPs in radiation-processed foods, and none were found.

The "Fact Sheet" continues, "Aflatoxin, a highly carcinogenic substance produced by molds, is produced in greater quantities in irradiated foods." As presented in Chapter 6, Sec. III, several studies have shown that no food irradiated and stored under conditions prevailing in practice would be at risk of increased formation of aflatoxin.

NCSFI further warns: "The doses are 100,000 to 3,000,000 times that of a chest X-ray." Irradiating man and irradiating man's food are entirely different things. Treatments with heat (boiling or frying), with smoke, or with freezer temperatures would all be deadly for man. That does not mean that cooked or smoked or frozen foods are not safe and wholesome.

"Irradiated food can become recontaminated if not sealed properly, undermining its primary purpose." This is also true of foods preserved by other methods, such as heat-sterilization, and is no valid argument against irradiation.

Sometimes the pamphlets published by anti–food irradiation groups support their claims of adverse effects of irradiated foods with references to the scientific literature. To the layperson or to the scientist not specializing in this field, such publications may look very scientific, very solid, very convincing. An example is Piccioni's journal article that was discussed in Chapter 5, sec. VIII.

As described in Chapter 5, Sec. VII, a number of studies seemed to demonstrate adverse effects of irradiated foods and caused considerable concern when they were first published. However, in each case, more extensive subsequent studies contradicted the earlier results and revealed weaknesses in the design, execution, and statistical evaluation of the earlier work. It is typical of the pseudo-scientific publications designed to warn against food irradiation that these earlier studies are quoted, but not the follow-up studies.

It is also typical of these publications that they quote scientific reports showing losses of vitamins, formation of off-flavors, effects on texture or color in foods treated with very high doses of radiation without making clear that the dose used for these experiments was high above that which is intended in practice.

The example of a German consumer information brochure illustrated with scary cartoons (Chapter 10, Fig. 1) has shown how innuendo is also used to influence the reader.

Authors opposing food irradiation rarely fail to make reference to the studies conducted by Industrial Biotest Labs (IBT), the firm convicted of performing fraudulent research (see Chapter 5). It is implied that the credibility of the scientific research on food irradiation must be questioned when "such things" could happen. In fact, IBT carried out only a small percentage of all the wholesomeness studies on irradiated foods. Plenty of solid information remains upon which FDA and other regulatory agencies have based their decisions.

Besides questioning the wholesomeness of irradiated foods, what are some of the other concerns? Opposition to food irradiation comes primarily from two groups. One is the group opposed to nuclear energy who think that food irradiation is part and parcel of the nuclear industry, of the "plutonium economy," and that approval of irradiated foods would strengthen the case of nuclear energy. The other group of opponents are advocates of natural foods, organic foods, biological foods, so-called alternative consumers, who are opposed to any kind of processing, and particularly to "high-tech" processing. Many opponents of food irradiation probably belong to or are sympathetic to both these camps.

Let us look first at some typical claims of the anti-nuclear movement:

1. Food irradiation is carried out with nuclear waste materials, and the nuclear industry wants food irradiation to get rid of their waste, with a profit.

Comment: Most of the irradiators now in operation in various parts of the world use ^{60}Co as a radiation source. This is not a nuclear waste material. Some irradiators, like the large wheat irradiation facility in the harbor of Odessa, are electron accelerators. These machines have nothing to do with the nuclear industry. In the United States and in the United Kingdom, spent-fuel rods from nuclear reactors have been used for experimental irradiation of food in the 1950s, but not since then. They are not suitable for commercial food irradiation, for reasons described in Chapter 2. Plans for pilot-scale irradiators using ^{137}Cs were proposed in the United States a few years ago, but these have been dropped. There is presently little or no available supply of ^{137}Cs, and there are many objections to its use. This isotope could only play a significant role as a radiation source if reprocessing plants were built. No such plant is now being built or planned in the United States. It is therefore not possible that ^{137}Cs-sources will play a major role in food irradiation in the near future (let us say until the year 2000) and not very likely that they will play that role in the longer-term future. The only country where the nuclear industry shows an interest in food irradiation is Canada. The overwhelming majority of those who have developed and promoted food irradiation since the 1940s came from the field of food science, not from nuclear science or nuclear industry. The main supporters of food irradiation in recent years have been the UN organizations FAO, IAEA, and WHO.

2. Food irradiation has been mostly sponsored by the Armed Forces, and it has been primarily developed in the interest of the military.

Comment: This claim is just as unfounded as the first one. It is true that the U.S. Army supported research and development on high-dose irradiation of foods between 1953 and 1980. This work has made important contributions, but it has been only a fraction of the worldwide efforts in this field. All practical applications of food irradiation now carried out anywhere in the world are low-dose or medium-dose applications. Outside the United States, the military has not sponsored any of the research and development work on irradiated foods. Furthermore, the mere fact that the Armed Forces are interested in a process is hardly a valid reason to be against that process. The Natick Laboratories of the U.S. Army have done outstanding work in the development of heat-sterilizable ("retortable") flexible pouches. Apparently nobody has ever considered this a reason to oppose the marketing of food in flexible pouches.

3. Commercial introduction of food irradiation would necessitate the transport of huge amounts of radioactive materials to the radiation facilities, and this would create the danger of radioactive spills in road accidents.

Comment: Huge amounts of radioactive materials have been transported for over 40 years and no serious accidents have occurred. Radiation sources for the sterilization of catgut, surgeons gloves, syringes, artificial joints, catheters, etc. and for other industrial uses such as radiation curing of plastics have been applied for over 20 years, all over the world, without problems. Food irradiation requires the same types of irradiators, transport containers and safety measures as these other industrial applications of radiation processing.

4. Workers employed in radiation facilities are not sufficiently protected. They face considerable risks in the event of malfunctioning equipment, leaking radioactive sources, or accidental exposure to the source.

Comment: Again, this assumption is not supported by decades of experience with megacurie radiation sources. Compared to other industries, the radiation processing industry has an excellent safety record. There are no good reasons to believe that this would change if food irradiation were carried out on a larger scale.

5. The permitted radioactive emission (from food irradiation facilities) are 20 times higher than (those of) nuclear power plants. The extremely high level of radiation involved will threaten workers and communities.

Comment: One wonders where NCSFI, from whose "Fact Sheet" this last claim is taken, gathered such nonsense. In contrast to nuclear power plants, no radioactivity is produced or emitted from any type of food irradiation facility.

What are the concerns of alternative consumers toward irradiated foods?

6. Irradiated foods are devitalized, denatured foods. How can such dead foods be wholesome?

Comments: There are no scientific grounds for assuming that man can assimilate certain vital forces from foods beyond the absorption of the known nutrients such as amino acids, vitamins, minerals, etc. From the nutritional viewpoint, there is no difference between living foods and dead foods as long as the nutrient content is the same. On the contrary, there are many examples that the raw (living) material is less well digested than the cooked (dead) material. For instance the starch in raw potatoes is practically indigestible. It would be most unreasonable to eat raw potatoes in the belief that they can convey some vital force.

Those who believe in the existence of such vital forces can be assured that irradiation is much less destructive than cooking. The low radiation doses applied to potatoes, onions, or fruits do not kill the plant cells. The tissue continues to consume oxygen, to produce carbon dioxide, and to maintain other metabolic reactions. Higher radiation doses, which lead to cell death, may be used for hygienization of spices, egg powder, meat, shellfish, etc. The only living cells in such products are those of microorganisms or possibly insects in spices or parasites in meat. Traditional preservation methods such as smoking, pickling, or heating also have the purpose of destroying organisms responsible for spoilage.

7. Irradiation permits food manufacturers and the food trade to make spoiled food marketable and thus to deceive the consumer. Irradiation is just a cosmetic treatment.

Comment: As discussed in Chapter 6, food spoilage is accompanied by physical and chemical changes in the food that cannot be reversed by irradiation. Spoiled meat or fish has a characteristic odor that does not disappear upon irradiation. Spoiled fruit is characterized by softening of the tissue and by changes in taste. Visible mold growth may be present. Again, none of these changes can be repaired by irradiation.

If irradiation reduces the number of microorganisms that cause spoilage of the food and possibly disease in man, it does the same as pasteurization of milk. Surely this cannot be interpreted as an attempt to deceive. Chemical fumigation is now used to prevent insect damage to stored grain or to prevent sprouting of potatoes or as a quarantine treatment of fruits to prevent the spreading of pests. Why should achieving the same purpose by a physical process, irradiation, be an attempt to deceive the consumer?

8. Irradiation causes "wet dog flavor" in meat, changes of color and texture in fruit, metallic off-flavors in other foods. The natural goodness of foods is destroyed by irradiation.

Comment: Like any other process, irradiation will maintain the good quality of food only if it is carried out under the right conditions. Irradiation with a dose that is too high for a particular food or irradiation with unsuitable packaging or at the wrong temperature can very well spoil a product. The fact that it is possible to burn a loaf of bread to ashes by overheating it is not a good argument that we should only eat raw grains. It is true that some foods, such as dairy products, cannot be preserved by irradiation because they acquire an off-flavor even from low doses of irradiation. But this is not a sufficient reason to condemn the process in general. Lettuce cannot be preserved well by freezing or canning. That does not mean that freezing and canning are bad processes.

Irradiated foods will have to compete with other foods in an open market. If consumers do not like the quality of irradiated foods, they will stop buying them.

9. Irradiation benefits only the food trade, because it keeps food in marketable condition for a longer period. The consumer will only have to pay the higher cost for the radiation treatment.

Comment: When irradiation is carried out to eliminate disease-causing microorganisms in foods, this is clearly in the interest of the consumer and it may well be worth a higher price. One can imagine, for instance, that pork with a label "Guaranteed trichina-safe by irradiation" or chicken with a label "Guaranteed salmonella-safe by irradiation" will be offered at a higher price than the unirradiated product, and that some consumers will be prepared to pay for such a guarantee. When the process is carried out to reduce spoilage losses, this may well lead to a lower price. After all, it is the consumer who pays for the trade's spoilage losses by paying a corresponding higher prince on that part which is sold.

At any rate, in an open market the future of food irradiation will depend on the quality and on the price of irradiated foods. If consumers have to pay a price that is not justified by the quality of the product, they will not continue to buy that product.

10. Irradiation of foods should not be permitted as long as reliable methods for identification of irradiated foods are not available. Without the means to test for food irradiation, labeling regulations cannot be enforced.

Comment: The lack of a reliable method for recognizing irradiated foods is not due to a lack of attempts to develop such a method. Scientists have tried for over 20 years to find suitable test methods. The fact that they have not been very successful is a good indication that the changes caused by irradiation are small and—what makes the task particularly difficult—they are nonspecific, i.e., they are not unique. Those who insist on such a method as a precondition for permitting the process practically say that they don't want food irradiation because it has so little effect on the composition of the food. With the constant improvement of analytical methods, it may eventually be possible to identify irradiated foods (first generation), but it is most unlikely that it will ever be possible to recognize irradiated ingredients in second- or third-generation products.

The way to enforce regulations of food irradiation is by controlling the irradiation facilities. In contrast to the use of additives, radiation processing cannot be carried out in somebody's back yard; it can only be done in government-approved and government-controlled irradiation plants. International agreements between countries permitting certain foods to be irradiated will allow enforcement of regulations in international trade. Those countries that permit no irradiated foods at all like the Federal Republic of Germany have the least control over imports of irradiated foods because identification of irradiated foods at the border is not possible and control of irradiation plants in exporting countries is not feasible for those governments which maintain a general prohibition of irradiated foods.

Adherents of Judaic or Islamic faith require meat from animals slaughtered in accordance with certain religious instructions. Science cannot provide a method capable of detecting any difference between meats fulfilling or not fulfilling these requirements. Nevertheless, there is worldwide trade with meat destined for consumption by these religious groups, and customers buying such meat can rely on its being in accordance with religious requirements. If this is possible by proper controls of production and trade, why should analogous controls not be possible for irradiated foods?

11. Irradiation up to a dose of 10 kGy may be safe. But dose control devices in the irradiation facility may fail and consumers might be harmed by overdosed foods.

Comment: In Chapter 5 studies were mentioned that showed no adverse effects even when radiation doses of 50–60 kGy were used. It should also be noticed that the sensory quality of all foods is affected by overdosing. As a rule, when a food is irradiated with a dose high enough to cause serious losses of vitamins or to cause other major changes in the chemical composition of the food, odor, taste, color, and/or texture will be seriously impaired, so that the consumer would reject such products. Food irradiation is thus a self-limiting process.

Some consumers, perhaps neither associated with the anti-nuclear movement nor having particular sympathies for alternative foods, have no objection to the irradiation of some more or less exotic products such as spices or mangoes or papayas, but they dislike the idea of "radiation facilities all over the country and all of our basic foods irradiated." They fear that with permission for spice irradiation or mango irradiation, the industry will "get its foot in the door, and the rest will follow."

Such a scenario is completely unrealistic. For economic and technical reasons, well-established processing methods such as canning or freezing or drying will not be replaced by irradiation. The new process requires high investments, and these will only be justified where irradiation offers decisive advantages over other processes. It is unlikely that the proportion of irradiated foods in the United States will exceed 1% by the year 2000, and even optimistic estimates do not foresee more than 5% in the long-range future. Irradiation has to be paid for, and it will therefore not be used where it is not needed. If spice irradiation is permitted (as it is in the United States), that does not mean all spices will be irradiated. Spices with low microbial contamination are needed as ingredients in the manufacture of certain highly perishable products, and it is only for that small section of the spice market that irradiated spices are needed. Similarly, there would be no point in irradiating the whole onion crop. Prevention of sprouting is only needed for those onions that are intended for long-term storage. If irradiation of poultry meat for salmonella clean-up is permitted or irradiation of pork for inactivation of trichina, this will not mean that all poultry and pork will be irradiated. The number of consumers willing to pay more for chicken bearing a label "Irradiated: guaranteed free of salmonella" or pork with a label "Trichina-safe by irradiation" will be limited.

Under these circumstances it is hard to understand why food irradiation has become such a hot political issue in some countries. Unfortunately, discussions of this topic in national parliaments, state legislatures, and city councils appear to be fueled more by ideologically motivated exaggerations than by a sober analysis of facts. How such discussions can be "snow jobbed" by "hand-picked local anti-nuclear-cum-anti-food irradiation activist agitators" has been described by Giddings [8].

Some studies mentioned in Sec. II have shown the existence of a population group who reject food irradiation because of personal beliefs, natural diets, or anti-nuclear concerns, and indicate that no educational campaign could change the attitude of these rejectors. The size of this group will differ in various geographical areas, and it will change in time. Events like the reactor accidents at Three Mile Island or Chernobyl (which hopefully will not be repeated) have strengthened the opposition to anything that was or seemed to be related to the nuclear industry. On the other hand, mass epidemics of food poisoning caused by salmonella species, campylobacter, shigella, or other pathogenic microorganisms (which undoubtedly will occur again and again) demonstrate the need for improving the hygienic quality of certain foods or food ingredients, and irradiation may be recognized as the most promising remedy.

The willingness of the remaining 85–95% of the population to purchase irradiated foods will depend very much on the availability of correct information about the process of food irradiation. The World Health Organization stated in its media service "In Point of Fact" no. 40/1987:

> Widespread information campaigns are still required for food irradiation to be fully accepted. WHO is concerned that rejection of the process, essentially based on emotional or ideological influences, may hamper its use in those countries which may benefit the most.

A Task Force on Irradiation Processing Wholesomeness Studies was established by the Third Conference for Food Protection, sponsored by the National Sanitation Foundation, meeting at Ann Arbor, Michigan, in August 1986. One of its recommendations stated [9]:

> Provide proper information and education to the public about the applications and limitations of food irradiation, and the form and meanings of labels on irradiated food, and about the basis for approving the process as safe and wholesome. Such education would minimize potential political exploitation of the understandable fears and disquiet of the public related to anything connected with nuclear energy and radiation.

Communications strategies for reaching consumers and the food industry with such information have been proposed by a task force meeting convened by the International Consultative Group on Food Irradiation [10].

IV. THE LABELING ISSUE

The FAO/IAEA/WHO Joint Expert Committee on Irradiated Food had made clear at its meeting in 1977 that "irradiation is a physical process for treating foods and as such it is comparable to the heating or freezing of foods for preservation" [11].

At its meeting in 1980, the Committee concluded that the irradiation of any food commodity up to an overall average dose of 10 kGy presents no toxicological hazard and no special nutritional or microbiological problems. Since irradiated foods would be subject to regulations covering foods generally, and to any specific food standards relating to individual foods, the Committee did not think it necessary on scientific grounds to envisage special requirements for labeling of irradiated foods [12].

However, labeling is as much a political issue as it is an issue to be discussed on scientific grounds. The conflict between a strictly scientific approach to this question and a consumer-oriented approach becomes particularly clear in the report of the British Advisory Committee on Irradiated and Novel Foods [13]. The Committee expressed its view on retail labeling thus:

> None of the considerations which, on health grounds alone, would require labelling to describe the process to which a food has been subjected applies to irradiated food.* There are no special microbiological consequences of food irradiation which mean that irradiated food must be handled in a special way by the consumer. Many items of food, irradiated or non-irradiated, should be stored under specified conditions and consumed within a set period of time, but this information can be conveyed to the consumer without any need to refer to the process used. We cannot

> foresee any special nutritional consequences of food irradiation which would necessitate retail labelling other than that already recommended for the nutritional content of foods in general. Finally, there is no evidence for any toxicological effects of food irradiation which would make it necessary for any groups to avoid or reduce their intake of irradiated food. We have therefore concluded that there are no health grounds for retail labelling of irradiated food.

Having said this, the report refers the reader to a report of the Food Advisory Committee, which states:

> . . . we felt strongly that there were real advantages in positive emphasis of irradiation as a beneficial treatment, and believe that the food industry could help by initiating a programme of education to promote the wider acceptance of the advantages of the process.
>
> We also recognised that despite any reassurance as to the safety and benefits of irradiation there were those members of the public who would wish to avoid irradiated food products. We could not ignore the fact that there was a significant emotive reaction among the public to any aspect of radiation which set this process apart from any other, and accordingly we took the view that failure to declare the fact of irradiation would be seen by the public not as a neutral act but as a positive concealment. As such it would be likely to exacerbate public fears and obscure the real benefits of irradiation.

Similar thoughts must have been on the FDA's mind when the agency demanded labeling of first-generation irradiated foods in its Final Rule of April 18, 1986 (see Chapter 10). The decision to require labeling [14]:

> depends not on the abstract worth of the information but on whether consumers view such information as important and whether the omission of label information may mislead the consumer. The large number of consumer comments requesting retail labeling attest to the significance placed on such labeling by consumers . . . . The agency emphasizes, however, that the labeling requirement is not based on any concern about the safety of the uses of radiation that are allowed under this final rule.

Consumer organizations that have formulated a policy toward food irradiation in different countries have invariably demanded retail labeling of irradiated foods. Governments cannot ignore these demands. But it must be recognized that this puts irradiated foods at a disadvantage in comparison to foods treated by other methods. If papayas have been quarantine-treated by the hot double-dip method, the consumer is not informed aobut this in any way. But he must be informed when the quarantine treatment was irradiation. The consumer will think that he has a choice between untreated papayas and irradiated papayas—and he will probably prefer the "untreated" ones. Chemical sprout inhibition of potatoes, onions, or garlic does not have to be labeled in most countries. Again, the consumer will assume that he can choose between untreated and irradiated produce, when he may actually be choosing between irradiated and chemically treated produce. Similarly, insect eradication in grain or dried fruits by chemical fumigation does not have to be labeled. Delay of senescence in fruits by controlled atmosphere storage does not require labeling. In all these cases, the labeled irradiated product will be at a disadvantage.

Obvious advantages of radiation processing combined with consumer education campaigns may overcome this handicap, for instance, in the case of irradiated papayas, the quality of which is clearly superior to that of double-dipped papayas. But the handicap for radiation processing becomes insurmountable in those countries where the labeling requirement also applies to irradiated ingredients or to the second- and third-generation products containing such ingredients. If sausages containing chemically fumigated spices do not have to be labeled while those containing irradiated spices have to be, every sausage manufacturer will prefer to use chemically treated spices.

As pointed out in Chapter 10, requiring the labeling of irradiated ingredients places an unreasonable burden on the food trade and also on the authorities charged with control of compliance. FDA has wisely limited its labeling requirement to first-generation irradiated foods. Some governments, for instance, that of the German Democratic Republic, require labeling of an irradiated ingredient only if it constitutes more than 10% of a food.

V. DEMANDS OF CONSUMER ORGANIZATIONS

Some consumer coalitions or self-appointed consumer advocates, like the California-based National Coalition to Stop Food Irradiation (NCSFI), have the declared purpose of preventing the use of this process. Others, such as the BEUC, the head organization of European Consumer Unions [15], are not opposed in principle, but demand retail labeling so that the consumer can make his own choice. The General Assembly of the International Organization of Consumer Unions, at its 12th World Congress held in Madrid, September 1987, demanded a worldwide moratorium on the further use and development of food irradiation "until there is a satisfactory resolution of issues of nutrition, safety, labelling and detection."

Of the comments and recommendations on food irradiation issued by various consumer organizations, those of the Australian Consumers' Association (ACA) are probably the most comprehensive [16]. In April 1987, ACA submitted the results of an inquiry and a list of 16 recommendations to the Commonwealth Minister of Health. Several of these relate to the specific Australian situation, where the current control of food irradiation is an ad hoc collection of Federal, State, and Territory laws and regulations. ACA demands that a Federal Food Irradiation Act be proclaimed to encompass all facets of the food irradiation industry, and that the responsibility to coordinate all matters under the Act be vested in a national body. Of broader interest outside of Australia are the following recommendations:

1. Blanket approval should not be granted to irradiated food. Approval should be granted on an item-by-item basis for specific foods, at a specific dose range (up to a maximum of 10 kGy), for a specific purpose, at an approved facility.

Comment: This is the path that has been followed in all of the countries where irradiation of foods is already allowed, with the exception that in a few instances (such as irradiation of spices in the United States) doses higher than 10 kGy are allowed.

2. Applications for approval should be accompanied by
   - a report assessing the toxicity aspects with reference to all relevant research
   - a report showing how the proposed application of irradiation fulfills a public health need and/or is the preferred technological option.

Comment: Again, this is the basis of approvals already granted in many countries.

3. Grains and ground nuts should be irradiated at a minimum level of 6 kGy.

Comment: It is rather amazing that a consumer organization would recommend a radiation dose more than 10 times higher than required. ACA is obviously under the mistaken impression that irradiation at a low dose stimulates aflatoxin production in grains and ground nuts and that therefore the dose should be high enough to kill aflatoxin-producing molds. As described in Chapter 6, the fear of radiation-stimulated aflatoxin production is based on some experiments carried out under completely unrealistic conditions. There are no indications that any food irradiated and stored under conditions prevailing in practice would be at risk of increased formation of mycotoxins. This ACA recommendation is clearly contrary to consumers' interests, and one can only hope that it will be quickly forgotten.

4. Federal, State, and Territory Departments of Health should keep up to date and accessible records of quantities of individual food items being irradiated.

Comment: The Recommended International Code of Practice for the Operation of Irradiation Facilities Used for the Treatment of Foods (see Annex II) contains under 4.3 the demand that irradiation facilities keep a record book that shows the nature and kind of the product being treated, its identifying marks if packed, or, if not, the shipping details, its bulk density, the type of source or electron machine, the dosimetry, the dosimeters used and details of their calibration, and the date of treatment. Why government agencies, in Australia even at three levels, should also keep such records is unclear. Government inspectors must have the right to inspect the records of the radiation facilities at any time. But why should they keep in triplicate what the company also has to keep?

5. Cesium-137 should be banned as a source of food irradiation.

Comment: At present ^{137}Cs is not used for commercial food irradiation anywhere. The demanded banning of this radioisotope would in no way threaten the practical use of this process.

6. Cobalt-60 should be purchased only under a contract that provides for transport, installation, maintenance, and return of cobalt after use.

Comment: This is normal practice.

7. Licenses should be granted conditional upon a full environmental impact study and an economic feasibility study including an appropriate sum of money for use in the event of an accident as a bond which covers public liability to people or to the environment.

Comment: Decades of experience have shown that irradiation facilities are not more damaging to the environment than any other industrial activity. Demands concerning environmental impact, liability, etc. should not be more severe than those demanded of other industrial enterprises. The US Food and Drug Administration found no significant environmental impact of permitting food irradiation and concluded in its final rule of April 18, 1986 that an environmental impact statement was not required.

Economic feasibility should be the problem of the investor, not the government.

8. Labeling regulations for all irradiated food including irradiated ingredients should be developed.

Comment: This topic was discussed in the previous section.

9. Approval to irradiate some food items should be conditional upon the use of specific additional labeling requirements, e.g., concerning the temperature at which the food should be stored.

Comment: No argument.

10. Consumer education undertaken or funded by governments should state the conditional safety and the conditional wholesomeness of irradiated foods.

Comment: The meaning of this recommendation is not clear. The safety of foods irradiated up to a dose of 10 kGy is not more or less conditional than the safety of foods treated in many other ways. If ACA means to say that demands for absolute safety can never be fulfilled, one can only agree. But this applies to all foods.

11. All costs associated with the establishment, operation, and safety of the industry should be charged to the industry.

Comment: This presumably means that none of these costs should be paid by the government. No argument, as long as this refers to the operation of commercial radiation facilities. On the other hand, many governments have spent large sums on research aimed at improving the sensory quality or the health safety or the economics of canned, frozen, fumigated, packaged, or otherwise processed foods. Government-sponsored research on irradiated foods is just at legitimate.

In summary, while some of these recommendations appear entirely reasonable, others would suffocate food irradiation under mountains of bureaucratic nonsense. ACA is not alone in exaggerating its safety demands to a point where practical use of the process becomes nearly impossible. In the last 30 years of its history, food irradiation has again and again been burdened with demands no other processing method has ever had to meet.

Some observers have voiced the opinion that much of the rigid opposition to food irradiation which has sprung up in some countries in the last few years is the result of intentional misinformation spread by small groups of political activists who consider the struggle over food irradiation as a good opportunity to gain influence and power. In the words of Jan Taylor, who has been active in consumer affairs for many years and who is now Commissioner for Consumer Affairs in Queensland, Australia [17]:

> I have yet to find any other consumer issue which measures up to the vitriol, the misinformation, the nosense, that has been stated about food irradiation . . . . While I am satisfied as to its safety and efficacy, quite patently many consumers are not. I would suggest that many of those consumers simply have not yet had the opportunity to learn about the process without the hysteria and emotionalism that seems to be such an intrinsic part of the current debate.

But she also had this to say about industry:

> Unfortunately there are still industry associations and individual corporations which demonstrate nothing but arrogance, condescension or paternalism in their dealings with the public.

And she urged industry to

> show their willingness not just to sell and promote the process but to *talk* about the process—to talk to consumers, government and the media, to acknowledge that this process is no more perfect than any other. They should show a preparedness for balanced discussion of food irradiation which shows it, warts and all, as yet another process to add to those already in existence . . . . One that doesn't meet every need but which is highly appropriate in certain circumstances and on certain foods, if only it was given the opportunity.

Surely this is sound advice—not only for Australia.

REFERENCES

1. Anonymous, Inform consumers about irradiated food, *Canadian Consumer*, 13(8):47 (1983).
2. van der Linde, H. J., and H. T. Brodrick, Commercial experience in introducing radurized foods to the South African market, in *Food Irradiation Processing*, Proceedings of a Symposium held in Washington, D.C., March 1985. Int. Atomic Energy Agency, Vienna, 1985, p. 137.
3. Bruhn, C. M., and J. W. Noell, Consumer in-store response to irradiated papayas, *Food Technology* (Chicago), 41(9):84 (1987).

4. Brand Group, Inc., Irradiated seafood products. A position paper for the seafood industry. Final report, Chicago, Ill., 1986.
5. Bruhn, C. M., H. G. Schutz, and R. Sommer, Attitude change toward food irradiation among conventional and alternative consumers, *Food Technology* (Chicago), 40(1):86 (1986).
6. Bruhn, C. M., R. Sommer, and H. G. Schutz, Effect of an educational pamphlet and posters on attitude toward food irradiation, *J. Indust. Irradiation Techn.*, *4*:1 (1986).
7. Wiese Research Associates, Inc., Consumer Reaction to the Irradiation Concept, Summary Report prepared for Albuquerque Operations Office, U.S. Dept. of Energy and National Pork Producers Council Contract No. DE-SC04-84AL 24460, March 1984.
8. Giddings, G. G., Update on N.J. irradiation bill, *Food Technology* (Chicago), *41*(9):38 (1987).
9. Elias, P. S., Task force on irradiation processing wholesomeness studies, in *Food Protection Technology*, C. W. Felix, ed. Lewis Publishers, Inc., Chelsea, Michigan, 1987, p. 349.
10. IAEA, Guidelines for Acceptance of Food Irradiation, IAEA-TECDOC-432, Vienna, 1987.
11. WHO, Wholesomeness of Irradiated Foods, WHO Technical Report Series 604, Geneva, 1977.
12. WHO, Wholesomeness of Irradiated Foods, WHO Technical Report Series, 659, Geneva, 1981.
13. HMSO, Report on the Safety and Wholesomeness of Irradiated Foods, Advisory Committee on Irradiated and Novel Foods, Her Majesty's Stationery Office, London, 1986.
14. FDA, Irradiation in the production, processing, and handling of food, *Federal Register*, *51*:13376 (1986).
15. BEUC, Discussion document on irradiated food, prepared for the 7th European Consumer Forum, Berlin, January 1983, AgV/BEUC/127/82.
16. ACA, Food Irradiation—An Inquiry by the Australian Consumers' Association, April 1987.
17. Taylor, J., Consumer views on acceptance of irradiated food, Presented at International Conference on Acceptance, Control of, and Trade in Irradiated Food, Geneva, Switzerland, December 1988.

SUGGESTED READING

Ford, N. J., and D. M. Rennie, Consumer understanding of food irradiation, *J. Consumer Studies and Home Econ.*, 11:305 (1987).

12
Outlook

Doubts about the safety of consumption of irradiated foods have held up the practical introduction of this process for many years. As described in Chapters 3 through 7, vast research programs have been carried out to test the wholesomeness of irradiated foods, and the processes occurring when foods are irradiated are now rather well understood. Committees of experts who have examined the available evidence at an international level and, in many countries, at a national level, have agreed that there is sufficient evidence to conclude that foods irradiated at a dose level of up to 10 kGy are safe and wholesome. Some experts, like those assembled in the CAST task force (Council for Agricultural Science and Technology) have avoided setting an upper dose limit and have concluded that foods exposed to ionizing radiation under the conditions proposed for commercial application are wholesome, that is, safe to eat.

Governments have not rushed into giving approvals for food irradiation on the basis of these committee recommendations. Not one government has given blanket approval for all irradiated foods. Where permissions have been granted, they have been limited to specific commodities or groups of commodities irradiated for specific purposes and at specific dose levels. These permissions have not led to an explosive growth of the radiation industry. In some countries, especially in Belgium and the Netherlands, growth has been steady in the last 10 years. In others, like the United States and Japan, developments have stagnated as a result of much adverse political activity. In a third group of countries, like France and the People's Republic of China, food irradiation on a commercial scale has been introduced so recently that it is too early to evaluate the progress. Finally, there are those countries where industrialization of food irradiation has not even been started, either because of strong political opposition, as in the Federal Republic of Germany, or because of lack of capital and/or infrastructure.

Of the many possible applications of food irradiation (see Chapter 9), only a limited number are now being used in commercial practice in those countries where permissions have been granted. They belong to the following categories:

Improvement of the hygienic quality of foods in cases where no other methods are available to achieve this purpose.
Replacement of chemical treatments.
Improvement of shelf life of certain fruits.
Improvement of sensory quality.

Representative of the first category is the irradiation of frozen deboned chicken meat (France), frozen shrimp and froglegs (Netherlands), and the irradiation of spices and dehydrated vegetables (Belgium and about a dozen other countries). The purpose of the treatment is destruction of salmonella, campylobacter, and other non-spore forming pathogenic microorganisms, and reduction in number of other microorganisms.

The second category comprises the irradiation of potatoes (Japan) and onions (German Democratic Republic) instead of using chemical sprout inhibitors, and irradiation of wheat (Soviet Union) in place of chemical fumigation for insect disinfestation.

To the third category belongs the irradiation of bananas, mangoes, strawberries, and some other fruits (Republic of South Africa).

In the fourth category is the irradiation of sweet potato brandy for improvement of taste (People's Republic of China).

At a time when mass epidemics of foodborne infections are of growing concern to public health authorities around the world, it appears safe to predict that the first of these applications will continue to grow, both in the number of countries where this is practiced and in the tonnage of food irradiated. The potential of irradiation for parasite disinfection in foods of animal origin has not yet been utilized anywhere. The fight against trichinosis, toxoplasmosis, cysticercosis, and other infections from parasitic protozoa and helminths, which are rampant in many tropical areas, could well be supported by low-dose irradiation of meat.

The future of the second application will depend very much on the public and regulatory attitude toward the remaining pesticides, fumigants, and sprout inhibitors. Several countries have restricted or prohibited the use of ethylene oxide for decontamination of spices and other dry food ingredients. The U.S. Environmental Protection Agency (EPA) has banned the use of the fumigant ethylene dibromide for insect disinfestation of fruits and vegetables, cereals, pulses, and nuts due to its suspected carcinogenicity. Similar fumigants, such as methyl bromide, are also suspect and are currently being closely studied. There is an urgent need to introduce alternative treatments to maintain international trade in various food commodities.

The prospects for the third application, the improvement of shelf life of fruits, are more uncertain, at least in those countries where governments are under a barrage of attacks from anti-irradiation movements. A minister of health who may be quite willing to publicly defend a permission for irradiation of poultry meat for salmonella control may be less inclined to do the same for a permission to prolong the shelf life of fresh fruits. Whereas he has a good chance of gaining applause for his action to protect the consumer against salmonellosis, he may risk his political future if he gives the media the opportunity to spread the activists' claim that "radioactive irradiation of fruits is only in the interest of the food trade and the nuclear industry." Consumers are, of course, interested in preventing quick spoilage of foods, as is demonstrated by every household refrigerator, but it is perhaps more difficult to bring this across than to point to the importance of salmonella control.

Applications of the fourth category, improvement of sensory quality, may well extend to products other than sweet potato brandy as the food industry gains experience with the process.

A potential fifth application, irradiation as a quarantine treatment, is only partly identical with the second, the replacement of chemicals. The example of papaya quarantine shows that formerly used fumigation treatments can be replaced by methods such as double dipping, but it also shows that the quality of the produce so treated leaves much to be desired. In that case superior quality of irradiated fruits may eventually lead to another commercial use of irradiation processing. Some of the miscellaneous other applications mentioned in Chapter 9 may find acceptance in the longer-range future.

The conclusion of the FAO/IAEA/WHO Joint Expert Committee on Irradiated Foods at its meeting in 1980 that irradiation of any food commodity up to an overall average dose of 10 kGy presents no toxicological hazard, and the issuance of the Codex Alimentarius Standard based on that conclusion, have set the stage for worldwide utilization of this process. However, as pointed out in Chapter 9, irradiation is not a miracle method. Its usefulness is limited to certain products and certain purposes. It will not provide competition to established processes such a canning, freezing and drying.

To what extent food irradiation will be commercially used in various countries does not depend on further research. It depends primarily on the public acceptance of the process. In the United States a number of groups have been formed under names such as

Coalition Against Food Irradiation. They have initiated letter-writing campaigns and contacted congressmen as well as state officials to seek legislation to delay the application of the FDA approvals for food irradiation or to ban the distribution of irradiated foods within the state. Similar activities exist in some European countries, expecially in the Federal Republic of Germany and in the United Kingdom. In Japan the opposition to food irradiation developed earlier than in these other countries and has apparently not waned in recent years. An International Network Against Food Irradiation was founded in 1987, supported by various consumer initiatives around the world.

In Chapter 11 an attempt was made to analyze some of the motives and arguments of groups opposed to food irradiation. An in-depth study of the anti-irradiation movement must be left to sociologists or political scientists. However, anyone who has observed developments in food technology and food legislation in recent years can see that this movement has been quite successful. It has delayed approvals for food irradiation in many countries and has completely prevented them in others. Where approvals have been granted, as in the United States, food manufacturers and food trade have been sufficiently disturbed by threats of picketing and consumer boycotts that in recent years, with the exception of some very limited test marketing, no company has dared to offer retail labeled irradiated foods for sale. In countries where a sizeable tonnage of foods is irradiated, like the Netherlands and Belgium, all irradiated products are used as ingredients by food manufacturers, and foods containing these ingredients are not labeled as irradiated. Where a permission for food irradiation has been coupled with a regulatory demand for labeling of second- and third-generation products, as in Denmark, the food industry has made no use of the permission (see Chapter 10). Test marketing in several countries has shown that a majority of consumers is willing to purchase foods labeled as irradiated. However, such test marketing situations, where special efforts were made to educate the public about the irradiation process and trained personnel were at hand to answer questions from consumers, may not have been typical of the normal marketing situation. How the general public will respond to foods labeled as irradiated remains to be seen.

As pointed out by the World Health Organization, information campaigns are required for food irradiation to be fully accepted by the public. The UN organizations, particularly FAO, IAEA, and WHO, have done much to achieve this. They have jointly established the International Consultative Group on Food Irradiation (ICGFI) in 1984. The major objectives of ICGFI are to evaluate global developments in the field of food irradiation and to provide a focal point of advice on the application of food irradiation to member states. At present, 26 governments are members of ICGFI and are contributing to its activities. ICGFI has assigned the highest priority in its program of work to the promotion of public information on food irradiation. This means discussing the advantages and limitations of irradiation in comparison with other preservation processes in an objective manner—and that may not always make very exciting reading. At a time when sensationalism seems to rule the media, it is difficult for those who insist on objective, balanced reporting to be noticed at all. The anti–food irradiation movement provides much more attractive headlines than the factual bulletins of a UN organization. One may expect, however, that in the long run the blatant disinformation spread by some of the opponents of food irradiation—examples of which have been cited in Chapters 5 and 11—will be counterproductive for the originators.

Looking at the future, the question is not "Will we eat irradiated food?" but "How much irradiated food will we eat?" Reliable methods of identifying irradiated commodities are not available, and identification of irradiated ingredients is even less likely. Even governments that have so far upheld a complete ban on irradiated foods cannot prevent the inflow of foods containing irradiated ingredients from neighboring countries where irradiation is allowed. This applies, for instance, to the Federal Republic of Germany, which imports much food from the Netherlands, France, and Belgium, where food irradiation is practiced on a growing scale. In the United States, even if state governments ban the sale of irradiated foods, it is unlikely that such bans could be effective in the face of FDA approvals for irradiation of a number of commodities and similar approvals in Canada and Mexico.

In Chapter 11 it was estimated that the proportion of irradiated foods in the United States will probably not exceed 1% by the end of the century, perhaps 5% in the long-range future. Similar figures appear realistic for the European Common Market. Continued political opposition will probably not reduce this to 0%, but it could greatly delay further steps towards commercial use of food irradiation. In some countries this opposition will assure a continuation of the present ban on food irradiation; in other countries it will convince regulatory agencies to make the conditions for premarket clearance and for practical use of the process so complicated and cumbersome that they become prohibitive.

Sceptics may ask, why should we not completely forget food irradiation, if it will apply to such a small percentage of the market? The answer is that food irradiation, even on a relatively small scale, promises to solve certain problems of food hygiene and quarantine which cannot be satisfactorily solved in any other way. And with regard to economics, it may be pointed out that 1% of the United States food market is a quantity of over 5000 tons per day or over two million tons per year, which would still be a sizeable economic activity.

Appendix I: Codex General Standard for Irradiated Foods

1. *SCOPE*

This standard applies to foods processed by irradiation. It does not apply to foods exposed to doses imparted by measuring instruments used for inspection purposes.

2. *GENERAL REQUIREMENTS FOR THE PROCESS*

2.1. *Radiation Sources*

The following types of ionizing radiation may be used:
(a) Gamma rays from the radionuclides ^{60}Co or ^{137}Cs;
(b) X-rays generated from machine sources operated at or below an energy level of 5 MeV.
(c) Electrons generated from machine sources operated at or below an energy level of 10 MeV.

2.2. *Absorbed Dose*

The oberall average dose absorbed by a food subjected to radiation processing should not exceed 10 kGy[1,2].

2.3. *Facilities and Control of the Process*

2.3.1. Radiation treatment of foods shall be carried out in facilities licensed and registered for this purpose by the competent national authority.

Source: Codex Alimentarius, vol. XV, 1st edit., 1984.
[1,2]See notes on last page of this appendix.

2.3.2. The facilities shall be designed to meet the requirements of safety, efficacy and good hygienic practices of food processing.
2.3.3. The facilities shall be staffed by adequate, trained and competent personnel.
2.3.4. Control of the process within the facility shall include the keeping of adequate records including quantitative dosimetry.
2.3.5. Premises and records shall be open to inspection by appropriate national authorities.
2.3.6. Control should be carried out in accordance with the Recommended International Code of Practice for the Operation of Radiation Facilities used for the Treatment of Foods.

3. *HYGIENE OF IRRADIATED FOODS*

3.1.

The food should comply with the provisions of the Recommended International Code of Practice—General Principles of Food Hygiene (Ref. No. CAC/RCP 1-1969, Rev. 1, 1979) and, where appropriate, with the Recommended International Code of Hygienic Practice of the Codex Alimentarius relative to a particular food.

3.2.

Any relevant national public health requirement affecting microbiological safety and nutritional adequacy applicable in the country in which the food is sold should be observed.

4. *TECHNOLOGICAL REQUIREMENTS*

4.1. Conditions for Irradiation

The irradiation of food is justified only when it fulfils a technological need or where it serves a food hygiene purpose[3] and should not be used as a substitute for good manufacturing practices.

4.2. Food Quality and Packaging Requirements

The doses applied shall be commensurate with the technological and public health purposes to be achieved and shall be in accordance with good radiation processing practice. Foods to be irradiated and their packaging materials shall be of suitable quality, acceptable hygienic condition and appropriate for this purpose and shall be handled, before and after irradiation, according to good manufacturing practices taking into account the particular requirements of the technology of the process.

---

[3]See note on last page of this appendix.

5. *RE-IRRADIATION*

5.1.

Except for foods with low moisture content (cereals, pulses, dehydrated foods and other such commodities) irradiated for the purpose of controlling insect reinfestation, foods irradiated in accordance with sections 2 and 4 of this standard shall not be re-irradiated.

5.2.

For the purpose of this standard food is not considered as having been re-irradiated when: (a) the food prepared from materials which have been irradiated at low dose levels e.g. about 1 kGy, is irradiated for another technological purpose; (b) the food, containing less than 5% of irradiated ingredient, is irradiated, or when (c) the full dose of ionizing radiation required to achieve the full dose of ionizing radiation required to achieve the desired effect is applied to the food in more than one installment as part of processing for a specific technological purpose.

5.3.

The cumulative overall average dose absorbed should not exceed 10 kGy as a result of re-irradiation.

6. *LABELLING*

6.1. *Inventory Control*

For irradiated foods, whether prepackaged or not, the relevant shipping documents shall give appropriate information to identify the registered facility which has irradiated the food, the date(s) of treatment and lot identification.

6.2. *Prepackaged foods intended for direct consumption*

The labelling of prepackaged irradiated foods shall be in accordance with the relevant provisions of the Codex General Standard for Labelling of Prepackaged Foods[4].

6.3. *Foods in bulk containers*

The declaration of the fact or irradiation shall be made clear on the relevant shipping documents.

---

[4]See note on last page of this appendix.

NOTES

(1)For measurement and calculation of overall average dose absorbed see Annex A of the Recommended International Code of Practice for the Operation of Radiation Facilities used for Treatment of foods.

(2)The wholesomeness of foods, irradiated so as to have absorbed an overall average dose of up to 10 kGy, is not impaired. In this context the term "wholesomeness" refers to safety for consumption of irradiated foods from the toxicological point of view. The irradiation of foods up to an overall average dose of 10 kGy introduces no special nutritional or microbiological problems (Wholesomeness of Irradiated Foods, Report of a Joint FAO/IAEA/WHO Expert Committee, Technical Report Series 659, WHO, Geneva, 1981).

(3)The utility of the irradiation process has been demonstrated for a number of food items listed in Annex B to the Recommended International Code of Practice for the Operation of Radiation Facilities used for the Treatment of Foods.

(4)Under revision by the Codex Committee on Food Labelling.

Appendix II. Recommended International Code of Practice for the Operation of Irradiation Facilities Used for the Treatment of Foods

1. *INTRODUCTION*

This code refers to the operation of irradiation facilities based on the use of either a radionuclide source (^{60}Co or ^{137}Cs) or X-rays and electrons generated from machine sources. The irradiation facility may be of two designs, either "continuous" or "batch" type. Control of the food irradiation process in all types of facility involves the use of accepted methods of measuring the absorbed radiation dose and of the monitoring of the physical parameters of the process. The operation of these facilities for the irradiation of food must comply with the Codex recommendations on food hygiene.

2. *IRRADIATION PLANTS*

2.1. *Parameters*

For all types of facility the doses absorbed by the product depend on the radiation parameter, the dwell time or the transportation speed of the product, and the bulk density of the material to be irradiated. Source-product geometry, especially distance of the product from the source and measures to increase the efficiency of radiation utilization, will influence the absorbed dose and the homogeneity of dose distribution.

2.1.1. *Radionuclide sources*

Radionuclides used for food irradiation emit photons of characteristic energies. The statement of the source material completely determines

Source: Codex Alimentarius, Vol. XV, 1st edit., 1984.

the penetration of the emitted radiation. The source activity is measured in Becquerel (Bq) and should be stated by the supplying organisation. The actual activity of the source (as well as any return or replenishment of radionuclide material) shall be recorded. The recorded activity should take into account the natural decay rate of the source and should be accompanied by a record of the date of measurement or recalculation. Radionuclide irradiators will usually have a well separated and shielded depository for the source elements and a treatment area which can be entered when the source is in the safe position. There should be a positive indication of the correct operational and of the correct safe position of the source which should be interlocked with the product movement system.

2.1.1. Machine sources

A beam of electrons generated by a suitable accelerator, or after being converted to X-rays, can be used. The penetration of the radiation is governed by the energy of the electrons. Average beam power shall be adequately recorded. There should be a positive indication of the correct setting of all machine parameters which should be interlocked with the product movement system. Usually a beam scanner or a scattering device (e.g., the converting target) is incorporated in a machine source to obtain an even distribution of the radiation over the surface of the product. The product movement, the width and speed of the scan and the beam pulse frequency (if applicable) should be adjusted to ensure a uniform surface dose.

2.2. Dosimetry and Process Control

Prior to the irradiation of any foodstuff certain dosimetry measurements[1] should be made, which demonstrate that the process will satisfy the regulatory requirements. Various techniques for dosimetry pertinent to radionuclide and machine sources are available for measuring absorbed dose in a quantitative manner[2].

Dosimetry commissioning measurements should be made for each new food, irradiation process and whenever modifications are made to source strength or type and to the source product geometry.

Routine dosimetry should be made during operation and records kept of such measurement. In addition, regular measurements of facility parameters governing the process, such as transportation speed, dwell time, source exposure time, machine beam parameters, can be made during the facility operation. The records of these measurements can be used as supporting evidence that the process satisfies the regulatory requirements.

---

[1]See Annex A to this Code.

[2]Detailed in the Manual of Food Irradiation Dosimetry, IAEA, Vienna, 1977, Technical Report Series No. 178.

3. *GOOD RADIATION PROCESSING PRACTICE*

Facility design should attempt to optimalize the dose uniformity ratio, to ensure appropriate dose rates and, where necessary, to permit temperature control during irradiation (e.g. for the treatment of frozen food) and also control of the atmosphere. It is also often necessary to minimize mechanical damage to the product during transportation irradiation and storage, and desirable to ensure the maximum efficiency in the use of the irradiator. Where the food to be irradiated is subject to special standards for hygiene or temperature control, the facility must permit compliance with these standards.

4. *PRODUCT AND INVENTORY CONTROL*

4.1.

The incoming product should be physically separated from the outgoing irradiated products.

4.2.

Where appropriate, a visual colour change radiation indicator should be affixed to each product pack for ready identification of irradiated and non-irradiated products.

4.3.

Records should be kept in the facility record book which show the nature and kind of the product being treated, its identifying marks if packed or, if not, the shipping details, its bulk density, the type of source or electron machine, the dosimetry, the dosimeters used and details of their calibration, and the date of treatment.

4.4.

All products shall be handled, before and after irradiation, according to accepted good manufacturing practices taking into account the particular requirements of the technology of the process[3]. Suitable facilities for refrigerated storage may be required.

---

[3]See Annex B to this Code.

ANNEX A

DOSIMETRY

1. *The overall average absorbed dose*

It can be assumed for the purpose of the determination of the wholesomeness of food treated with an overall average dose of 10 kGy or less, that all radiation chemical effects in that particular dose range are proportional to dose.

The overall average dose, $\overline{D}$, is defined by the following integral over the total volume of the goods

$$\overline{D} = \frac{1}{M} \int \rho\,(x, y, z) \cdot d\,(x, y, z) \cdot dV$$

where

M = the total mass of the treated sample

ρ = the local density at the point (x, y, z)

d = the local absorbed dose at the point (x, y, z)

dV = dx dy dz the infinitesimal volume element which in real cases is represented by the volume fractions.

The overall average absorbed dose can be determined directly for homogeneous products or for bulk goods of homogeneous bulk density by distributing an adequate number of dose meters strategically and at random throughout the volume of the goods. From the dose distribution determined in this manner an average can be calculated which is the overall average absorbed dose.

If the shape of the dose distribution curve through the product is well determined the positions of minimum and maximum dose are known. Measurements of the distribution of dose in these two positions in a series of samples of the product can be used to give an estimate of the overall average dose. In some cases the mean value of the average values of the minimum ($\overline{D}_{min}$) and maximum ($\overline{D}_{max}$) dose will be a good estimate of the overall average dose. i.e. in these cases

$$\text{overall average dose} \approx \frac{\overline{D}_{max} + \overline{D}_{min}}{2}$$

2. Effective and limiting dose values

Some effective treatment e.g. the elimination of harmful microorganisms, or a particular shelflife extension, or a disinfestation requires a minimum absorbed dose. For other applications too high an absorbed dose may cause undesirable effects or an impairment of the quality of the product.

The design of the facility and the operational parameters have to take into account minimum and maximum dose values required by the process. In some low dose applications it will be possible within the terms of section 3 on Good Radiation Processing Practice to allow a ratio of maximum to minimum dose of greater than 3.

With regards to the maximum dose value under acceptable wholesomeness considerations and because of the statistical distribution of the dose a mass fraction of product of at least 97.5% should receive an absorbed dose of less than 15 kGy when the overall average dose is 10 kGy.

3. Routine Dosimetry

Measurements of the dose in a reference position can be made occasionally throughout the process. The association between the dose in the reference position and the overall average dose must be known. These measurements should be used to ensure the correct operation of the process. A recognized and calibrated system of dosimetry should be used.

A complete record of all dosimetry measurements including calibration must be kept.

4. Process Control

In the case of a continuous radionuclide facility it will be possible to make automatically a record of transportation speed or dwell time together with indications of source and product positioning. These measurements can be used to provide a continuous control of the process in support of routine dosimetry measurements.

In a batch operated radionuclide facility automatic recording of source exposure time can be made and a record of product movement and placement can be kept to provide a control of the process in support of routine dosimetry measurements.

In a machine facility a continuous record of beam parameters, e.g. voltage, current, scan speed, scan width, pulse repetition and a record of transportation speed through the beam can be used to provide a continuous control of the process in support of the routine dosimetry measurements.

ANNEX B

EXAMPLES OF TECHNOLOGICAL CONDITIONS FOR THE IRRADIATION OF SOME INDIVIDUAL FOOD ITEMS SPECIFICALLY EXAMINED BY THE JOINT FAO/IAEA/WHO EXPERT COMMITTEE

This information is taken from the Reports of the Joint FAO/IAEA/WHO Expert Committees on Food Irradiation (WHO Technical Report Series, 604, 1977 and 659, 1981) and illustrates the utility of the irradiation process. It also describes the technological conditions for achieving the purpose of the irradiation process safely and economically.

1. *CHICKEN (Gallus domesticus)*

1.1. *Pruposes of the Process*

The purposes of irradiating chicken are:
(a) to prolong storage life
and/or
(b) to reduce the number of certain pathogenic microorganisms, such as *Salmonella* from eviscerated chicken.

1.2. *Specific Requirements*

Average dose: for (a) and (b), up to 7 kGy

2. *COCOA BEANS (Theobroma cacao)*

2.1. *Purposes of the Process*

The purposes of irradiating cocoa beams are:
(a) to control insect infestation in storage
(b) to reduce microbial load of fermented beans with or without heat treatment.

2.2. *Specific Requirements*

2.2.1. Average dose: for (a) up to 1 kGy, for (b) up to 5 kGy.
2.2.2. Prevention of Reinfestation: Cocoa beans whether prepackaged or handled in bulk, should be stored as far as possible, under such conditions as will prevent reinfestation and microbial recontamination and spoilage.

3. *DATES (Phoenix dactylifera)*

3.1. Purpose of the Process

The purpose of irradiating prepackaged dried dates is to control insect infestation during storage.

3.2. Specific Requirements

3.2.1. Average dose: up to 1 kGy
3.2.2. Prevention of Reinfestation: Prepackaged dried dates should be stored under such conditions as will prevent reinfestation.

4. *MANGOES (Mangifera indica)*

4.1. Purposes of the Process

The purposes of irradiating mangoes are:
(a) to control insect infestation
(b) to improve keeping quality by delaying ripening
(c) to reduce microbial load by combining irradiation and heat treatment.

4.2. Specific Requirements

Average dose: up to 1 kGy

5. *ONIONS (Allium cepa)*

5.1. Purpose of the Process

The purpose of irradiating onions is to inhibit sprouting during storage.

5.2. Specific Requirement

Average dose: up to 0.15 kGy

6. *PAPAYA (Carica papaya L.)*

6.1. Purpose of the Process

The purpose of irradiating papaya is to control insect infestation and to improve its keeping quality by delaying ripening.

6.2. Specific Requirements

6.2.1. Average dose: up to 1 kGy
6.2.2. Source of Radiation: The source of radiation should be such as will provide adequate penetration.

7. *POTATOES (Solanum tuberosum L.)*

7.1. *Purpose of the Process*

The purpose of irradiating potatoes is to inhibit sprouting during storage.

7.2. *Specific Requirement*

Average dose: up to 0.15 kGy

8. *PULSES*

8.1. *Purpose of the Process*

The purpose of irradiating pulses is to control insect infestation in storage.

8.2. *Specific Requirement*

Average dose: up to 1 kGy

9. *RICE (Oryza species)*

9.1. *Purpose of the Process*

The purpose of irradiating rice is to control insect infestation in storage.

9.2. *Specific Requirements*

9.2.1. Average dose: up to 1 kGy
9.2.2. Prevention of Reinfestation: Rice, whether prepackaged or handled in bulk, should be stored as far as possible, under such conditions as will prevent reinfestation.

10. *SPICES AND CONDIMENTS, DEHYDRATED ONIONS, ONION POWDER*

10.1. *Purposes of the Process*

The purposes of irradiating spices, condiments, dehydrated onions and onion powder are:
(a) to control insect infestation
(b) to reduce microbial load
(c) to reduce the number of pathogenic microorganisms.

10.2. *Specific Requirement*

Average dose: for (a) up to 1 kGy, for (b) and (c) up to 10 kGy.

11. STRAWBERRY (Fragaria species)

11.1. Purpose of the Process

The purpose of irradiating fresh strawberries is to prolong the storage life by partial elimination of spoilage organisms.

11.2. Specific Requirement

Average dose: up to 3 kGy

12. TELEOST FISH AND FISH PRODUCTS

12.1. Purposes of the Process

The purposes of irradiating teleost fish and fish products are:
(a) to control insect infestation of dried fish during storage and marketing
(b) to reduce microbial load of the packaged or unpackaged fish and fish products
(c) to reduce the number of certain pathogenic microorganisms in packaged or unpackaged fish and fish products.

12.2. Specific Requirements

12.2.1. Average dose: for (a) up to 1 kGy, for (b) and (c) up to 2.2 kGy.
12.2.2. Temperature Requirement: During irradiation and storage the fish and fish products referred to in (b) and (c) should be kept at the temperature of melting ice.

13. WHEAT AND GROUND WHEAT PRODUCTS (Triticum species)

13.1. Purpose of the Process

The purpose of irradiating wheat and ground wheat products is to control insect infestation in the stored product.

13.2. Specific Requirements

13.2.1. Average dose: up to 1 kGy
13.2.2. Prevention of Reinfestation: These products, whether prepackaged or handled in bulk, should be stored as far as possible under such conditions as will prevent reinfestation.

Appendix III: Glossary of Terms and Abbreviations

accelerator A device for accelerating atomic particles (e.g., **electrons**, **protons**) to high speed, imparting large amounts of kinetic energy to these particles.

alpha particle Helium nucleus, consisting of 2 **protons** and 2 **neutrons**, emitted from the nucleus of a **radionuclide**.

aqueous electron, e^-_{aq} The hydrated or solvated **electron**, a product of the **radiolysis** of water.

atomic number The number of **protons** in an atomic nucleus. The atomic number is commonly written as a subscript before the symbol of the element (e.g., $_{92}U$ for uranium).

a_w See **water activity**.

becquerel (Bq) The SI unit of **radioactivity**, being one radioactive disintegration per second of time. It has the dimension of s^{-1}, and its relationship to the formerly used unit, the **curie (Ci)**, is: 1 Bq = 2.7×10^{-11} Ci.

beta particle High-speed **electron** (β^-) or **positron** (β^+) emitted from a nucleus during radioactive decay.

carcinogen A substance that causes cancer.

carcinogenicity The power or property of causing cancer.

---

*Cross reference to other items in the alphabetical list is provided by boldface printing.

Codex Alimentarius A collection of internationally adopted food standards (codes of practice, guidelines, etc.) resulting from the activities of the Joint FAO/WHO Food Standards Program as implemented by the Codex Alimentarius Commission, on which 130 states were represented in 1988.

curie (Ci) The formerly used unit of **radioactivity**, originally defined as the radioactivity of 1 gram of radium. The SI unit **becquerel** (Bq) should now be used. 1 Ci = 3.7×10^{10} Bq.

D_{min}, D_{max} Minimum and maximum absorbed **radiation doses** in the product.

D_{10} The required **radiation dose** to reduce the number of microorganisms by a factor of 10 or one log cycle.

disinfestation Control of the proliferation of insect and other pests in grain, cereal products, dried fruit, spices, etc.

dose In **radiation** technology this refers to the **radiation dose**. In toxicology the word dose refers to the quantity of a substance fed to test animals or consumed by man, usually indicated in mg per kg of body weight per day.

dose distribution The spatial variation in **radiation dose** throughout the product, the dose having the extreme values D_{max} and D_{min}.

dose meter A device, instrument, or system having a reproducible and measurable response to **radiation** that can be used to measure or evaluate the quantity termed absorbed dose, exposure, or similar radiation quantity. The word dosimeter has been replaced by dose meter as standard terminology (dose meter, but dosimetry, dosimetric).

dose uniformity or dose uniformity ratio (U) The ratio of maximum to minimum **radiation dose** in the product, i.e., $U = D_{max}/D_{min}$.

dosimetry The measurement of **radiation** quantities, specifically absorbed **radiation dose** and **radiation dose rate**.

dominant lethal assay An in vivo test for **mutagenicity**. The incidence of dominant lethal mutations is assessed by exposing male animals to the test material. They are then mated at various times with groups of females which have not been exposed to the test materials and the pregnant females are dissected at the mid-point of their pregnancy to determine the numbers of live and dead embryos in the uterus. By comparison with data from females mated with untreated

dominant lethal mutations Changes in the genetic material of the germ cells which give rise to abnormalities so severe that death occurs at an early stage after fertilisation. Dominant mutations are those which are not compensated for by the existence of a normal gene in the germ cell or opposite sex.

males it is possible to determine any increase in fetal deaths. If an increase is observed it is probably associated with genetic factors in the embryo transmitted by the spermatozoa and is good evidence that the test material has induced lethal mutations in the developing spermatozoa of the male.

dwell time The time the product remains in **irradiation** position in an **irradiator**.

electromagnetic spectrum The electromagnetic **radiations** arranged on a scale according to their energy, wavelength, or frequency, ranging from the low-frequency, high-wavelength radio waves to the high-frequency, low-wavelength cosmic radiation.

electron A negatively charged particle that is a constituent of all atoms.

electron accelerator A device for imparting large amounts of kinetic energy to electrons.

electron beam An essentially monodirectional stream of (negative) electrons which have usually been accelerated electrically or electromagnetically to high energy.

electronvolt (eV) A unit of energy. One electronvolt is the kinetic energy acquired by an **electron** in passing through a potential difference of one volt in a vacuum. One eV corresponds approximately to 1.60×10^{-19} **joule**. A commonly used multiple is the mega-electron volt, MeV (1 MeV = 10^6 eV).

erg Formerly used unit of energy. The SI unit **joule** (J) should now be used. 10^7 erg = 1 J.

excitation Attainment of a higher than normal energy level of an atom or molecule, as a consequence of absorption of energy, e.g., by heating or **irradiation**.

FAO Food and Agriculture Organization of the United Nations, Rome.

free radical An atom or molecule with an unpaired **electron**. A dot placed in a manner such as shown in H• (a hydrogen radical) desinnates a free radical.

gamma radiation One of the types of **radiation** emitted from radioactive **isotopes**. Like visible light or radio waves, gamma rays form a part of the **electromagnetic spectrum**.

gamma source Radioactive material emitting **gamma radiation**, in food **irradiation** usually cobalt-60.

gray (Gy) The SI unit of the **radiation dose**, being equal to one **joule** of energy absorbed per kilogram of matter being irradiated. It has the dimension of $J \cdot kg^{-1}$, and its relationship to the formerly used unit, the **rad**, is: 1 Gy = 1 J · kg^{-1} = 100 rad. A commonly used multiple is the kilogray, kGy (1 kGy = 10^3 Gy).

G-value The **radiation** yield of chemical changes in an irradiated substance in terms of the number of specified chemical changes produced per 100 **eV** or per **joule** of energy absorbed from **ionizing radiation**. Examples of such chemical changes are production of particular molecules, **free radicals**, or **ions**. At present, most data are given as number of chemical changes per 100 eV. To convert to SI units, transform to per-eV value and divided by 1.602×10^{-19} to obtain data in per-joule value, e.g.,

$$15.6 \text{ per } 100 \text{ eV} \rightarrow 15.6 \times 10^{-2} \text{ eV}^{-1} \rightarrow \frac{15.6 \times 10^{-2}}{1.602 \times 10^{-19}} \text{ J}^{-1}$$

$$\rightarrow 9.74 \times 10^{17} \text{ J}^{-1}$$

half-life The time in which half of any starting amount of a **radionuclide** decays.

IAEA International Atomic Energy Agency, Vienna.

induced radioactivity **Radioactivity** resulting when a material is irradiated with **radiation** of sufficiently high energy to cause nuclear reactions.

in vitro "In glass." Biological, biochemical, or other studies carried out on cells or tissues surviving in the test tube are in vitro studies.

in vivo "In life." Biological, biochemical, or other studies carried out on living animals or plants are in vivo studies.

ion An atom or molecule bearing a positive (cation) or negative electrical charge (anion) caused by a deficiency or an excess of electrons.

ionization Production of **ion** pairs, one of which is usually an **electron**, the other a positively charged atom or molecule (cation).

ionizing radiation Any **radiation** causing **ionization** of the irradiated material.

irradiation treatment with **radiation**.

irradiator That part of a **radiation** facility that houses the **gamma source**, **electron accelerator**, or **X-ray** machine, i.e., the radiation chamber inside the radiation protection shield.

isotopes **Nuclides** having the same **atomic number** but different mass **number**.

JECFI Joint FAO/IAEA/WHO Expert Committee on Irradiated Foods.

joule (J) SI unit of energy, replacing the formerly used unit **erg**. 1 J = 10^7 erg.

mass number The total number of **protons** and **neutrons** or an atom. The mass number is the whole number of integer nearest in value to the atomic mass when the latter is expressed in atomic mass units. The mass number is commonly written as a superscript before the symbol of the element (e.g., ^{60}Co) or after the name of the element (e.g., cobalt-60).

metabolism The processes by which a living organism utilizes nutrients to support its functions, producing energy and waste products is doing so.

micronucleus test an in vivo test for **mutagenicity**. The genetic apparatus of each cell is normally enclosed in a membrane and constitutes a body called the nucleus. Agents that interfere with the genetic material can cause the separation of fragments from the main body of the nucleus which appear as independent micro-nuclei. The test involves the examination of developing blood cells in the bone marrow for the appearance of such bodies after treatment of the animal with the test material. This test is quicker and simpler than methods of detecting changes in the chromosomes themselves.

mutagen A substance that causes changes (mutations) in the hereditary (genetic) traits transmitted from parent to offspring.

mutagenicity The power or property of causing mutations.

neutron An electrically neutral elementary particle; a component of the nucleus of an atom (except the hydrogen atom).

nuclide Any atomic form of an element. Nuclides are characterized by the number of **protons** and **neutrons** in the nucleus.

polyploidy Occurrence, as detected under the microscope, of cells containing twice or more the normal number of chromosomes. Human cells have 46 chromosomes. If they are polyploid they have 92 chromosomes and occasionally 138 or even more.

positron An elementary particle having a positive charge and the same mass as an **electron**.

post mortem "After death." The autopsy is a post mortem examination of the body.

proton An elementary particle having a positive charge and the mass **number** 1. The proton is the nucleus of the hydrogen atom and is also one of the constituents of every atomic nucleus. The number of protons in a nucleus is indicated by the **atomic number**.

rad The formerly used unit of **radiation dose**, now superseded by the **gray (Gy)**. 1 rad = 0.01 Gy = 0.01 J/kg. Commonly used multiples were the kilorad (krad) and the megarad (Mrad). 1 Mrad = 10^3 krad = 10^6 rad.

radappertization The application to foods of doses of **ionizing radiation** sufficient to reduce the number and/or activity and viable mciroorganisms to such an extent that very few, if any, are detectable in the treated food by any recognized method (viruses being excepted). In the absence of postprocessing contamination, no microbial spoilage or toxicity should become detectable with presently available methods, no matter how long or under what conditions the food is stored. Doses in the range of 25 to 45 kGy may be required to achieve radappertization.

radiation In the context of this book, radiation means **ionizing radiation**, or specifically **gamma radiation**, **X-rays**, or **electron beams**.

radiation dose The absorbed dose is the amount of energy absorbed per unit mass of irradiated matter at a point in the region of interest. The SI unit of absorbed dose is the **gray (Gy)**; the unit **rad** was formerly used.

radiation dose rate The increment of absorbed dose in a particular medium during a given time interval. The SI units is Gy $\cdot$ s^{-1}.

radiation source A radioactive substance or a machine (**electron accelerator** or **X-ray** machine) delivering **ionizing radiation**.

radicidation The application to foods of a dose of **ionizing radiation** sufficient to reduce the number of viable specific non–spore-forming pathogenic microorganisms to such a level that none are detectable when the treated food is examined by any recognized method. The required dose is in the range of 2 to 8 kGy. The term may also be applied to the destruction of parasites like tapeworms and trichinae in meat, in which case the required dose is in the range of 0.1 to 1 kGy. When the process is used specifically for destroying enteropathogenic and enterotoxinogenic organisms belonging to the genus *Salmonella*, it is referred to as "Salmonella radicidation."

radioactivity Spontaneous disintegration of the atomic nucleus of a **radionuclide**, associated with emission of **ionizing radiation**. Radioactivity is measured in **becquerel**, formerly in **curie**.

radioisotope Formerly used term for a radioactive **nuclide**, now superseded by the term **radionuclide**.

radiolysis Chemical changes caused by **ionizing radiation**.

radiolytic products Substances produced by the process of **radiolysis**, e.g., in a food when that food is irradiated.

radionuclide A radioactive **nuclide.** The term supersedes the older usage, **radioisotope.**

radurization The application to foods of doses of **ionizing radiation** sufficient to enhance keeping quality (usually at refrigeration temperature) by causing a substantial decrease in numbers of viable specific spoilage microorganisms. The dose required is in the range of 0.4 to 10 kGy.

Salmonella typhimurium reverse mutation test (Ames test) A very widely used in vitro test for mutagenicity in bacteria which has been developed by Bruce Ames of the University of California. Strains of the bacterium *Salmonella typhimurium* have been derived which lack the ability to grow in the absence of the amino acid histidine. When treated with mutagens, genetic changes occur in the bacterial population and some of thes emutated cells can survive without histidine. The appearance of growing cells in medium lacking this amino acid is therefore an indication of mutagenesis.

sister-chromatid exchange test (SCE test) The normal chromosome consists of a filament containing the genetic material, DNA. This filament is duplicated during cell growth so that at division, the chromosome consists of two filaments called chromatids, joined at one point. The two chromatids separate at division and move to the two daughter cells. Damage to DNA, or interference with its reproduction can lead to disturbance of the distribution of the newly synthesized material; parts of the DNA molecules become exchanged between the two developing (sister) chromatids. Staining methods have been devised which show up such exchanges. The method is a sensitive indicator of genetic damage used with cells in vivo or in culture (in vitro).

SI system The International System of Units (SI, from the French "système international"), the worldwide official system of units since 1986. The adoption of the SI system requires, for instance, the use of **joules** instead of calories or of the dose unit **gray (Gy)** instead of **rad.** The SI units are described in detail in a useful booklet from the World Health Organization (*The SI for the Health Professions.* Prepared at the request of the Thirtieth World Health Assembly. WHO, Geneva 1977. See also Lundberg, G. D., C. Iverson, G. Radulescu. Now read this: The SI units are here, *J. Am. Med. Assoc., 255*:2329 (1986)).

toxicology The science that deals with poisons (toxic substances) and their effects.

uniformity ratio See **dose uniformity.**

water activity, a_w The availability of water for microbial, enzymic or chemical activity in a food (or other material) is characterized by this term. Water activity is defined as the ratio of the vapor pressure of water in a food to the saturated vapor pressure of water at the same temperature. In pure water $a_w = 1$; in very dry materials a_w approaches zero.

WHO World Health Organization, Geneva.

wholesomeness Safety for consumption. With regard to irradiated foods, considerations of wholesomeness involve 1) radiological safety, 2) toxicological safety, 3) microbiological safety, and 4) nutritional adequacy.

X-rays A type of **ionizing radiation** produced by the impact of high-energy **electrons** on matter; a part of the **electromagnetic spectrum**. In their physical properties X-rays are very similar to the **gamma rays** emitted from certain **radionuclides**. Also named roentgen rays after W. C. Rontgen, who discovered them in 1895.

Index

D

E

F

N

O

P

Q

R

S